CMP

W
DE
G

Acknowledgement is made to G. Comini and O. Saro for the use of Figures 4a and c on p.33, which appear on the front cover of this book.

Computational Modelling of Free and Moving Boundary Problems

Vol. 2 Heat Transfer

Proceedings of the First International Conference, held 2-4 July 1991, Southampton, U.K.

Editors: L.C. Wrobel
C.A. Brebbia

Computational Mechanics Publications
Southampton Boston

Co-published with

Walter de Gruyter
Berlin New York

L.C. Wrobel
Wessex Institute of Technology
Computational Mechanics Institute
Ashurst Lodge, Ashurst
Southampton SO4 2AA

C.A. Brebbia
Wessex Institute of Technology
Computational Mechanics Institute
Ashurst Lodge, Ashurst
Southampton S04 2AA

A Catalogue record for this book is available from the British Library

ISBN 1-85312-158-4 Computational Mechanics Publications, Southampton
ISBN 1-56252-085-7 Computational Mechanics Publications, Boston
ISBN 3-11-013173-0 de Gruyter, Berlin
ISBN 0-89925-909-X de Gruyter, New York

set

ISBN 1-85312-142-8 Computational Mechanics Publications, Southampton
ISBN 1-56252-070-9 Computational Mechanics Publications, Boston
ISBN 3-11-013174-9 de Gruyter, Berlin
ISBN 0-89925-910-3 de Gruyter, New York

Library of Congress Catalog Card Number 91 72243

Die Deutsche Bibliothek - Cataloging-in-Publication Data

Computational modelling of free and moving boundary problems:
proceedings of the ... international conference ...-Southampton ; Boston :
Computational Mechanics Publ. ; Berlin ; New York : de Gruyter

1. Held 2-4 July 1991, Southampton, U.K.
Vol.2. Heat Transfer.-1991
ISBN 3-11-013173-0 (de Gruyter, Berlin)
ISBN 3-11-013174-9 (de Gruyter, Berlin, 1, Vol. 1 und 2)
ISBN 0-89925-909-X (de Gruyter, New York)
ISBN 0-89925-910-3 (de Gruyter, New York, 1, Vol. 1 und 2)
ISBN 1-85312-158-4 (Computational Mechanics Publ., Southampton)
ISBN 1-56252-085-7 (Computational Mechanics Publ., Boston)
ISBN 1-85312-142-8 (Computational Mechanics Publ., Southampton, 1, Vol. 1 und 2)
ISBN 1-56252-070-9 (Computational Mechanics Publ., Boston, 1, Vol. 1 und 2)

Printed in Great Britain by The Alden Press, Ltd, Oxford

PREFACE

The classification "free boundary problem" or "moving boundary problem" can be applied to a large variety of physical situations. The common feature is the presence, in their mathematical modelling, of an initially unknown (free) boundary or a boundary which moves throughout the analysis, the determination of which is an important part of the solution procedure. Practical examples are ingot solidification and scrap melting in metallurgy, the re-entry problem of hypersonic missiles in aeronautical sciences, the cooling, freezing and cold storage of foodstuffs in the industry, to name but a few. Although completely different in nature, these problems have several common features regarding their mathematical modelling and computer methods of solution.

The present book and its companion volume contain edited versions of the papers presented at the First International Conference on Computational Modelling of Free and Moving Boundary Problems, held in Southampton, England, in July 1991. The purpose of the conference was to promote the interaction between engineers, applied mathematicians and numerical analysts involved in the creation, development and application of computational methods to free and moving boundary problems. In particular, the conference intended to provide a useful connection between scientists working in different areas of application but using similar numerical techniques.

Although the papers have been typographically edited they have been reproduced directly from material submitted by the authors, and their content is a reflection of the authors' research and opinion.

The editors are grateful for the assistance of the staff of Computational Mechanics Publications in the preparation of the volumes.

Southampton, July 1991
L.C. Wrobel and C.A. Brebbia

CONTENTS

SECTION 1: SOLIDIFICATION AND MELTING

FIRST INTERNATIONAL CONFERENCE
ON
COMPUTATIONAL MODELLING OF FREE AND MOVING BOUNDARY PROBLEMS 91

SCIENTIFIC COMMITTEE

SECTION 1: SOLIDIFICATION AND MELTING

Recent Developments in the Modelling of Solidification Processes

V.R. Voller

Department of Civil and Mineral Engineering, University of Minnesota, Minnesota, Minneapolis, MN 55455, USA

ABSTRACT

A central theme in this paper is the need to deal with disparate length scales in modelling solidification systems. This theme is illustrated in outlining two solidification models recently reported in the literature. The first is a macroscopic model of heat and fluid flow in a casting. The second is a microscopic model of mass and heat diffusion in the two phase mushy zone.

INTRODUCTION

Solidification processes and the related phenomena are highly complex systems and a variety of modelling approaches have been developed [1,2]. The major problem facing a modeler is in dealing with the multiple length scales that occur in a solidification process (ranging from microns in the solid/liquid mushy zone to meters on the full scale of the process). In practice it is not possible to separate these length scales. Many macroscopic phenomena are intimately linked to micro-structural considerations. On the other hand, microscopic phenomena are influenced by macroscopic events such as fluid flow and heat transfer. In this paper both the macro and micro modelling approaches will be explored. In particular:

1. a full macroscopic model of solidification which attempts to account for the key heat and mass transfer phenomena, and

2. a microscopic model of the solidification in the columnar dendritic structure of the two-phase solid/liquid mushy zone.

will be investigated.

THE MACROSCOPIC MODEL

An area that has attracted much recent attention is the development of comprehensive macroscopic equations describing the transport of solute, heat and momentum [3-5]. In this section a review, of sorts, is given on presenting a slightly simplified version of equations previously derived by Voller et al [5].

Consider the solidification of a binary material in a two dimensional cavity insulated on three sides and cooled on the left boundary, Figure 1. On making appropriate assumptions (described below) the macroscopic behavior of the solidification can be described by basic transport equations

Transient Term + Convective Term = Diffusive Term + Source

for heat, solute, and momentum. In arriving at the appropriate equations there are a number of important phenomena that need to be accounted for. As the solidification proceeds the solute phase in the binary material is rejected (partition ratio k<1) into the liquid phase. This rejection leads to a differential solidification and a so-called "mushy zone" forms in which solid and liquid co-exist. There are some key solidification features associated with the mushy zone.

Morphology: Depending on the cooling conditions the morphology of the mushy region can, in general, take two forms

(i) a columnar dendritic crystal structure which attaches to the fully solid region

(ii) an equi-axed dendritic structure of free floating grains

These two morphologies are schematically illustrated in Figure 1. Note that, in most casting geometries both fixed columnar and free equi-axed grain morphologies can occur.

Thermodynamic Coupling : It is important to recognize that, in the mushy region, the temperature and concentration field are coupled via thermodynamic considerations. The exact form of this coupling depends on the nature of the solute diffusion on the local microscopic scale. There are a number of assumptions that can be made by which this microscopic behavior can be "captured" at the macroscopic scale. The most straight forward is to assume that local thermodynamic equilibrium is maintained throughout the mushy region (the lever assumption). This equilibrium assumption will mean that at any point in the mush

$$T_s = T_l = T \; ; \qquad and \qquad C_s = k\, C_l \tag{1}$$

where C is the solute concentration, T is the temperature and

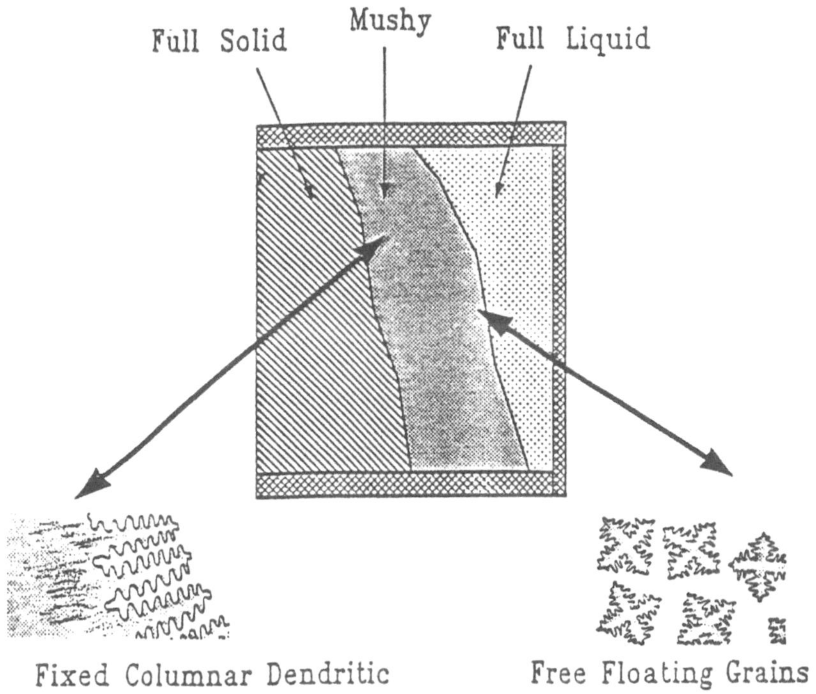

Figure 1. Test geometry showing schematic of mushy zone morphologies.

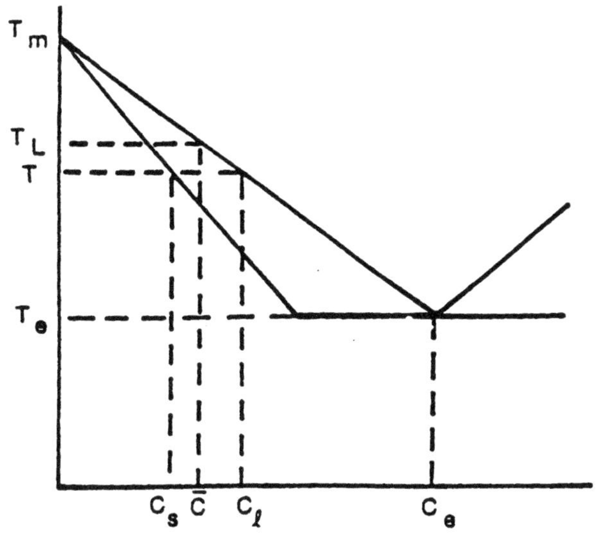

Figure 2. The equilibrium phase diagram.

the subscripts s and l refer to the solid and liquid phases respectively. The temperature and solute fields are coupled by the liquids line in the phase diagram, Figure 2.

$$T = T_m + M_l C_l \qquad (2)$$

where T_m is the fusion temperature of the pure material and M_l is the slope of the liquids line.

<u>Double Diffusive Phenomena:</u> In the current work it will be assumed that the solute phase of the binary material is lighter than the solvent. As a result the rejection of the solute phase in the mushy region will induce a buoyancy driven convection. In early stages of the solidification this convection will compete with the thermal buoyancy convection. These opposing convection flows will give rise to a double diffusive layer at the mush/liquid interface.

All of the above aspects effect the progress of the solidification, in particular, the assumption of the mushy zone morphology takes a critical role in determining the form of the governing equations. Following Voller et al [5] the effect of the mushy region morphology can best be illustrated on considering two extreme models, with opposite morphology assumptions.

1. Fixed columnar model : The mushy region consists of fixed columnar dendrites (i.e., the solid velocity $\mathbf{u}_s = 0$).

2. Free floating equi-axed grain model : The mushy region consists of free floating equi-axed grains that move with the velocity of the fluid (i.e., the solid and liquid velocities are equal, $\mathbf{u}_s = \mathbf{u}_l = \mathbf{u}$).

The objective in obtaining governing equations, related to the above models, is to derive equations which are simultaneously applicable to the whole domain, solid, mush and liquid (Bennon and Incropera [3] refer to this as the continuum approach). In these equations, however, it is important to account for the multi-component nature of the problem. Voller et al [5], achieve this on starting the derivation from a two-phase modelling perspective. This approach leads to heat and solute transfer continuum transport equations which are consistent with the two-phase mushy region. It is only in the derivation of the momentum transport equations that some what ad-hoc assumptions need to be made. In the fixed columnar model an analog to flow in porous media is made and a 'Darcy' source term, based on the local liquid fraction is added into the momentum equations. In the free floating grain model the mixture viscosity associated with the mushy zone is enhanced as the liquid fraction decreases. The net effect of both of these approaches is to reduce the velocity across the mushy zone from the full velocity in the liquid phase to a zero velocity as the solid phase is approached. On making the assumption of constant specific heat and density in each phase a general set of continuum transport equations (applicable to both the chosen models) are

Mass:

$$\nabla.(\rho \mathbf{u}^{+}) = 0 \tag{3}$$

where $\mathbf{u}^{+}=(u,v)$ is the system velocity (see Figure 3).

Momentum:

$$\frac{\partial(\rho u)}{\partial t} + \nabla.(\rho \mathbf{u}^{+} u) = -\frac{\partial P}{\partial x} + \nabla.(\mu^{+}\nabla u) + F_x \tag{4}$$

$$\frac{\partial(\rho v)}{\partial t} + \nabla.(\rho \mathbf{u}^{+} v) = -\frac{\partial P}{\partial y} + \nabla.(\mu^{+}\nabla v) + F_y \tag{5}$$

$$- \rho_{ref} g_r [\beta_T (T - T_e) + \beta_s (C_l - C_e)]$$

where the Boussinesq approximation has been used to replace the body force term with a solutal and thermal buoyancy term.

Temperature:

$$\frac{\partial(\rho c T)}{\partial t} + \nabla.(\rho \mathbf{u} c T) = \nabla.(K^{*}\nabla T) - \rho L \frac{\partial g}{\partial t} + S_T \tag{6}$$

where K^{*} is the mixture conductivity $(gK_l + (1-g)K_s)$ and g is the local liquid fraction in the mushy zone.

Solute:

$$A\frac{\partial \rho C_l}{\partial t} + \nabla.(\rho \mathbf{u}^{+} C_l) = \nabla.(\rho D^{*}\nabla C_l) + (k-1)\rho C_l \frac{\partial g}{\partial t} + S_c \tag{7}$$

where and D^{*} is the mixture diffusivity $(gD_l+(1-g)kD_s)$ and $A=[1+(k-1)(1-g)]$.

In these equations the definitions of the source terms, F and S, and the values marked with + superscripts depend upon the mushy region model, the exact definitions are graphically illustrated in Figure 3.

The governing equations need to be solved numerically. One approach, based on a control volume algorithm [6], is to use the local liquid fraction as an iterative field in achieving the temperature and solute thermodynamic coupling[7]. In a time step, for a given choice of local liquid fraction and pressure field, the momentum (equations 4 and 5), heat (equation 6) and solute (equation 7) are solved. In general, at the end of an iteration,

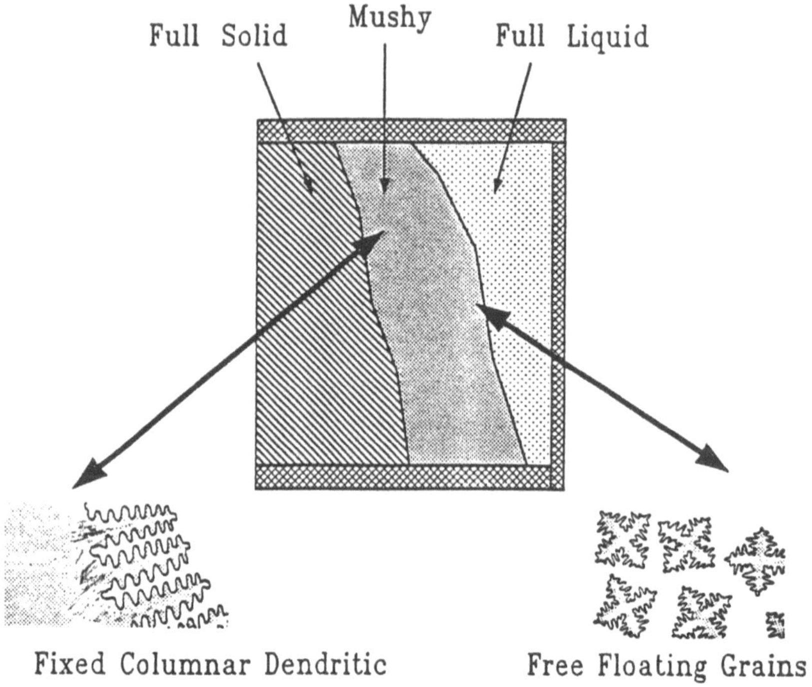

Fixed Columnar Dendritic

System Velocity :

$$u^{+} = g\, u_l$$

Viscosity :

$$\mu^{+} = \mu_l$$

$$S_c = 0$$

$$S_h = 0$$

Darcy Source:

$$F_x = \frac{K_o\,[1-g]^2}{g^3}\, u$$

$$F_y = \frac{K_o\,[1-g]^2}{g^3}\, v$$

Free Floating Grains

System Velocity :

$$u^{+} = u_s = u_l$$

Viscosity :

$$\mu^{+} = \mu_{eff} = g\mu_l + (1-g)\,\mu_s \quad [\,\mu_s = 10^4]$$

$$S_c = -\,(k-1)\,[\nabla.(\rho u (1-g)\, c_l)]$$

$$S_h = -\,\nabla.(\rho\, g\, u\, L)$$

Darcy Source :

$$F_x = 0$$

$$F_y = 0$$

Figure 3. Definitions of sources and terms related to the morphology.

(i) the flow solution will not satisfy the mass conservation and (ii) the solute and temperature field in the mushy zone will not be consistent with the coupling dictated by thermodynamic equilibrium (equations 1 and 2). Before the next iteration step (i) the pressure and velocity fields are updated in order to achieve a continuity satisfying flow field (the well known SIMPLE algorithm [6]) and (ii) in a similar manner, the liquid fraction, solute and temperature fields are updated so that thermodynamic equilibrium is maintained. These iterations steps continue to convergence. Full details of appropriate numerical solution approaches can be found in Prakash and Voller [7] and Bennon and Incropera [8].

More general equations which account for different values of density and specific heat in each phase can be found in reference [5]. Alternative forms of the governing equations for similar systems can be found in the work of Bennon and Incropera [3] and Beckermann and co-workers [4,9]. Equations and models for other systems have been proposed by Rappaz and Voller [10], who consider non-equilibrium conditions in the mushy zone, and Flood et. al. [11] who consider the case of a continuous casting geometry.

A key feature in the governing equations shown in Figure 3 is the nature of the solute and enthalpy source terms. In the free floating grain model solid is transported with the flow and this results in latent heat and solute advection source terms, S_T and S_C, which are not present in the fixed columnar model. These source terms can lead to major difference in predictions between the mushy region models. Figure 4, taken from the work of Voller et al [5], shows predicted solute composition fields (iso-concentrate contours) on the final solidification of an ammonium chloride solution in a small square cavity (see Table 1 for the thermal and physical data). Initially, in the liquid state, the super heated liquid phase is at a uniform solute composition $C_l=0.1$. During the solidification the rejection of solute in the mushy region and its subsequent advection by thermal and solutal natural convection fields leads to a non-uniform distribution composition, C_s, in the solid (macro-segregation). In areas of positive segregation, shown shaded in Figure 4, $C_s>0.1$ and in areas of negative segregation $C_s<0.1$. In the early stages of solidification concentration gradients are small and the flow in the cavity is driven by thermal natural convection in an anti-clockwise direction. Rejection of the lighter solute in the mushy region, however, leads to a clockwise convection cell. As solidification continues the solute gradients increase in size whereas, due to the dissipation of the superheat, the thermal gradients decay. In the fixed columnar model the solute driven convection cell becomes dominant at an early stage of the solidification. Thus the fluid flow produces a clockwise "washing" of the solute from the mushy region resulting in areas of positive segregation on the right and top walls. In the free floating grain model concentration gradients created by solute

Table 1. Data for test problem. ammonium chloride solution

Initial and boundary conditions		
Cavity dimensions	w =	0.025 m
Initial liquid concentration	C_{int} =	0.1 kg m^{-3}
Initial temperature	T_{cold} =	400 K
Thermophysical properties		
Specific heat	c =	3000 J kg^{-1} K^{-1}
Thermal conductivity	K =	0.4 W m^{-1} K^{-1}
Density	ρ =	1000 kg m^{-3}
Liquid viscosity	μ_l =	1.0 x 10^{-3} kg m^{-1} s^{-1}
Solid viscosity	μ_s =	1 x 10^{4} kg m^{-1} s^{-1}
Species diffusion coefficient	D =	4.8 x 10^{-9} m^{2} s^{-1}
Latent heat	L =	3 x 10^{5} J kg^{-1}
Permeability coefficient	K_0 =	2 x 10^{6} kg m^{-3} s^{-1}
Thermal expansion coefficient	β_T =	4 x 10^{-5} K^{-1}
Solutal expansion coefficient	β_s =	0.025
Eutectic temperature	T_e =	250 K
Eutectic concentration (mass fraction)	C_e =	0.8
Melting point of pure material	T_m =	630 K
Equilibrium partition ratio	k =	0.3

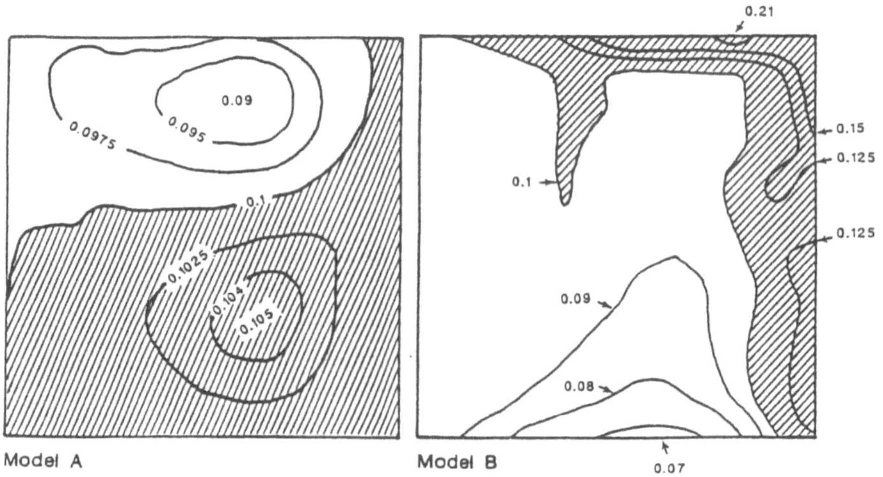

Figure 4. Solid solute composition predictions taken from reference [5]. Model A = free floating grain model, Model B = fixed columnar model.

rejection in the mushy region are smoothed out by the migration of the solid with the fluid motion. Hence an anti-clockwise flow is maintained throughout the solidification resulting in a totally different shape for the final solute field in which segregation is reduced.

In summary equations 3-8 with the definitions of Figure 3. provide reasonable limits on the required system of equations to describe mass, momentum, heat and solute transport in a static casting under the assumption of local thermodynamic equilibrium. The results in Figure 4. clearly show the impact of the mushy region morphology assumption on the transport of the solute during solidification. Elsewhere Voller et al. [12] have shown an equally dramatic effect of mushy morphology on the transport of heat. It is worth realizing, however, that in the general case both fixed columnar and free grain will be present and further the grains may not travel identically with the fluid. The treatment of such cases is a major challenge in solidification modelling.

THE MICROSCOPIC MODEL

Next we turn attention to a microscopic solidification model. A major objective of such modelling is the prediction of the micro-structure and micro-composition (micro-segregation) within the dendritic mushy zone of a binary alloy. As an example we consider a recent micro-model [13] involving coupled heat and mass diffusion in the mushy zone. The domain of interest is the region $0 < x < X(t)$ between the secondary dendritic arms in the mushy zone (see top half of figure 5). In analyzing the heat and mass transfer in this domain the following assumptions are made

1. The domain of interest $0 < x < X(t)$ is one dimensional.

2. Mass and heat transfer is controlled by diffusion alone, i.e., no flow effects arising from shrinkage and bulk flow are considered.

3. The interface between the solid and the liquid is sharp, i.e., effects of constitutional undercooling leading to morphological instability [14] are ignored.

4. Thermodynamic equilibrium holds at the solid-liquid interface, i.e., kinetic undercooling effects are neglected (Figure 2 and equations 1 and 2 above)

5. No solute enters or leaves the domain, ie., the integrated composition in $0 < x < X(t)$ remains at the initial composition C_o.

6. In the case of an expanding domain (see below) the material to the right of X(t) is assumed to be at the initial concentration and temperature, i.e.,

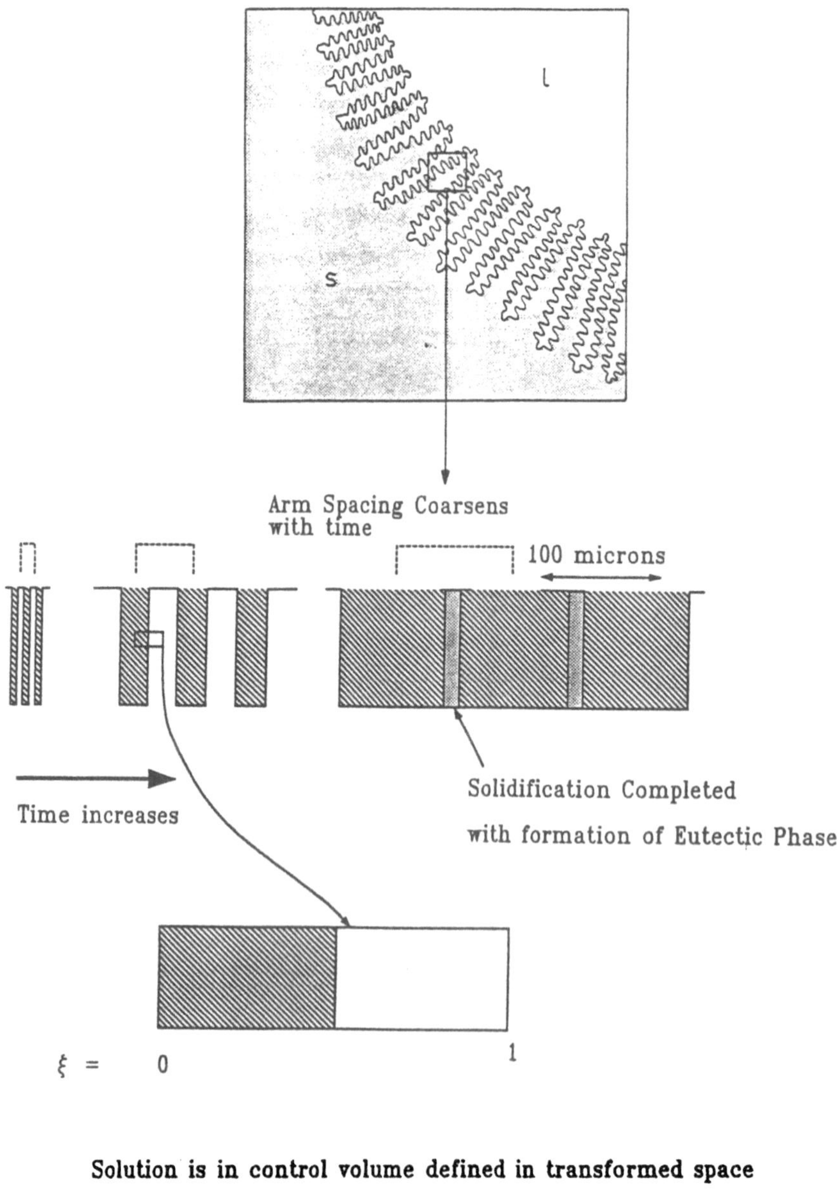

Figure 5. Micro-segregation in coarsening secondary dendrite arms.

in $x > X(t)$ $\quad C = C_o$ and $T = T_{liquids}$

where $T_{liquids}$ is the temperature corresponding to C_o on the liquids line of the phase diagram. This assumption ensures that the mass is conserved as the system expands.

As given this problem is closely related to the well known double diffusive "binary alloy problem" [14,15], in the current case a convenient and descriptive label is the "micro-segregation" model. The "extra spice" in the micro-segregation model is that the solution domain between the secondary dendrite arms is expanding as the solidification proceeds (this is schematically illustrated in the middle section of Figure 5). The expansion of the domain is caused by the coarsening of the secondary dendritic arms. Coarsening occurs by "melting back" of some of the arms or when arms break off by the process of "necking". This coarsening behavior can be described by an empirical relationship of the form [16]

$$\lambda_2 = \beta \, t^n \tag{8}$$

where λ_2 is the secondary arm spacing (μ m), $X(t) = \dfrac{\lambda_2}{2}$.

At the start of the process the arm spacing is zero. As solidification proceeds the arm spacing grows as does the solid fraction within the spacing. New material enters the liquid end of the arm spacing at the initial composition and temperature. Solidification is complete when the composition in the liquid fraction of the arm spacing reaches the eutectic composition. At this point the remainder of the solidification occurs at a constant interface temperature, T_e, a constant composition, C_e , and a constant arm spacing. The object of the modelling analysis is to predict the microstructure on solidification (defined by the final secondary arm spacing) and the micro-segregation (characterized by the mass fraction of eutectic composition in the solidified arm spacing).

Earlier micro-segregation models [16,17] assume complete solute mixing in the liquid phase and uniform temperature throughout the domain at any given time. Recently, Battle and Pehlke [18] have presented a solution based on the method of lines/invariant imbedding approach [19]. Battle and Pehlke not only solve the problem in an expanding domain but also consider the heat and mass transfer in the solid phase. Although, this model produces excellent agreement with a range of experiments a major drawback is the "excessive" number of grid points, on the order of 100,000, required to maintain solute mass conservation. These grid sizes result in a large computational cost (1 hour to several days on an Apollo domain DN 4000 work station). The model presented by Suresh and Voller [13] also solves for the heat and mass transfer in both liquid and solid in the expanding domain. In this work, however, no mass conservation problems are encountered and grid independent solutions can be obtained with relatively few nodes (100) and low CPU times (seconds on an IBM 386 PC).

In finding an appropriate solution route to the micro-segregation problem there are two major numerical hurdles to be overcome, both associated with moving boundaries.

1. the solid liquid interface, x=s(t), at which mass and solute balances must be met, and

2. the secondary arm spacing x=X(t).

Following previous workers [20] the first of these problems is overcome on writing the governing equations in a conservation form (i.e., in terms of enthalpy H and solute concentration C). The result is single domain equations for the heat and mass diffusion, i.e.,

$$\frac{\partial H}{\partial t} = \frac{\partial}{\partial x}\left(K\frac{\partial T}{\partial x}\right) \tag{9}$$

$$\frac{\partial C}{\partial t} = \frac{\partial}{\partial x}\left(D^m\frac{\partial V}{\partial x}\right) \tag{10}$$

where $H = cT + gL$, $V = M_iC_i$ is the chemical activity and $D^m = D_i/M_i$ (i=s or l and M_i are the slopes in the phase diagram).

The boundary conditions are:

At x = 0

$$-K\frac{\partial T}{\partial x} = q \quad (a\ prescribed\ flux), \qquad \frac{\partial C}{\partial x} = 0$$

At x = X(t)

$$\frac{\partial T}{\partial x} = 0\,, \quad \frac{\partial C}{\partial x} = 0$$

The moving domain boundary, X(t) is dealt with via a Landau [21] transformation. A new coordinate system is defined as

$$\xi = \frac{x}{X(t)} \tag{11}$$

and the expanding domain $0 < x < X(t)$ is transformed into a fixed domain

$$0 < \xi < 1$$

In the transformed coordinates the governing conservation equations become:

$$\frac{\partial H}{\partial t}\Big|_{\xi} = \frac{1}{X^2}\frac{\partial}{\partial \xi}\left[K\frac{\partial T}{\partial \xi}\right] + \frac{1}{X}\frac{dX}{dt}\left[\frac{\partial(\xi H)}{\partial \xi} - H\right] \quad (12)$$

$$\frac{\partial C}{\partial t}\Big|_{\xi} = \frac{1}{X^2}\frac{\partial}{\partial \xi}\left[D^m\frac{\partial V}{\partial \xi}\right] + \frac{1}{X}\frac{dX}{dt}\left[\frac{\partial(\xi C)}{\partial \xi} - C\right] \quad (13)$$

The advantage of equations (12) and (13) is that a numerical solution can be carried out using a fixed grid. The solution approach used by Suresh and Voller [13] is based on the node jumping solution method previously proposed by Voller [20]. In a control volume discretization of the fixed transformed domain the operation of the node jumping solution involves the selection of a variable time step such that the planar solid liquid interface "jumps" form one node of the grid to the next in subsequent time steps. In jumping from one node to the next the numerical source terms and coefficients are fully defined and a linear tri-diagonal system results which is readily solved. The key in the solution is to choose the required time step. This is achieved on iteratively choosing a set of time steps. Convergence is declared when the current choice of time step results in a temperature, T, and chemical composition, V, fields that satisfy (or nearly satisfy) the condition of thermodynamic equilibrium at the node point coinciding with the solid liquid interface. Full details of this approach can be found in references [13,20].

The above model has been numerically validated on solving a binary alloy problem in a semi-infinite domain. Results are in excellent agreement with the Rubinstein analytical solution [15], in particular sharp jumps in concentration and enthalpy fields at the solid/liquid interface are fully picked up. Further testing across a range of problems indicates that the model and associated numerical solution can produce grid independent and mass conserved results with grids of 100 nodes or less.

In this paper results related to the solidification of a aluminum copper systems (thermal and physical data given in Table 2) will be presented. Michael and Bever [22] conducted experiments on a this system in which the initial composition and cooling rate was varied. A selection of the conditions used by Michael and Bever are given in Table 3. Comparisons between the experimental and the numerical results for the final secondary arm spacing and the mass fraction of eutectic are given in Figure 6 (points that fall on the 45 degree line indicate a match between experiment and model) . These results clearly show the excellent utility of the model in predicting microscopic solidification phenomena.

Table 2.- Aluminum-copper Data

Property	Symbols	Values	Units
Thermal Conductivity	K	100	W/m-°K
Specific Heat	c	800	J/kg-°K
Latent Heat	L	$4x10^5$	J/kg
Density	ρ	2500	kg/m^3
Liquid Mass Diffusivity	D_l	$5x10^{-9}$	m^2/sec
Solid Mass Diffusivity	D_s	$0.29 X 10^{-4}$ exp(-15610/T)	m^2/sec
Partition Coefficient (Based on mole %)	k	0.145	---
Eutectic Composition	C_e	17.4	mole %
Eutectic Temperature	T_e	821.2	°K
Fusion Temperature	T_m	933.54	°K
Liquidus Temperature	$T_{liquidus}$	T_m - 6.57 C_o	°K
Coarsening Equation Coefficient	β	19.70	---
Coarsening Equation Exponent	n	0.33	---

Table 3. Experimental Conditions of Michael and Bever [22]

Sample	C_o (Wt %)	Cooling Rate (° C/sec)	Prescribed Flux (J/m^2-sec)
A	1.96	1.4e-2	23
B	2.81	1.6e-2	22
C	2.07	53.0	5610
D	2.94	16.2	2320
E	3.96	5.0	915
F	3.93	49.0	4300
G	4.71	15.7	18

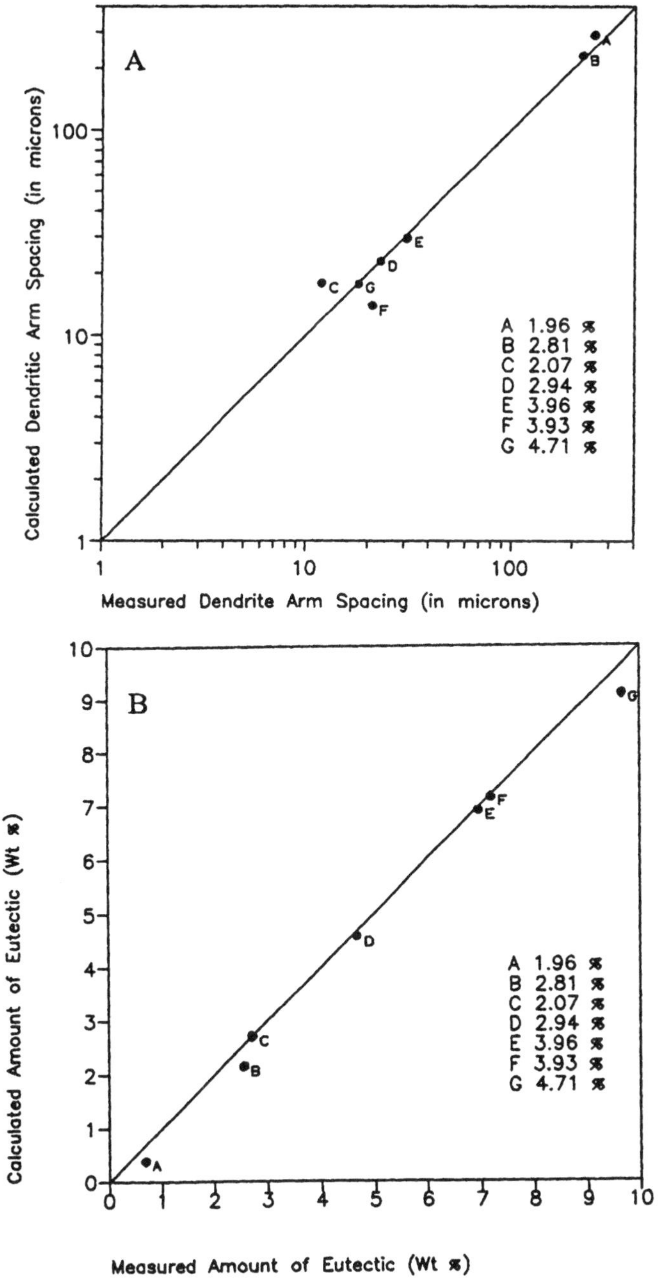

Figure 6. Comparison between experimental and numerical micro-segregation results.

CONCLUSIONS

The objective of this paper was to outline current and recent developments in the modelling of solidification systems. A central thesis of the paper is that the major challenge in modelling solidification systems is the need to account for the multiple length scales associated with solidification phenomena. This point was illustrated on outlining two models at different extremes of the length scale. The macroscopic model provided insight into the role of the mushy region morphology in determining the solidification transport phenomena. The microscopic model represented an attempt at modelling the local scale behavior in a mushy zone. Both of these models can be deemed successful in that the macroscopic model gave qualitative insight and the microscopic model predicted results in close agreement with experiment. There is still much room for improvement, however. The macroscopic model would be improved by a better accounting of the thermodynamic processes in the mushy zone (processes driven by the solute mass diffusion). Likewise the microscopic model would be improved by accounting for the advection of the solute phase (macro-segregation). The central requirement is to identifying ways in which the micro and macro models can be effectively and efficiently coupled together.

ACKNOWLEDGEMENTS

The author would like to acknowledge the contributions made by his students Dr. Allon Brent, now at BHP in Australia, and S. Suresh who were responsible for much of the "leg work" involved in the numerical solutions of the macro and micro models respectively. Research support from the Army High Performance Computing Research Center at the University of Minnesota, the Minnesota Supercomputer Institute and BHP Research and Technology in Australia is also acknowledged.

REFERENCES

1. Voller, V.R. An overview of the modelling of heat and fluid flow in solidification systems, in Modelling of Casting, Welding and Advanced Solidification Systems, TMS, 1991.

2. Rappaz, M. Modelling of Microstructure Formation in Solidification Processes, Int. Mat. Review, 34, pp. 93-123, 1989.

3. Bennon, W.D and Incropera, F.P. The Evolution of Macro-segregation in Statically Cast Binary Ingots, Met. Trans. B, 18B pp 611-612, 1987.

4. Beckermann, C. and Viskanta, R. Double-diffusive Convection during Dendritic Solidification of a Binary Mixture, PCH, 10 pp 195-213, 1988.

5. Voller, V.R., Brent, A.D. and Prakash, C. The Modelling of Heat, Mass and Solute Transport in Solidification Systems, Int. J. Heat Mass Transf., 32, pp1719-1731, 1989.

6. Patankar, S.V. Numerical Heat Transfer and Fluid Flow, McGraw Hill, New York, 1980.

7. Prakash, C. and Voller, V.R. On the numerical solution of continuum mixture model equations describing binary solid-liquid phase change, Numer. Heat Trans B, 15,pp 171-189, 1989.

8. Bennon, W.D. and Incropera, F.P Numerical analysis of binary solid-liquid phase change using a continuum model, Numer. Heat Trans. 13, pp277-296, 1988.

9. Ni, J. and Beckermann, C. A volume averaged two-phase model for transport phenomena during solidification (paper presented at 5th AIAA/ASME Thermophysics and heat transfer Conference, Seattle, June 1990).

10. Rappaz, M. and Voller, V. Modeling of Micro-Macrosegregation in solidification processes, Met Trans A, 21A, 749-753, 1990.

11. Flood, S.C., Katgerman, L and Voller, V.R. The calculation of macro-segregation and heat and fluid flows in the D.C. casting of aluminum alloys, in Modelling of Casting, Welding and Advanced Solidification Systems, TMS, 1991.

12. Voller, V.R., Brent, A.D. and Prakash, C. Modelling the mushy region in a binary alloy, Applied Mathematical Modelling, 14,pp 320-326, 1990.

13. Suresh, S and Voller, V.R. Numerical modelling of the binary alloy problem in an expanding domain, submitted to Int. J. Heat Mass Transfer, 1991.

14. Wilson, D.G., Solomon, A.D. and Alexiades, V. A. Shortcoming of the explicit solution for the binary alloy solidification problem, Letters in Heat and Mass transfer, 9, pp421-428, 1982.

15. Rubinstein, L.I. The Stefan Problem , Translations of Mathematical Monographs, 27, American Mathematical Society, 1971.

16. Roosz,A., Halder, E., Exner, H.E. Numerical calculation of micro-segregation in coarsened dendritic micro-structure, Mat. Sci. and Tech., 2, pp1149-1155, 1986.

17. Ogilvy, A.J.W. and Kirkwood, D.H. A model for the numerical computation of micro-segregation in alloys, App.Sci.Res.,44, pp 43-49, 1987.

18. Battle, T.P and Pehlke, R.D. Mathematical modeling of micro-segregation in binary metallic alloys, Met.Trans.B, 21B, pp357-375, 1990.

19. Meyer, G.H. Initial Value Methods for Boundary Value Problems, Academic Press, New York, NY, 1973.

20. Voller, V.R. An implicit enthalpy solution for phase change problems with application to a binary alloy solidification, Appl. Math. Model., 11, pp. 110-116, 1987.

21. Landau, H.G. Heat conduction in a melting solid, Q. Appl. Math., 8, pp 81-94, 1950.

22. Michael, A.B. and Bever, M.B. Solidification of Aluminum-rich Aluminum-Copper Alloys, Journal of Metals, Trans., ACME, pp 47-56, 1954.

Numerical Modelling of Freezing Processes in Foodstuffs

G. Comini, O. Saro
Istituto di Fisica Tecnica e di Tecnologie Industriali, Università di Udine, Viale Ungheria 43, 33100 Udine, Italy

ABSTRACT

Freezing times and temperature distributions in food materials, are among the most important factors to consider when designing a freezing process. Analytical solutions to heat conduction problems in foods undergoing freezing are available, but they are usually based on restrictive assumptions, or are limited to specific processes. Consequently, numerical methods are now considered to be the most accurate and cost effective way to model freezing processes of practical interest to food technology. In this paper, a complete mathematical model for freezing processes in foodstuffs is developed, on the basis of the physical aspects of the problem. The equivalent heat capacity approach and a finite element discretization are utilized in the numerical solutions.

KEYWORDS: Foodstuffs freezing, Finite element method, Thermal properties.

INTRODUCTION

In the food industry, freezing is commonly used to retard deterioration which occurs during storage and transportation of foodstuffs because of irreversible biochemical reactions and undesirable bacterial proliferations [1, 2]. On the other hand, the manner in which freezing itself takes place is of prime importance for the final quality of the products. At a slow cooling rate, few crystallization centers are formed and the ice crystals can grow to a relatively large average size. These large crystals can damage the texture of biological materials and can be responsible for undesirable effects like an increase of "drip" on thawing. As the cooling rate increases, the number of crystallization centers grows, while the average size of ice crystals becomes smaller. This way, texture and flavour of the products are, in general, better preserved. Besides, faster freezing methods lower the temperature of the products quickly to values where the rate of moisture evaporation from the external surface is small and adverse physico-chemical changes are negligible. As a consequence, rapid freezing yields, usually, higher quality and lower weight losses than slow freezing. However, an exceedingly high rate of freezing may also have negative effects, since it can lead to a pronounced toughening of muscle tissues and to the build up of excessive internal stresses, causing breaking and shattering [1-7].

Because of their influence on the quality of the final product, freezing times and rates of cooling are among the most important factors to be accounted for in the design of industrial freezing equipment. Many analytical procedures are available for estimating heat conduction that occurs during phase change of food materials and excellent reviews of approximate prediction methods can be found in the literature (see, for example [8, 9]). However, prediction formulae are limited to geometrically regular shapes and/or to particular freezing processes. Instead, numerical methods can always be relied upon to model any freezing process of technical interest in a reasonably accurate and very cost effective way [10]. Besides, numerical methods are being increasingly used, as an alternative to experiments, in the production of data for the development of predictive formulae through regression analyses [11-15].

In recent years, most numerical calculations of freezing processes in foodstuffs have been based on the definition of an equivalent heat capacity, which includes the latent heat of fusion. With this use, temperature distributions in the products can be obtained immediately by solving the resulting nonlinear heat conduction equation in a single homogeneous domain [16-19]. Thermal properties utilized in the mathematical model can be estimated by means of simple formulae, which take into account the physical aspects of freezing processes [20-23]. This way, any reliable code for the analysis of nonlinear heat conduction can be used for the numerical solution of phase change problems in foodstuffs. However, the numerical solutions described in this paper are formulated in terms of finite element discretizations, both for the sake of concreteness and because of the Authors' familiarity with these methods.

GOVERNING EQUATION AND BOUNDARY CONDITIONS

During freezing of water or solidification of pure metals, the phase change region consists of distinct solid and liquid phases, separated by a smooth front and the phase transformation occurs at a constant temperature. In such a case, the system can be characterized by the position of the interface and "front tracking" methods can be considered in the numerical solution [24, 25]. However, when the solid/liquid interface is not distinct, as it happens with food products, front tracking becomes computationally expensive, if not impossible. In such instances, the characterization of the phase change is best achieved by alternative approaches, which can be classified as "fixed grid" methods [24]. The essential feature of these methods is that the latent heat evolution is accounted for in the governing equation instead of being introduced as a boundary condition. Consequently, the numerical solution can be carried out on a space grid that remains fixed throughout the calculations, without having to worry about the position of the freezing front. The choice of one particular fixed grid method depends on the personal preference of the investigator, but the physics of the process must also be considered.

The physics of freezing processes in foods is well known. When foods are cooled at temperatures below freezing, ice formation occurs, since water is the major component of food materials. Ice formation starts at an initial freezing temperature, between 0°C and -3°C, which depends on the molar concentration of soluble cell electrolytes and on the texture of biological components. As the temperature is progressively reduced, more and more water is turned into ice and the latent heat of ice formation adds to the sensible heat involved in cooling both ice and the unfrozen part. This leads to large variations in heat capacities, while thermal conductivities too change considerably, mainly because the thermal conductivity coefficient of ice is about four times greater than that of water. For most food materials, the largest part of the freezing process takes place in a temperature interval between -1°C and -10°C, while the largest variation in heat capacity occurs between -1°C and -3°C. However, only at temperatures ranging from -20°C to -40°C, and below, there is no measurable change with temperature in the amount of ice present and the remaining water, if any, can be considered as non-freezable [2].

Thus, for freezing processes in foodstuffs, the physics itself suggests a mathematical model where an equivalent, volumetric heat capacity C, defined as the derivative of enthalpy H with respect to temperature T:

$$C = dH/dT \tag{1}$$

includes the latent heat of solidification. Obviously, in order to yield a single valued function of temperature $C(T)$ and two related variables $H(T)$ and T, this definition assumes that local equilibrium conditions are satisfied. Happily, this is almost invariably the case, since industrial freezing processes are not accompanied, in general, by significant degrees of supercooling or, at least, temperature distributions during these processes do not seem to be significantly influenced by non-equilibrium effects such as supercooling [2, 26, 27].

Starting from definition (1), the governing heat conduction equation can be written as:

$$C\frac{\partial T}{\partial t} = \nabla(k\nabla T) \tag{2}$$

where k is the thermal conductivity. In Eq. (2) both C and k are temperature dependent properties, which can be estimated by means of the techniques illustrated in the next Sections.

The governing equation (2) is identical in form to the basic, non-linear Fourier equation. As a consequence, this approach can be easily incorporated into existing codes for the analysis of nonlinear heat conduction, even if some care must be taken in the evaluation of the equivalent heat capacity. In fact, the equivalent heat capacity, defined by Eq. (1), is highly dependent on both space and time and, as it will be shown in the next Sections, this requires the development of effective means of numerical approximation.

A most complete set of boundary conditions for Eq. (2) is:

$$T = T_b \tag{3}$$

on part of the boundary Γ_1, and

$$k\frac{\partial T}{\partial n} + q + q_c + q_r = 0 \tag{4}$$

on part of the boundary Γ_2. In Eqs. (3) and (4), T_b and q are prescribed values of temperature and heat flux, $\partial T / \partial n$ denotes the derivative in the direction of the outward normal, while the usual relationships can be written for convection and radiation fluxes:

$$q_c = \alpha_c(T - T_c) \tag{5}$$

$$q_r = \frac{\sigma F(T^4 - T_r^4)}{T - T_r}(T - T_r) = \alpha_r(T - T_r) \tag{6}$$

In Eqs. (5) and (6) α_c is the convective heat transfer coefficient, σ is the Stefan-Boltzmann constant, F modifies the equation for perfect radiators to account for the emissivities and relative geometries of the actual bodies, and T_c and T_r are the equilibrium temperatures for which, respectively, no convection or no radiation occurs.

ESTIMATION OF THERMAL PROPERTIES

For many food materials, curves of heat capacity and thermal conductivity vs. temperature can be found in the literature [2] or can be estimated, with great accuracy, on the basis of well known thermodynamic relationships [22, 23]. However, accurate representations of $C(T)$ and $k(T)$ are not really needed, since it has been shown that even rough estimates of thermal properties do not lead to significant errors in the numerical solutions if [20, 21]:

- thermal conductivities are interpolated linearly between "liquid" and "solid" values and
- the "approximate" heat capacities retain enthalpy changes across the freezing interval.

Therefore, for practical purposes, a lower limit to the phase change interval can be defined on the basis of a ratio of ice to total water content of, say, 90%. This way, the phase change is assumed to take place between two well defined temperatures, T_i and T_f while, for reasons of simplicity, heat capacity and thermal conductivity, above and below, are approximated by constant values. In the phase change zone instead, thermal conductivity is interpolated by a straight line, while heat capacity is approximated by a triangle, with the peak value at a temperature T_p closer to T_i than to T_f [20, 21].

With reference to a test substance such as Tylose, a methyl-cellulose gel (23% solid and 77% water in mass) whose properties are well known [28], this procedure leads to the schematization shown in Fig.1. The reliability of this schematization has been evaluated experimentally, by means of several tests on one dimensional Tylose samples. The center and the external surface temperatures of the samples have been recorded during freezing experiments. Then, with reference to measured initial and boundary conditions, temperature distributions have been computed numerically. The agreement between measured and computed time-temperature curves has been of the order of 1.5% of the total temperature change when experimental values of thermal properties have been employed and of the order of 2.5% when approximate values have been used [21].

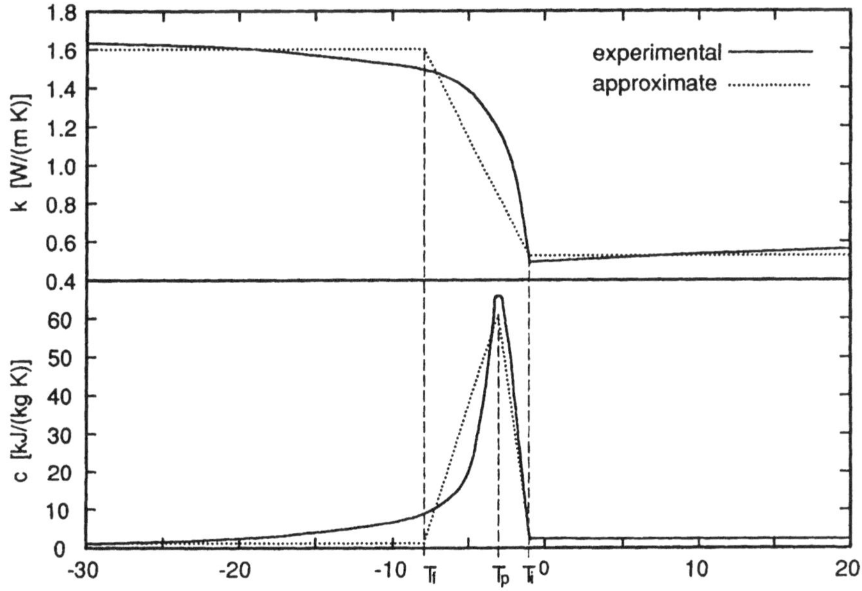

Figure 1. Thermal conductivity k and specific heat capacity of Tylose, a cellulose gel (77% moisture) whose thermal properties are similar to those of lean beef [17].

This approximation procedure has then been extended to substances whose thermal properties are not known from the literature. In fact, for homogeneous food materials, approximate values of thermal properties, above and below freezing, can be obtained from the formulae [21]:

$$c_l = m\,c_{wl} + (1 - m)\,c_d \tag{7}$$

$$k_l = m\,k_{wl} + (1 - m)\,k_d \tag{8}$$

$$c_s = m\,c_{ws} + (1 - m)\,c_d \tag{9}$$

and

$$k_s = m\,k_{ws} + (1 - m)\,k_d \tag{10}$$

where

$$c_d \cong 1.26 \quad \text{kJ/kg K and } k_d \cong 0.26 \quad \text{W/m K}$$

represent specific heat capacity and thermal conductivity of the dry solid fraction, while

$$c_{wl} \cong 4.19 \ \text{kJ/kg K}\,; \qquad k_{wl} \cong 0.59 \ \text{W/m K}$$

and

$$c_{ws} \cong 2.10 \ \text{kJ/kg K}\,; \qquad k_{ws} \cong 2.44 \ \text{W/m K}$$

represent thermal properties of (liquid) water and (solid) ice. Instead, the moisture content m, i.e. the mass fraction of H_2O referred to in Eqs. (7)-(10), can be easily determined experimentally, if not already known from the literature. In fact, we have obviously:

$$m = 1 - m_d \tag{11}$$

where the dry solid fraction m_d is the residual left after completely drying the product.

The latent heat effect, represented by the dashed area in Fig. 1, is estimated as:

$$L = m\,L_w \tag{12}$$

where $L_w \cong 335$ kJ/kg is the heat of fusion/solidification of water. Therefore, we only need simple algebra to calculate the peak value of the specific heat capacity if T_i, T_p and T_f are known in advance. With most food materials, a fairly safe course is to assume:

$$T_i = -1\ °\text{C}\,; \qquad\qquad T_p = -3\ °\text{C}$$

and to estimate T_f as the value that gives the best fit between experimentally determined and computed time-temperature curves [21]. Finally, volumetric heat capacities C can be estimated from their definition:

$$C = \rho c \tag{13}$$

where specific heat capacities c are given by formulae (7) and (9), while densities ρ can be easily measured, if not already known from the literature.

Alternative approaches are also possible, since heat capacity and thermal conductivity, below the initial freezing point, can be expressed directly, as functions of temperature, by means of suitable empirical formulae [11, 15, 29]. However, for many calculations of technical interest, the simpler procedure described here can be used confidently.

For porous materials, the void fraction ε must be taken into account, for example by assuming:

$$k_e \cong (1 - \varepsilon)\, k \tag{14}$$

where k_e is the equivalent thermal conductivity, while the thermal conductivity of the material k is computed from formulae (2) and (4), respectively for temperatures above and below the freezing interval [21]. Instead, for the equivalent heat capacity C_e it seems quite reasonable to assume:

$$C_e = \rho_a c \tag{15}$$

where ρ_a is the apparent bulk density of the samples and c is computed from formulae (7) and (9), and from the knowledge of the latent heat effect L [21].

The reliability of these approximations has also been evaluated experimentally for several food materials, with reference to thermal properties estimated from the parameters reported in Table 1.

Table 1. Values of parameters used for the estimation of thermal properties of food materials.

Food material	m [%]	T_i [°C]	T_p [°C]	T_f [°C]	ρ [kg/m^3]	ε
Wreck fish	80	-1	-3	-9	1006	0
Cod	76	-1	-3	-10	1039	0
Sole	82	-1	-3	-10	994	0
Shrimps	83	-1	-3	-10	1041	0
Shortbread	16	-1	-3	-40	1145	0
Peas	74	-1	-3	-8	890	0.11
Green Beans	89	-1	-3	-8	872	0.13
Spinach	93	-1	-3	-7	954	0.05

In all freezing tests, the agreement between measured and computed time-temperature curves has always been of the order of a few percent of the total temperature change when approximate values of thermal properties have been used [21]. Therefore, the accuracy reached can be considered to be well within the limits imposed for thermal design.

Because of its simplicity and reliability, the approximation procedure described in this Section has gained wide acceptance over the years [15], and it has been used extensively, with minor modifications, by several Authors (see, for example, [11, 30, 31]).

FINITE ELEMENT FORMULATION

The discretized finite element analog of Eq. (2), subjected to boundary conditions (3) and (4), can be written in matrix form as:

$$\mathbf{C}\dot{\mathbf{T}} + \mathbf{K}\mathbf{T} + \mathbf{Q} = 0 \tag{16}$$

where the matrices **C**, **K** and the vector **Q** are strongly dependent on T. In Eq. (16) $\dot{\mathbf{T}} = \partial \mathbf{T}/\partial t$ represents the time derivative and the definitions of the various matrices and vectors can be found in References 18 and 19. On the other hand, the analogy with equations (2)-(4) is apparent.

The set of ordinary differential equations (16), which defines the discretized problem, can be solved using one of the many recurrence schemes for time integration. If it is assumed that the temperature varies linearly in the small time interval $2\Delta t$, between time instants $(n-1)\Delta t$ and $(n+1)\Delta t$, Eq. (6) can be approximated as:

$$\mathbf{C}^n(\mathbf{T}^{n+1} - \mathbf{T}^{n-1})/(2\Delta t) + \mathbf{K}^n(\mathbf{T}^{n+1} + \mathbf{T}^n + \mathbf{T}^{n-1})/3 + \mathbf{Q}^n = 0 \tag{17}$$

where superscripts $n-1$, n and $n+1$ refer to time instants $(n-1)\Delta t$, $n\Delta t$ and $(n-1)\Delta t$. This, after some algebra, results in the following recurrence formula for final integration:

$$\begin{aligned} \mathbf{T}^{n+1} = & -\left[\mathbf{C}^n/(2\Delta t) + \mathbf{K}^n/3\right]^{-1} \cdot \\ & \cdot \left[\mathbf{K}^n\left(\mathbf{T}^n + \mathbf{T}^{n-1}\right)/3 + \mathbf{Q}^n - \mathbf{C}^n\mathbf{T}^{n-1}/(2\Delta t)\right] \end{aligned} \tag{18}$$

where iterations are avoided since only central values of **K**, **C** and **Q** occur. This scheme is apparently not self starting but, as we shall always refer to known initial conditions, two starting values will be certainly available.

Algorithm (18) has been found to be unconditionally stable [16, 18]. Oscillations arising from the noise inherent in the numerical approximations can be greatly reduced if, at each time step after the first one, instead of letting simply:

$$\mathbf{T}^{n-1} \Leftarrow \mathbf{T}^n \tag{19}$$

the vector $\mathbf{T}^{n-1}$ is redefined as:

$$\mathbf{T}^{n-1} \Leftarrow \left(\mathbf{T}^{n+1} + \mathbf{T}^{n} + \mathbf{T}^{n-1}\right)/3 \tag{20}$$

before starting again the calculations [32].

The program for implementing algorithm (18) follows the usual pattern [18, 19], even if some care has to be taken in the evaluation of matrices $\mathbf{K}^n$ and $\mathbf{C}^n$. Past experience has shown that a direct evaluation of thermal conductivities at the time level n, in correspondence with the integration points, always yields a satisfactory approximation for the matrix $\mathbf{K}^n$. Instead, the direct use of heat capacities, evaluated at integration points, can lead to errors, if the chosen time step allows the latent heat effect to skip, partially or completely, a few integration points.

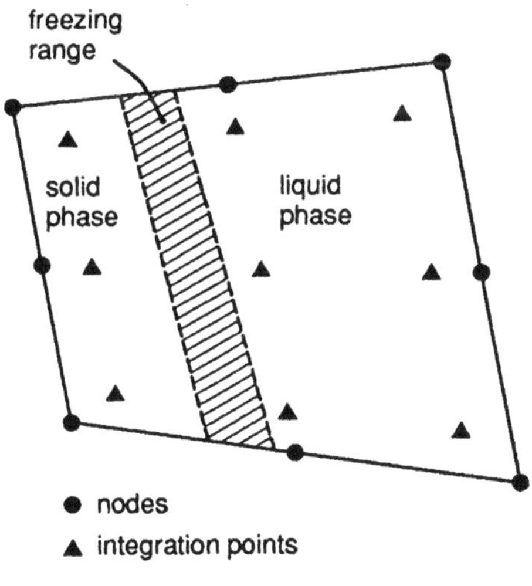

Figure 2. A possible location of the freezing zone in a typical element [13].

These errors can be easily explained by referring to phase change processes like the one represented in Fig. 2, where the freezing zone falls between integration points. In such situations, if no precaution is taken, the contributions of the element to the capacitance matrix would be underestimated. In fact, in the situation of Fig. 2, the heat capacity at the integration points would assume values corresponding either to the "solid" or to the "liquid" phase, even if a freezing zone is passing through the element. As a general rule, decreasing the time step and the size of the elements will reduce the risk of missing the latent heat effect. On the other hand, this might require the use of too many steps and elements, even if such a situation may be alleviated by adopting, for equivalent heat capacities, suitable numerical approximations falling in one of two possible general classes defined below [24, 33, 34].

In the first class we have approximations which lead to volumetric averages of heat capacities over each element. According to these "volumetric" approximations, the equivalent heat capacity is expressed, at the time level n, in terms of enthalpy and temperature gradients as [19]:

$$C^n = \frac{\nabla H^n \cdot \nabla T^n}{\nabla T^n \cdot \nabla T^n} \tag{21}$$

or in the form [35]:

$$C^n = \sqrt{\frac{\nabla H^n \cdot \nabla H^n}{\nabla T^n \cdot \nabla T^n}} \tag{22}$$

which yields the same estimates of C^n whenever an exact evaluation of ∇H^n is available. Obviously, if temperature is uniform within an element, recourse is made to the direct evaluation heat capacities.

In the second class, we have approximations based on temporal averaging:

$$C^n = \left(\frac{dH}{dt}\right)^n = \left(\frac{\partial H}{\partial t} / \frac{\partial T}{\partial t}\right)^n \cong \frac{H^n - H^{n-1}}{T^n - T^{n-1}} \tag{23}$$

where heat capacities are obtained directly at integration points by means of a backward difference representation [36]. These "temporal" evaluations are explicit and still allow the use of Eq. (18) without iteration, even if formula (23) represents a departure from the central representations of the three-time-level method of integration. Of course, if the temperature is constant, i.e. $T^{n+1} = T^n$, a direct evaluation of heat capacities is required.

In general, use of the finite element method, in connection with one of the numerical approximations (21) to (23), will lead to efficient and accurate solutions for freezing processes in foodstuffs, even with reasonably large time steps and element sizes (see, for example, [19, 29, 37-39]).

EXAMPLES OF APPLICATION

In the past twenty years, numerical methods have been used extensively in the simulation of freezing processes in foodstuffs. The results obtained have been favourably compared with experimental data and have provided reliable information for the design of industrial equipment. Therefore, we will only need to present here a few demonstrative examples to illustrate the accuracy of the predictions made and the capabilities of the finite element method in thermal design.

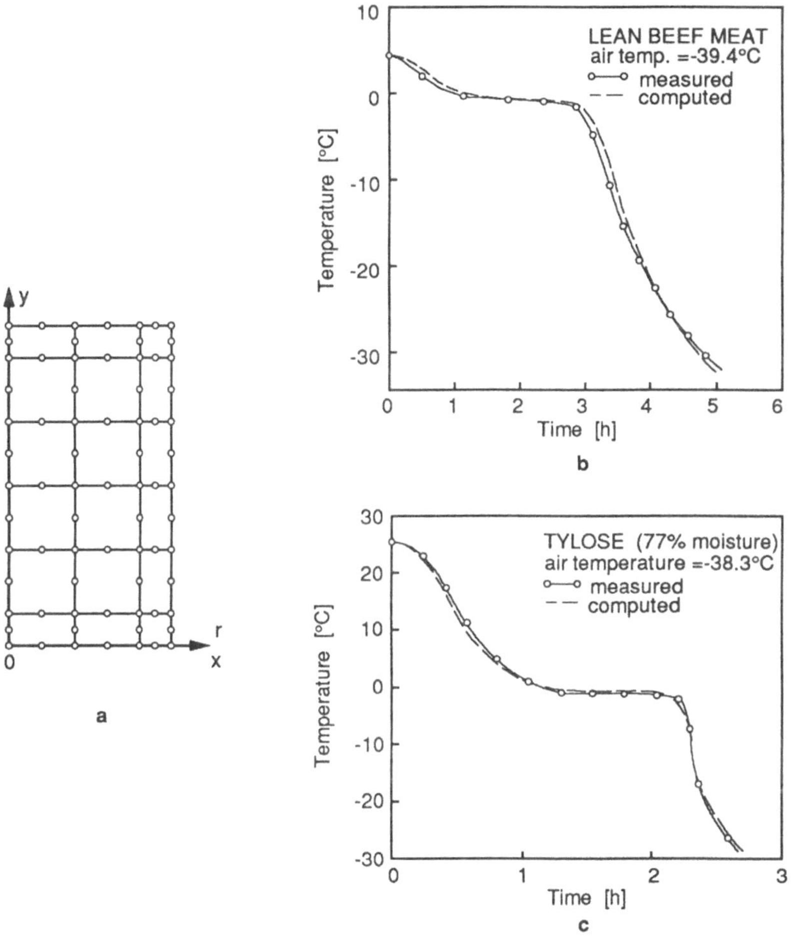

Figure 3. Freezing test with lean beef meat and with a cellulose gel 77% moisture (Tylose).

a) Isoparametric elements and nodal points used for rectangular (x, y) and cylindrical (r, y) geometries. Because of the symmetry only one-fourth of the domain is considered.

b) Measured and computed temperature histories of rectangularly trimmed lean beef. The side dimensions of meat are 7.7 cm and 7.0 cm respectively. The average dimensionless value of the convective coefficient is estimated as $Bi = hl/k = 0.361$ from the experimental runs.

c) Measured and computed temperature histories of cylindrically shaped cellulose gel 77% moisture (Tylose), whose radius and height are respectively equal to 3.3 cm and 7.6 cm. The average dimensionless value of the convective coefficient is estimated as $Bi = hr/k = 0.167$ from the experimental runs.

Comparison with experiments

Samples were prepared from lean beef meat and cellulose gel (Tylose), since the physical properties of these materials are available in the published literature [2, 15, 28]. The samples, shaped as rectangular bricks and finite cylinders, were subjected to freezing processes in a low-temperature, forced-air chamber, starting from a uniform initial temperature. Two opposite surfaces of the rectangular bricks were insulated to obtain two-dimensional thermal fields, while axial symmetry was taken into account to deal with cylindrical geometries. The convective heat transfer coefficients were obtained by finding the values which gave the best fit between measured and computed temperatures [29].

The finite element mesh used is illustrated in Fig. 3a, while comparisons between experimentally obtained and computed time-temperature curves are reported in Fig. 3b and in Fig. 3c, respectively for rectangular and cylindrical shapes [29].

Industrial freezing processes

The finite element method has been applied to the computation of temperature distributions in foodstuffs of irregular shape during industrial freezing processes. Freezing of a beef side has thus been examined. To avoid toughening of the meat, the air temperature is maintained at above freezing values during the first period of chilling. Afterwards, the air temperature is suddenly reduced to the final value [19, 37].

A cross section of the hind leg was considered for the finite element analysis and the mesh utilized is illustrated in Fig. 4a. Thermal properties of beef meat and bone are known from the literature [2, 3], while convective boundary conditions, with $T_c = -40°C$ and $\alpha_c = 18$ W/m^{2}K, were assumed. The initial temperature value was taken to be uniform and equal to 1°C. Isotherm fields, computed at different times, are reported in Fig. 4b to 4d.

CONCLUSIONS

Numerical modelling of freezing processes in foodstuffs is a viable area of research and of industrial applications. Many successful approaches are available for the numerical modelling and the choice of any one model depends on the problem at hand and on the preference of the investigator. This paper has attempted to describe the equivalent heat capacity approach and a widely used calculation procedure, based on the finite element method.

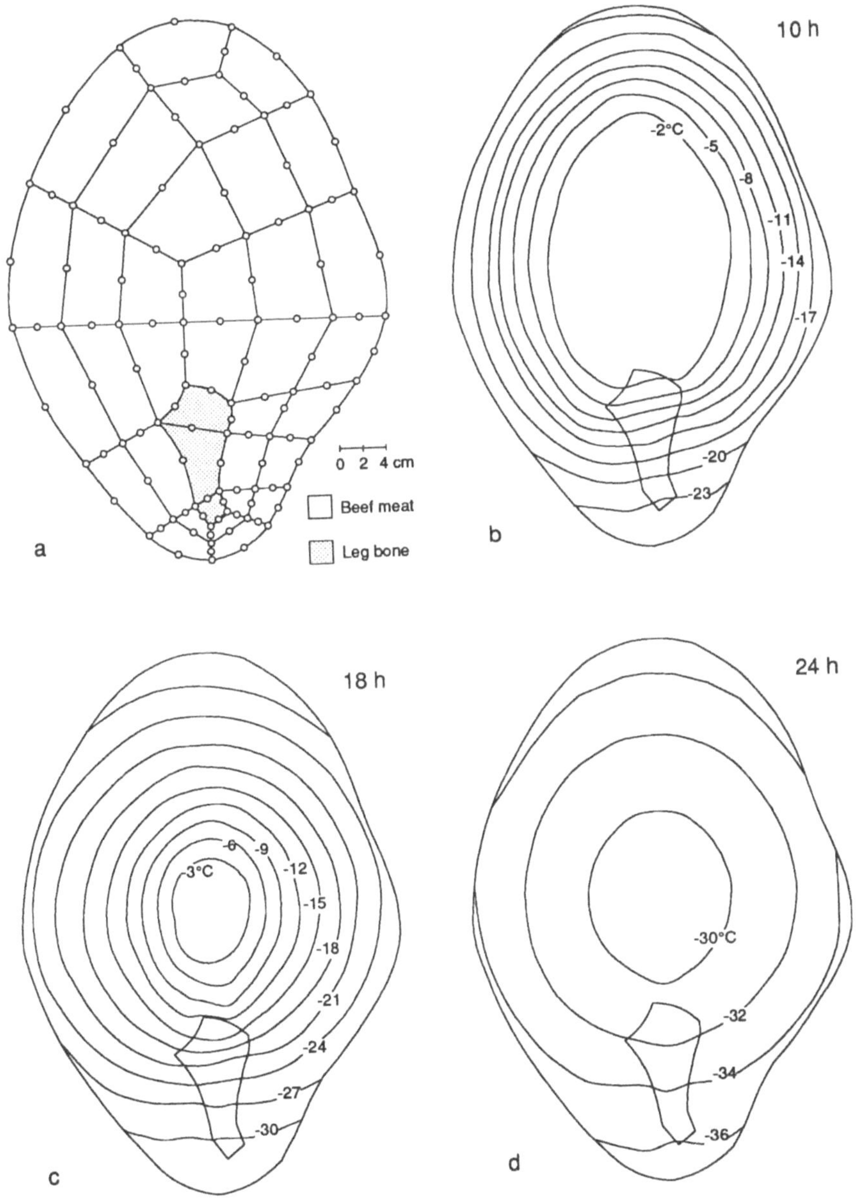

Figure 4. Air blast freezing of a hind leg of beef. Convective boundary conditions with $T_c = -40°C$ and $\alpha_c = 18\ W/m^2 K$ are assumed, together with a uniform initial temperature equal to 1°C.
a) Finite element mesh.
b) Isotherm field after 10 hours.
c) Isotherm field after 18 hours.
d) Isotherm field after 24 hours.

NOMENCLATURE

Bi dimensionless convective coefficient (Biot number)
c specific heat capacity [J/kg K]
C heat capacity [J/m^3 K]
$\mathbf{C}$ capacity matrix, defined in Eq. (16)
H enthalpy per unit volume [J/m^3]
k thermal conductivity [W/m K]
$\mathbf{K}$ conductivity matrix, defined in Eq. (16)
l reference length [m]
L latent heat effect [J/kg]
m moisture mass fraction [$kg_{water}/kg_{material}$]
n outward normal direction
q heat flux [W/m^2]
$\mathbf{Q}$ heat load vector, defined in Eq. (16)
r spatial radial coordinate [m]
t time [s]
T temperature [°C]
$\mathbf{T}$ temperature vector, defined in Eq. (16)
$\dot{\mathbf{T}}$ vector of temperature rate, defined in Eq. (16)
x spatial cartesian coordinate [m]
y cartesian and cylindrical spatial coordinate [m]
α heat transfer coefficient [W/m^2 K]
ε void fraction
ρ density [kg/m^3]
σ Stefan-Boltzmann constant [W/m^2 K^4]

Subscripts and superscripts

a apparent
b boundary
c convective
d referred to the dry fraction
e equivalent
f final
i initial
l referred to the liquid phase
n at the time instant $n\Delta t$
p peak
r radiative
s referred to the solid phase
w referred to water

REFERENCES

1. ASHRAE Handbook - Refrigeration Systems and Applications, Chapter 9, *Commercial freezing methods*, pp. 7-8, ASHRAE, Atlanta, 1986.
2. IIR, *Recommendations for the Processing and Handling of Frozen Foods*, 2nd ed., Annexe au Bulletin IIF/IIR, Paris, 1972.
3. Fleming, A.K. *Physical Aspects of Meat Cooling*, Annexe au Bull. IIF/IIR 1970-3, , pp. 151-160, Paris, 1970.
4. Bonacina, C. and Comini, G. *Mass and Heat Transfer During Cooling, Freezing and Cold Storage of Foodstuffs*, Bulletin IIF/IIR, Vol. 51, pp. 1539-1558, 1971.
5. Singh, R.P. and Heldman, D.R. *Introduction to Food Engineering*, Academic Press, New York, 1984.
6. Ogawa, Y. *Studies on Gape and Heave of Foodstuffs Due to Internal Pressure During Freezing: Freezing of Tuna by Brine Spray Systems*, International Journal of Refrigeration, Vol. 12, pp. 95-106, 1989.
7. Hanenian, R., Mittal, G.S. and Usborne, W.R. *Effects of Pre- Chilling, Freezing Rate and Storage Time on Beef Patty Quality*, Journal of Food Science, Vol. 54, pp. 532-535, 1989.
8. Cleland, A.C and Earle, R.L. *Assessment of Freezing Time Prediction Formulae*, Journal of Food Science, Vol. 49, pp. 1034- 1042, 1984.
9. Le Blanc, D.I., Kok, R. and Timbers, G.E. *Freezing of a Parallelipiped Food Product. Part 2: Comparison of Experimental and Calculated Results*, International Journal of Refrigeration, Vol. 13, pp. 379-392, 1990.
10. Comini, G. and Bonacina, C. *Application of Computer Codes to Phase-Change Problems in Food Engineering*, Annexe au Bulletin IIF/IIR 1974-3, pp. 15-27, 1974.
11. Hayakawa, K.I., Nonino, C. and Succar, J. *Two Dimensional Heat Conduction in Food Undergoing Freezing: Predicting Freezing Time of Rectangular or Finitely Cylindrical Food*, Journal of Food Science, Vol. 48, pp.1841-1848, 1983.
12. Pham, Q.T. *Simplified Equation for Predicting the Freezing Time of Foodstuffs*, Journal of Food Technology, Vol. 21, pp. 209- 219, 1986.
13. Cleland, D.J., Cleland, A.C. and Earle, R.L. *Prediction of Freezing and Thawing Times for Multi-Dimensional Shapes by Simple Formulae. Part 1: Regular Shapes*, International Journal of Refrigeration, Vol. 10, pp. 156-164, 1987.
14. Cleland, D.J., Cleland, A.C. and Earle, R.L. *Prediction of Freezing and Thawing Times for Multi-Dimensional Shapes by Simple Formulae. Part 2: Irregular Shapes*, International Journal of Refrigeration, Vol. 10, pp. 234-240, 1987.
15. ASHRAE Handbook - Fundamentals, Chapter 29, *Cooling and Freezing Times of Foods*, pp. 1-16, ASHRAE, Atlanta, 1989.
16. Bonacina, C. and Comini, G. *On the Solution of the Nonlinear Heat Conduction Equation by Numerical Methods*, International Journal of Heat and Mass Transfer, Vol. 16, pp. 581-589, 1973.

17. Bonacina, C., Comini, G., Fasano, A. and Primicerio, M. *Numerical Solution of Phase-Change Problems*, International Journal of Heat and Mass Transfer, Vol. 16, pp. 1825-1832,1973.
18. Comini, G., Del Giudice, S., Lewis, R.W. and Zienkiewicz, O.C. *Finite Element Solution of Nonlinear Heat Conduction Problems With Special Reference to Phase Change*, International Journal for Numerical Methods in Engineering, Vol. 8, pp. 613-624, 1974.
19. Rebellato, L., Del Giudice, S. and Comini, G. *Finite Element Analysis of Freezing Processes in Foodstuffs*, Journal of Food Science, Vol. 43, pp. 239-243, 1978.
20. Bonacina, C., Comini, G., Fasano, A. and Primicerio, M. *On the Estimation of Thermophysical Properties in Nonlinear Heat Conduction Problems*, International Journal of Heat and Mass Transfer, Vol. 17, pp. 861-867,1974.
21. Comini, G., Bonacina, C. and Barina, S. *Thermal Properties of Foodstuffs*, Annexe au Bulletin IIF/IIF 1974-3, pp. 163-172, 1974.
22. Heldman, D.R. *Predicting the Relationship Between Unfrozen Water Fraction and Temperature During Food Freezing Using Freezing Point Depression*, ASAE Transactions, Vol. 17, pp.63-66, 1974.
23. Schwartzberg, H.G. *Effective Heat Capacities for the Freezing and Thawing of Foods*, Annexe au Bulletin IIF/IIR 1977-1, pp. 303- 310, 1977.
24. Voller, V.R. and Swaminathan, C.R. *Fixed Grid Techniques for Phase Change Problems: a Review*, International Journal for Numerical Methods in Engineering, Vol. 30, pp. 875-898, 1990.
25. Crank, J. *Free and Moving Boundary Problems*, Clarendon Press, Oxford, 1984.
26. Pham, Q.T. *Effect of Supercooling on Freezing Time Due to Dendritic Growth of Ice Crystals*, International Journal Of Refrigeration, Vol. 12, pp. 295-300, 1989.
27. Rubinsky, B. *The Energy Equation for Freezing of Biological Tissue*, Transactions of the ASME, Series C, Journal of Heat Transfer, Vol. 111, pp. 988-997, 1989.
28. Riedel, L. *A Test Substance for Freezing Experiments* (in German), Kältetechnik, Vol.12, pp. 222-225, 1960.
29. Hayakawa, K.I., Nonino, C., Succar, J., Comini, G. and Del Giudice, S. *Two Dimensional Heat Conduction in Food Undergoing Freezing: Development of Computerized Model*, Journal of Food Science, Vol. 48, pp.1849-1853, 1983.
30. Ansari, F.A., Charan, V. and Varma, H.K. *Measurement of Thermophysical Properties and Analysis of Heat Transfer During Freezing of Slab-Shaped Food Commodities*, International Journal of Refrigeration, Vol. 8, pp. 85-90, 1985.
31. Le Blanc, D.I., Kok, R. and Timbers, G.E. *Freezing of a Parallelipiped Food Product. Part 1: Experimental Determination*, International Journal of Refrigeration, Vol. 13, pp. 371-378, 1990.

32. Wood, W.L. and Lewis, R.W. *A Comparison of Time Marching Schemes for the Transient Heat Conduction Equation*, International Journal for Numerical Methods in Engineering, Vol. 9, pp. 679-689 ,1975.
33. Comini, G., Del Giudice, S. and Saro, O. *A Conservative Algorithm for Multidimensional Conduction Phase Change*, International Journal for Numerical Methods in Engineering, Vol. 30, pp. 697-709, 1990.
34. Comini, G., Nonino, C. and Saro, O. *Performance of Enthalpy- Based Algorithms for Isothermal Phase Change*, in Advanced Computational Methods in Heat Transfer, Vol. 3: Phase Change and Combustion Simulation (Eds. Wrobel, L.C., Brebbia, C.A. and Nowak, A.J.), pp. 3-13, Proceedings of the First International Conference, Portsmouth, UK, 1990. Springer-Verlag, Berlin, 1990.
35. Lemmon, E.C. *Phase Change Techniques for Finite Element Codes*, in Numerical Methods in Thermal Problems (Eds. Lewis, R.W. and Morgan, K.), pp. 149-158, Proceedings of the First International Conference, Swansea, UK, 1979.
36. Morgan, K., Lewis, R.W. and Zienkiewicz, O.C. *An Improved Algorithm for Heat Conduction Problems with Phase Change*, International Journal for Numerical Methods in Engineering, Vol. 12, pp. 1191-1195, 1978.
37. Comini, G., Del Giudice, S., Strada, M. and Rebellato, G. *The Finite Element Method in Refrigeration Engineering*, International Journal of Refrigeration, Vol. 1, pp. 113-119, 1978.
38. Cleland, D.J, Cleland, A.C., Earle, R.L. and Byrne, S.J. *Prediction of Rates of Freezing, Thawing and Cooling in Solids od Arbitrary Shape Using the Finite Element Method*, International Journal of Refrigeration, Vol. 7, pp. 6-13, 1984.
39. Cleland, D.J, Cleland, A.C., Earle, R.L. and Byrne, S.J. *Prediction of Freezing and Thawing Times for Multi-Dimensional Shapes by Numerical Methods*, International Journal of Refrigeration, Vol. 10, pp. 32-39, 1987.

Numerical Solution of a Stefan Problem Arising in the Thermoelectrical Modelling of Aluminium Electrolytic Cells

A. Bermúdez, M.C. Muñiz, P. Quintela

Department of Applied Mathematics, University of Santiago de Compostela, 15706 Santiago de Compostela, Spain

ABSTRACT

In this paper we solve a free boundary problem related to the numerical simulation of the thermoelectrical behaviour of an aluminium electrolytic cell. We introduce a nonlinear weak formulation for which we prove existence of solution. A numerical method is given and numerical results for real electrolytic cells are presented.

INTRODUCTION

In industry , aluminium is produced by electrolytic reduction of alumina (Al_2O_3) dissolved in a bath based on molten cryolite (Na_3AlF_6) (see for instance Grjotheim and Kvande [1], Grjotheim, Krohn, Malinovsky, Matiasovsky and Thonstad [2]). The metal is formed around 960°C at the molted aluminium cathode and oxygen is discharged at the carbon anode where reacts to form carbon dioxide. In addition some of the metal is dissolved in the electrolyte and transported so that it can be reoxidized by the CO_2 leading to a decrease in efficiency.

This process is called Hall-Heróult after its inventors C.M. Hall (USA) and P.L.T. Héroult (France) who, in 1886, applied for the registration of a patent.

The cell voltage is usually between 4 and 5 volts and the amperage is in the range from 100 to 300 kA. The energy consumed is about 15 Kwh per Kg Al. Thus the energy efficiency of the process is only around 40% and a large amount of the energy input is lost in the form of heat to the surroundings.

An aluminium electrolytic cell consists of a rectangular steel shell thermally insulated with refractory materials. Inside of this there is an lining of prebaked carbon cathode blocs with inner steel current collector bars. On the inner walls of the cell a protective coating of electrolyte, the so called ledge, is frozen.

On the other hand magnetic effects produce the bath and the liquid metal move. This causes difference in level at the interface of the liquid an important factor that influences cell operation efficiency. One of the objectives of cell sidewall design is to promote the formation of a good ledge profile to give stable, efficient cell operation and long sidewall life. This ledge plays an important role to help keep the metal still and thereby maintain stable cell operation. Furthermore it ensures that the permanent sidewall material is not exposed to erosion by bath and metal pad circulation and prevents infiltrations to the cathodic block.

Numerical simulation of the behaviour of the cell is being extensively used to aid the design of new ones or the modification of existing ones in order to increase the sidewall life and cell operation efficiency (see Bruggeman and Danka [3], Arita, Urata and Ikeuchi [4], Utne [5] , Sulmot and Hudault [6]).

In these papers the form of the ledge is fitted along an iterative procedure. This makesit neccessary to modify the mesh at each iteration leading to difficulties in computer programming.

In the present paper we introduce a finite element method in a fixed domain. As a first step, this paper only deals with the thermoelectrical problem. Thus electromagnetic effects are not considered.

The model is stated in section 2. It consists of a system of partial differential equations taking place in a unknown domain. Therefore it is a free boundary problem, more precisely, the thermal submodel is a Stefan problem with source at the free boundary very similar to the so called ablation problem (Saguez [7], Fasano and Primicerio [8], Elliott and Ockendon [9], Bossavit, Damlamiam and Frémond [10], Hoffmann and Sprekels [11] and Wrobel, Brebbia and Nowak [12]).

In section 3 we propose a weak formulation in a fixed domain . The objective is to use numerical discretizations by finite element methods associated to a fixed mesh thus avoiding front tracking algorithms. We prove existence of solution of this weak formulation by regularization.

Section 4 is devoted to numerical solution. The above mentioned weak formulation is dicretized by finite elements of degree one. Then an iterative algorithm is proposed for solving the discretized problem.

Numerical results are presented in section 5. They correspond to an academic example with an analytical solution which is used to validate the proposed methodology and to a real aluminium reduction cell.

2. STATEMENT OF THE MODEL

The basic equation used to calculate the electric potential and current in the cell is the continuity equation for electric charge:

$$\nabla . J(x) = 0 \tag{1}$$

where J is the current density vector which is related to the potential by

$$J(x) = \sigma(x, T) \nabla V(x) \tag{2}$$

σ being the electrical conductivity supposed to be dependent on temperature.

Heat production by Joule effect causes increasing of temperature which is governed by the nonlinear partial differential equation

$$-\nabla . (k(x, T) \nabla T(x)) = \sigma(x, T) |\nabla V(x)|^2 \tag{3}$$

where k denotes the thermal conductivity also depending on temperature.

In addition we have boundary conditions on the surface of the cell. They consists of prescribing intensity for the electrical submodel and radiative and convective heat exchange with surroundings for the thermal one.

On the surface of the frozen bath (the ledge), which is a free boundary, two thermal boundary conditions are given: temperature is fixed to the eutectic temperature of the bath and heat flux is prescribed.

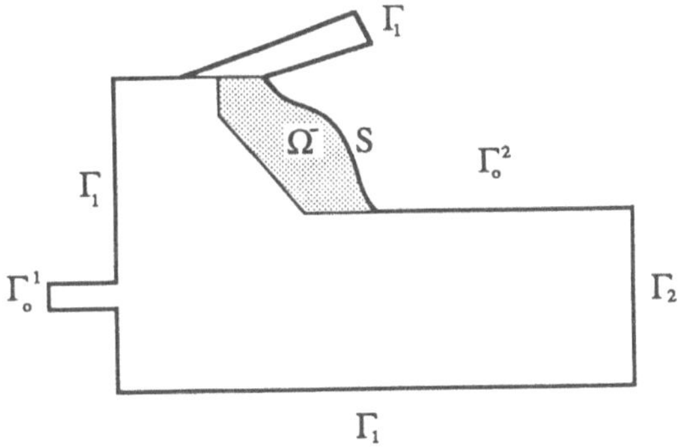

Figure 1

To be more precise we refer to figure 1 where a half cross section of the cathode of an aluminium reduction cell is shown. Let n be an unit outward normal vector on the boundary. Then boundary conditions are the following:

Electric problem

$$J.n = 0 \quad on \quad \Gamma_1 \cup \Gamma_2 \cup S \tag{4}$$

$$J.n = j \quad on \quad \Gamma_0^1 \tag{5}$$

$$V = 0 \quad on \quad \Gamma_0^2 \tag{6}$$

Thermal problem

$$k\frac{\partial T}{\partial n} = \alpha(T_c - T) + \beta(T_r^4 - T^4) \quad on \quad \Gamma_1 \tag{7}$$

$$k\frac{\partial T}{\partial n} = 0 \quad on \quad \Gamma_2 \tag{8}$$

$$T = T_b \quad on \quad \Gamma_0^1 \tag{9}$$

$$T = T_{op} \quad on \quad \Gamma_0^2 \tag{10}$$

$$T = T_{eu} \quad on \quad S \tag{11}$$

$$k\frac{\partial T}{\partial n} = \varphi \ on \ S \tag{12}$$

where the following notations have been used:

j : is the density current.

T_c : convective temperature of surroundings.

T_r : radiative temperature of surroundings.

T_b : temperature of the cathodic bar.

T_{eu} : eutectic temperature of the bath.

T_{op} : operation temperature.

φ : Heat flux through the free boundary.

3. MATHEMATICAL ANALYSIS OF THE THERMAL PROBLEM

In this section we study a weak formulation of the thermal problem in a fixed domain. More precisely, as a first step, we consider here the free boundary problem taking place in the region occupied by the metal and the bath shown in figure 2.

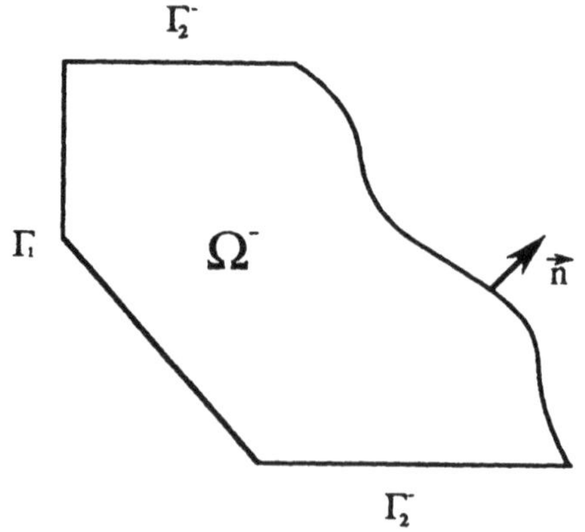

Figure 2

Let Ω^- be the frozen bath (ledge). Denote by $\Gamma_1 \cup \Gamma_2^-$ the part of its boundary different from the free boundary S (see figure 2). Consider the following heat transfer problem

$$-\nabla.(k(x,T))\nabla T) = 0 \quad in \quad \Omega^- \tag{13}$$

$$k\frac{\partial T}{\partial n} + \alpha(T - T_e) = 0 \quad on \quad \Gamma_1 \tag{14}$$

$$k\frac{\partial T}{\partial n} = 0 \quad on \quad \Gamma_2^- \tag{15}$$

$$T = T_{eu} \quad on \quad S \tag{16}$$

$$k\frac{\partial T}{\partial n} = hn_1 \quad on \quad S \tag{17}$$

where h is a nonnegative function that only depends on x_2.

The presence of the x_1 component of the normal vector in the boundary condition (17) is reasonable because the gradient in the liquid phase is almost parallel to the x_1 direction. Function h depends on the composition of the electrolyte, temperature of operation, etc. In practice it should be identified for each reduction cell.

For numerical purposes as well as for theoretical ones it is interesting to embed the problem (13) -(17), taking place in a unknown domain Ω^-, into another one defined in a fixed domain.

For that consider the set $\Omega = \Omega^- \cup S \cup \Omega^0$ given in figure 3.

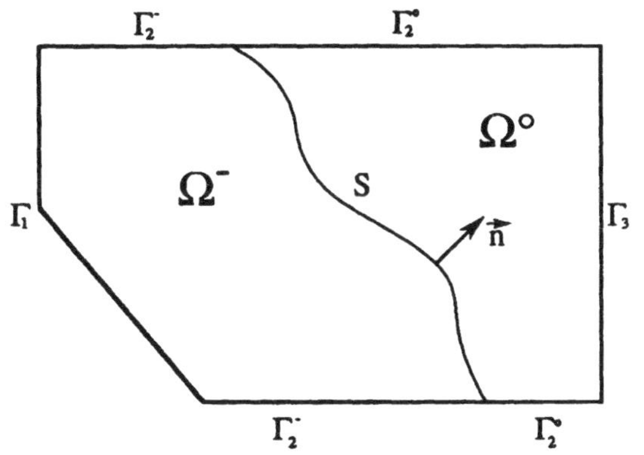

Figure 3

Let Γ be the boundary of Ω. Then $\Gamma = \Gamma_1 \cup \Gamma_2 \cup \Gamma_3$ with $\Gamma_2 = \Gamma_2^- \cup \Gamma_2^0$.

We set the following problem:

Find (T,q) defined in Ω such that

$$-\nabla.(k(x)\nabla T(x)) = 0 \quad in \quad \Omega^- \tag{18}$$

$$T < T_{eu}, \quad q = 0 \quad in \quad \Omega^- \tag{19}$$

$$-h\frac{\partial q}{\partial x_1} = 0 \quad in \quad \Omega^0 \tag{20}$$

$$T = T_{eu}, \quad 0 \leq q \leq 1 \quad in \quad \Omega^0 \tag{21}$$

$$T = T_{eu} \quad on \quad S \tag{22}$$

$$k\frac{\partial T}{\partial n} = hqn_1 \quad on \quad S \tag{23}$$

$$k\frac{\partial T}{\partial n} + \alpha(T - T_c) = 0 \quad on \quad \Gamma_1 \tag{24}$$

$$k\frac{\partial T}{\partial n} = 0 \quad on \quad \Gamma_2 \tag{25}$$

$$hq = h \quad on \quad \Gamma_3 \tag{26}$$

where in (23) q represents the trace of $q_{/\Omega^0}$ on S. Notice that if Ω^0 is connected and $\Gamma_3 \subset \partial\Omega^0$ (20) and (26) imply hq = h in Ω^0 and then (17) follows.

We do the following assumptions on the data appearing in (18)- (26)

$(H1)$ $\quad k(x) \in L^\infty(\Omega)$ *and there exists* $k_0 > 0$ *such that* $k(x) \geq k_0$ *a.e. in* Ω.

$(H2)$ $\quad T_{eu}$ *is a positive constant.*

$(H3)$ $\quad h \in L^2(o, a)$ *and* $f \in L^2(\Omega)$ *are nonegative,*

where a is the height of the domain.

$(H4)$ $\quad \alpha \in L^\infty(\Gamma_1)$, $T_c \in L^2(\Gamma_1)$ *with* $\alpha \geq \alpha_0 > 0$, $0 \leq T_c \leq T_{eu}$ *a.e. on* Γ_1.

$(H5)$ $\quad -h(x_2)n_1(x_2) \leq \alpha(T_{eu} - T_c(x_2))$ *a.e.* $x_2 \in [0, a]$,

where n_1 denotes the first component of the outward unit normal vector to Γ_1. Assume that $-1 \leq n_1 \leq 0$.

We introduce the following variational problem:

$$T \in H^1(\Omega) \quad and \quad q \in L^\infty(\Omega) \tag{27}$$

such that

$$\int_\Omega k(x)\nabla T.\nabla z dx + \int_\Omega hq\frac{\partial z}{\partial x_1}dx + \int_{\Gamma_1} \alpha(T - T_c)zd\Gamma =$$

$$\int_{\Gamma_3} hzd\Gamma, \forall z \in H^1(\Omega) \tag{28}$$

$$q \in H(T - T_{eu}), \tag{29}$$

where H denotes the multivalued Heaviside function given by

$$H(s) = \begin{cases} 0 & \text{if} \quad s \leq 0 \\ [0,1] & \text{if} \quad s = 0 \\ 1 & \text{if} \quad s > 0 \end{cases} \tag{30}$$

Let the sets Ω^-, S, Ω^0, Γ_1^- and Γ_1^0 be defined by

$$\Omega^- = \{x \in \Omega : T(x) < T_{eu}\} \tag{31}$$

$$S = \partial\Omega^- \cap \Omega \tag{32}$$

$$\Omega^0 = \Omega/(\Omega^- \cup S) \tag{33}$$

$$\Gamma_2^- = \Gamma_2 \cap \partial\Omega^-, \Gamma_2^0 = \Gamma_2 \cap \partial\Omega^0 \tag{34}$$

Notice that q is equal to zero in Ω^- from (29).

We have the following:

Proposition 1 .- If there exists a regular solution of problem P such that T $\leq T_{eu}$ a.e. in Ω and if Ω^- and Ω^0 are open sets with a Lipschitz boundary then T and q satisfy (18)-(26).

Proof : Taking z $\in D(\Omega^-)$ (resp. z $\in D(\Omega^0)$) we get equation (18) (resp. (20)) because in Ω^0 T is constant and equal to T_{eu}.

Let z be any function in $H^1(\Omega)$. Multiplying (18) and (20) by z, integrating in Ω^- and Ω^0 respectively, using a Green formula and taking into account that $T = T_{eu}$ in Ω^0 we obtain:

$$-\int_{\Gamma_1 \cup S \cup \Gamma_2} k(x)\frac{\partial T}{\partial n}zd\Gamma + \int_\Omega k(x)\nabla T.\nabla z dx$$

$$+\int_S hqn_1 z d\Gamma + \int_\Omega hq\frac{\partial z}{\partial x_1}dx - \int_{\Gamma_3} hqn_1 z d\Gamma = 0 \tag{35}$$

Combining this equality with (28) we easily obtain (23), (24), (25) and (26). #

Next we show the existence of solution for problem P. Proof is related to the ones given in Brézis,Kinderlehrer and Stampacchia [13] or Bayada-Chambat [14] for the dam problem and the lubrication with cavitation problem respectively. However we have not Dirichlet boundary conditions and then we do not automatically have $T \leq T_{eu}$ for a solution of problem P. In order to get this property assumption H5 will be considered. As in the above mentioned papers we use a regularization technique.

Let H $_\epsilon$ he Yosida regularization of H (see figure 4) defined by

$$H_\epsilon(s) = \begin{cases} 0 & \text{if} \quad s \leq 0 \\ \frac{s}{\epsilon} & \text{if} \quad 0 \leq s \leq \epsilon \\ 1 & \text{if} \quad s \geq \epsilon \end{cases} \tag{36}$$

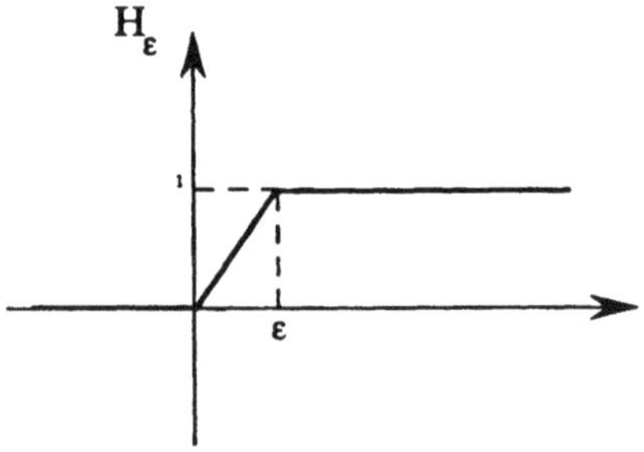

Figure 4

We introduce the following problem (P $_\epsilon$). Find $T_\epsilon \in H^1(\Omega)$ such that

$$\int_\Omega k(x)\nabla T_\epsilon.\nabla z dx + \int_\Omega hH_\epsilon(T_\epsilon - T_{eu})\frac{\partial z}{\partial x_1}dx$$
$$+\int_{\Gamma_1} \alpha(T_\epsilon - T_c)z d\Gamma = \int_{\Gamma_3} hz d\Gamma \quad \forall z \in H^1(\Omega) \tag{37}$$

Proposition 2 .-Under the assumptions H1 to H4 problem P_ϵ has unique solution.

Proof : Let L_ϵ be the operator defined by

$$L_\epsilon : u \in H^1(\Omega) \rightarrow L_\epsilon(u) = u_\epsilon \in H^1(\Omega)$$

where u_ϵ is the unique solution of the linear problem

$$\int_{\Omega} k(x)\nabla u_\epsilon.\nabla z dx + \int_{\Gamma_1} \alpha(u_\epsilon - T_c) z d\Gamma = -\int_{\Omega} hH_\epsilon(u - T_{eu})\frac{\partial z}{\partial x_1} dx$$

$$+\int_{\Gamma_3} hzd\Gamma \quad ,\forall z \in H^1(\Omega) \tag{38}$$

It is easy to prove that L_ϵ is a well defined compact operator. On the other hand u_ϵ belongs to a bounded set of $H^1(\Omega)$ independent of ϵ. These facts imply the existence of a solution for P_ϵ from the Schauder fixed point theorem.

Let T_ϵ^1 and T_ϵ^2 two solutions of (37). Subtracting (37) for T_ϵ equal T_ϵ^1 and T_ϵ^2 and taking

$$z = \frac{(T_\epsilon^1 - T_\epsilon^2 - \delta)^+}{T_\epsilon^1 - T_\epsilon^2}$$

we obtain for $q_\epsilon = T_\epsilon^1 - T_\epsilon^2$

$$\delta \int_{\Omega} k(x)\frac{|\nabla(q_\epsilon - \delta)^+|^2}{q_\epsilon^2} dx + \delta \int_{\Omega} h(H_\epsilon(T_\epsilon^1 - T_{eu}) - H_\epsilon(T_\epsilon^2 - T_{eu}))\frac{\frac{\partial(q_\epsilon - \delta)^+}{\partial x_1}}{q_\epsilon^2} dx$$

$$= -\int_{\Gamma_1} \alpha(q_\epsilon - \delta)^+ d\Gamma \leq 0 \tag{39}$$

As in Brézis, Kinderlehrer and Stampacchia [13] we can deduce from (39)

$$\frac{\alpha_0}{\delta}\int_{\Gamma_1} (q_\epsilon - \delta)^+ d\Gamma + \int_{\Omega} |\nabla log(1 + \frac{(q_\epsilon - \delta)^+}{\delta})|^2 dx \leq C_\epsilon \tag{40}$$

where C_ϵ is independent *of* δ.

Letting δ go to zero we obtain $q_\epsilon \leq 0$ a. e. in Ω. Interchanging T_ϵ^1 and T_ϵ^2 it follows $q_\epsilon = 0$ and then uniqueness. #

Proposition 3 .- Assume hypothesis H1 to H5. Then we have

$$T_\epsilon \leq T_{eu} + \epsilon \quad a.e. \quad in \quad \Omega \tag{41}$$

Proff : Taking z= $(T_\epsilon - T_{eu} - \epsilon)^+$ in (37) we obtain

$$\int_{\Omega} k(x)|\nabla(T_\epsilon - T_{eu} - \epsilon)^+|^2 dx + \int_{\Gamma_1} h(T_\epsilon - T_{eu} - \epsilon)^+ n_1 d\Gamma +$$

$$\int_{\Gamma_1} \alpha(T_\epsilon - T_c)(T_\epsilon - T_{eu} - \epsilon)^+ d\Gamma = 0 \tag{42}$$

because $H_\epsilon(T_\epsilon - T_{eu}) = 1$ if $(T_\epsilon - T_{eu} - \epsilon)^+ > 0$. On the order hand, assumption H5 implies

$$0 \le \epsilon \int_{\Gamma_1} \alpha(T - T_{eu} - \epsilon)^+ d\Gamma \le \int_{\Gamma_1} h(x_2)(T_\epsilon - T_{eu} - \epsilon)^+ n_1 d\Gamma +$$

$$\int_{\Gamma_1} \alpha(T_\epsilon - T_c)(T_\epsilon - T_{eu} - \epsilon)^+ d\Gamma \qquad (43)$$

and using (42) we get

$$\epsilon \alpha_0 \int_{\Gamma_1} (T_\epsilon - T_{eu} - \epsilon)^+ d\Gamma + k_0 \int_\Omega |\nabla (T_\epsilon - T_{eu} - \epsilon)^+|^2 dx = 0 \qquad (44)$$

from which (41) follows. #

Proposition 4 .- Assume H1 to H5 hold. Then there exists $T \in H^1(\Omega)$ solution of problem P and such that $T \le T_{eu}$ a.e. in Ω.

Proof : The sequence T_ϵ and $H_\epsilon(T_\epsilon - T_{eu})$ being bounded we can extract subsequences still denoted T_ϵ and $H_\epsilon(T_\epsilon - T_{eu})$ such that

$$T_\epsilon \rightharpoonup T \quad in \quad H^1(\Omega) \quad weakly \qquad (45)$$

and then in $L^2(\Omega)$ strongly and a.e. in Ω,

$$H_\epsilon(T_\epsilon - T_{eu}) \rightharpoonup q \quad in \quad L^2(\Omega) \quad weakly \qquad (46)$$

On the other hand (41) implies $T \le T_{eu}$.

To finish the proof we only have to show that q=0 a.e. in $\{x \in \Omega : T(x) < T_{eu}\}$. For this notice that if $T(x) < T_{eu}$ we have $T_\epsilon(x) < T_{eu}$ for ϵ small enough and *then* $H_\epsilon(T_\epsilon(x) - T_{eu}) = 0$. #

4.NUMERICAL ALGORITHM

In order to solve the problem P we discretize (27)-(29) by a finite element method. Thus we obtain a nonlinear problem for which an iterative algorithm will be proposed.

Let V_h the internal approximation of the space $H^1(\Omega)$ given by

$$V_h = \{z_h \in C(\Omega) : z_{h/K} \in P^1, \quad \forall K \in \sum_h\} \qquad (47)$$

where $\sum_h$ is a given family of triangulations of Ω. In other words V_h is the triangular Lagrange finite element space of degree one. Associated to it consider the following discretized problem

Find T_h , $q_h \in V_h$ such that

$$\int_\Omega k(x) \nabla T_h . \nabla z_h dx + \int_\Omega h q_h \frac{\partial z_h}{\partial x_1} dx +$$

$$\int_{\Gamma_1} \alpha(T_h - T_c) z_h d\Gamma = \int_{\Gamma_3} h z_h dx \quad \forall \quad z_h \in V_h \tag{48}$$

$$q_h(p) \in H(T_h(p) - T_{eu}) \quad for \quad all \quad vertex \quad p \tag{49}$$

In order to solve this problem we can use the following iterative algorithm (Bermúdez and Moreno [15], Bermúdez and Durany [16]).

Let r_h^0 arbitrarily given in V_h. At the iteration n we first compute T_h^n as the solution of the linear problem

$$\int_\Omega k(x)\nabla T_h^n.\nabla z_h dx + \omega \int_\Omega h T_h^n \frac{\partial z_h}{\partial x_1} dx + \int_{\Gamma_1} \alpha(T_h^n - T_c) z_h d\Gamma$$

$$= -\int_\Omega h r_h^n \frac{\partial z_h}{\partial x_1} dx + \int_{\Gamma_3} h z_h d\Gamma, \quad \forall \quad z_h \in V_h \tag{50}$$

and then r_h^n is updated by

$$r_h^{n+1}(p) = \rho H_\lambda^\omega (T^n(p) - T_{eu} + \lambda r_h^n(p)) + (1-\rho) r_h^n(p) \tag{51}$$

where H_λ^ω denotes the Yosida regularization of $H - \omega I$, i.e.

$$H_\lambda^\omega(s) = \begin{cases} -\omega\frac{s}{1-\lambda\omega} \text{ if } s \leq 0 \\ \frac{s}{\lambda} \text{ if } 0 \leq s \leq \lambda \\ \frac{1-\omega s}{1-\lambda\omega} \text{ if } \lambda \leq s \end{cases} \tag{52}$$

which is a Lipschitz continuous operator for $\lambda\omega \leq 0.5$.

In Bermúdez- Moreno [15] it is shown that $r \in H(u) - \omega u$ if and only if $r = H_\lambda^\omega(u + \lambda r)$. Therefore every limit point of the sequence r_h^n is a solution of (48)-(49).

5. NUMERICAL RESULTS

In this section numerical results for real aliminium reduction cells obtained with the methods described in this paper are shown.

In a first part we solve a one dimensional version of problem P having an analytical solution which allows to know the behaviour of the numerical algorithm. More precisely consider the following problem:

$$\int_0^1 \frac{dT}{dx}\frac{dz}{dx} dx + \int_0^1 hq \frac{dz}{dx} dx + \alpha(T(0) - T_c) z(0) = hz(1) \quad \forall z \in H^1(0,1) \tag{53}$$

Its solution is as follows:

i) For $h \leq \alpha\frac{(T_{eu} - T_c)}{\alpha + 1}$

$$T(x) = hx + \frac{\alpha T_c + h}{\alpha} \tag{54}$$

$$q(x) = 0 \tag{55}$$

Thus no free boundary exists.

ii)For $\alpha\frac{(T_{eu}-T_c)}{\alpha+1} < h \leq \alpha(T_{eu} - T_c)$

$$T(x) = \begin{cases} hx + \frac{\alpha T_c + h}{\alpha} & \text{if} \quad x \leq \gamma \\ T_{eu} & \text{if} \quad x > \gamma \end{cases} \tag{56}$$

$$q(x) = \begin{cases} 0 & \text{if} \quad x < \gamma \\ 1 & \text{if} \quad x > \gamma \end{cases} \tag{57}$$

where γ, the position of the free boundary, is given by

$$\gamma = \frac{\alpha(T_{eu} - T_c) - h}{\alpha h} \tag{58}$$

iii) For $h > \alpha(T_{eu} - T_c)$

$$T(x) = \frac{h + \alpha T_c}{\alpha} \tag{59}$$

$$q(x) = 1 \tag{60}$$

and then condition T $\leq T_{eu}$ does not hold.

Figures (5)-(10) show numerical and exact solutions when h is chosen satisfying i),ii) and iii) successively.

In all cases we have taken

$$\alpha = 6, \quad T_{eu} = 2.5, \quad T_c = 0.5 \tag{61}$$

and used a uniform mesh of the interval (0,1) with 51 points ($\Delta x = 0.02$).

Velocity of convergence of algorithm (50)-(51) depends on values of λ,ω and ρ. For

$$\lambda = 0.1 \quad \omega = 5 \quad \rho = 0.9$$

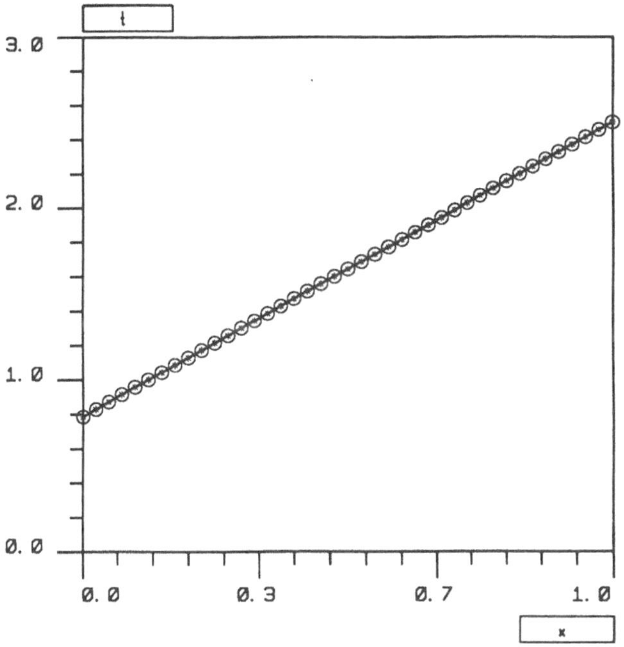

Figure 5 : discretized and exact temperature for h=1.71.

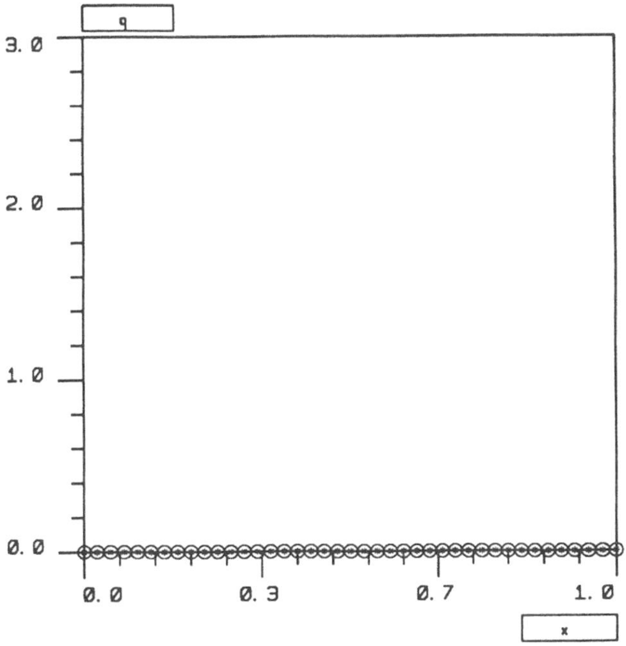

Figure 6 : discretized and exact function q for h=1.71.

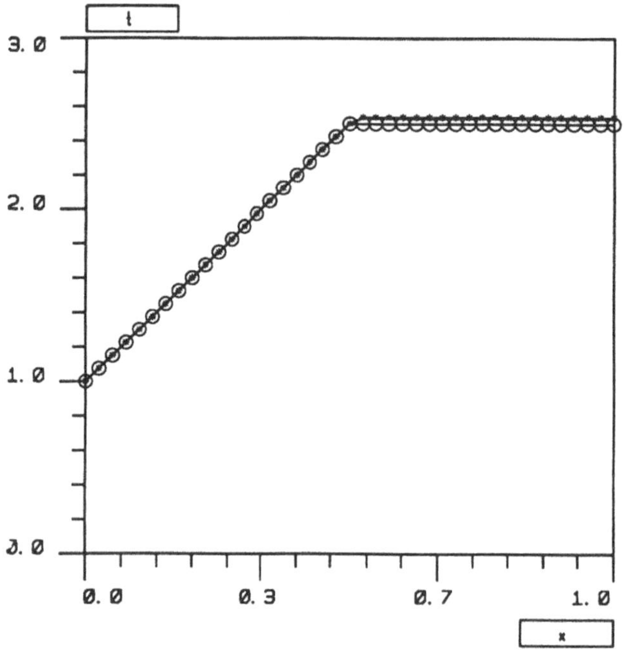

Figure 7 : discretized and exact temperature for h=3.

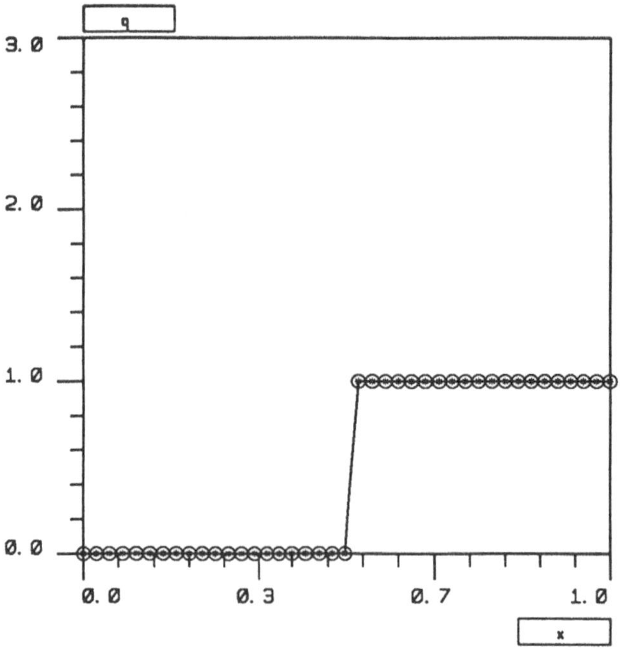

Figure 8 : discretized and exact function q for h=3.

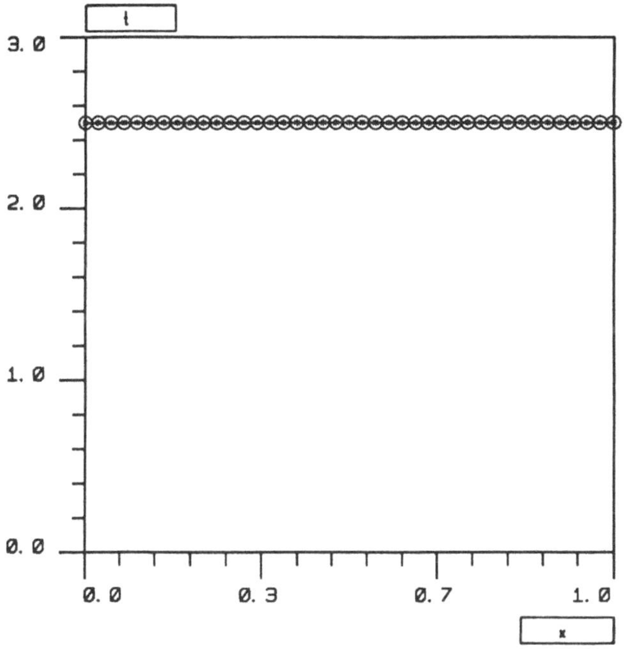

Figure 9 : discretized and exact temperature for h=12.

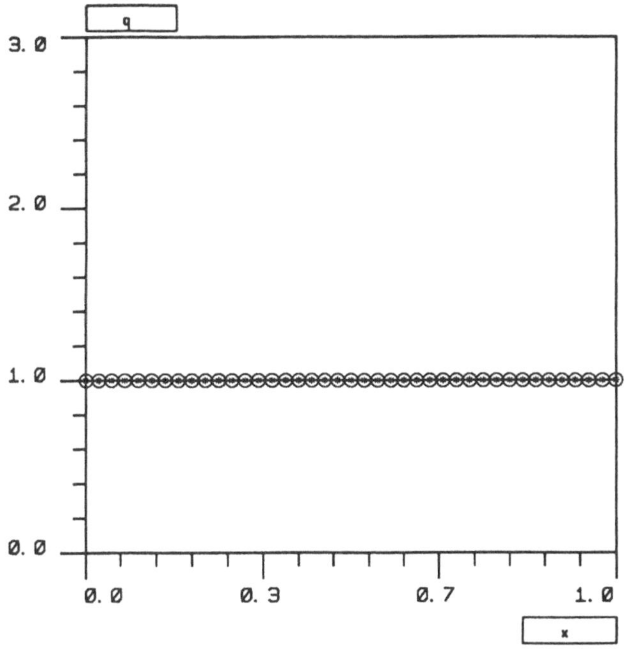

Figure 10 : discretized and exact function q for h=12.

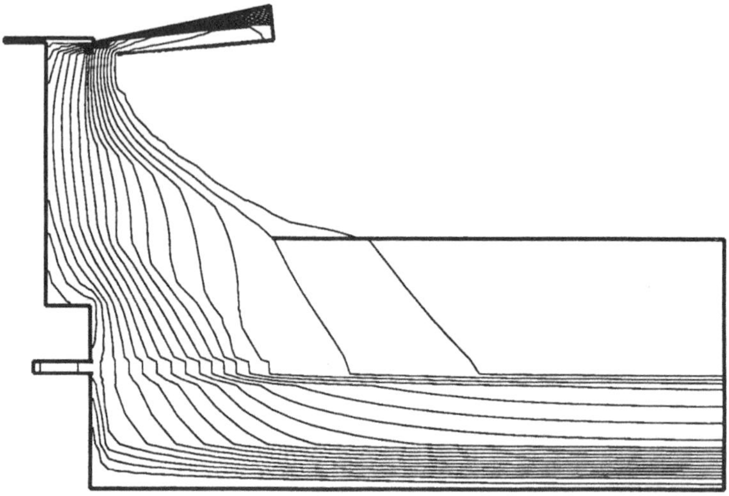

Figure 11 : isolines for temperature for intensity=108000 A.

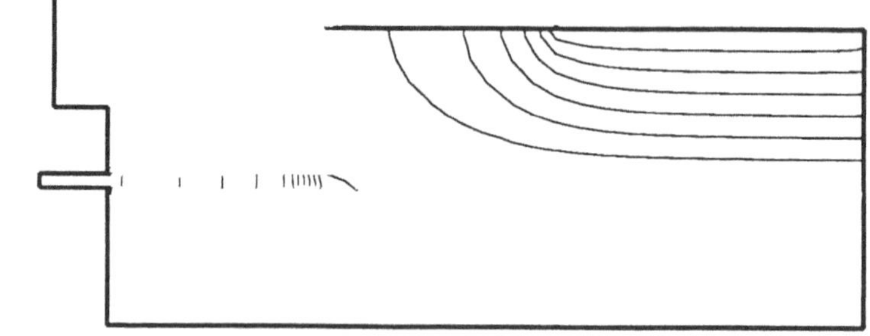

Figure 12 : isolines for electric potential for intensity=108000 A.

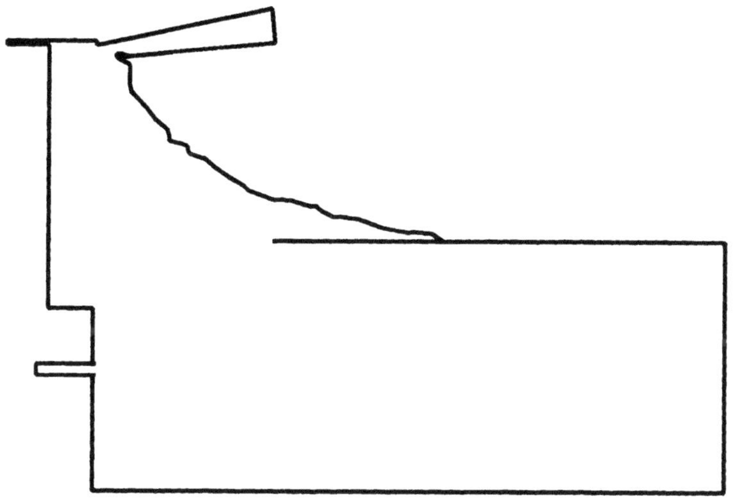

Figure 13 : free boundary for intensity=108000 A.

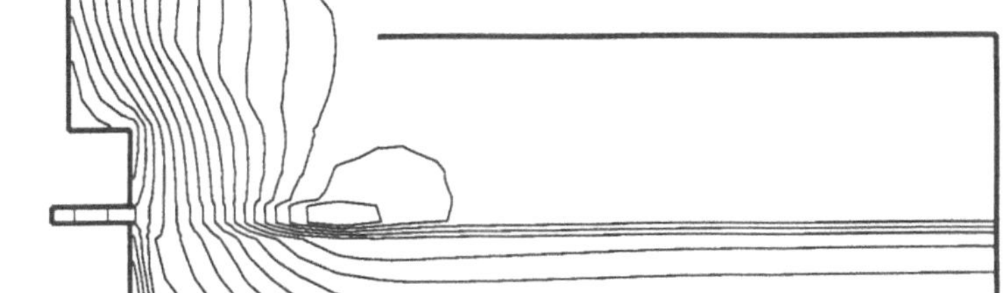

Figure 14 : isolines for temperature for intensity=128000 A.

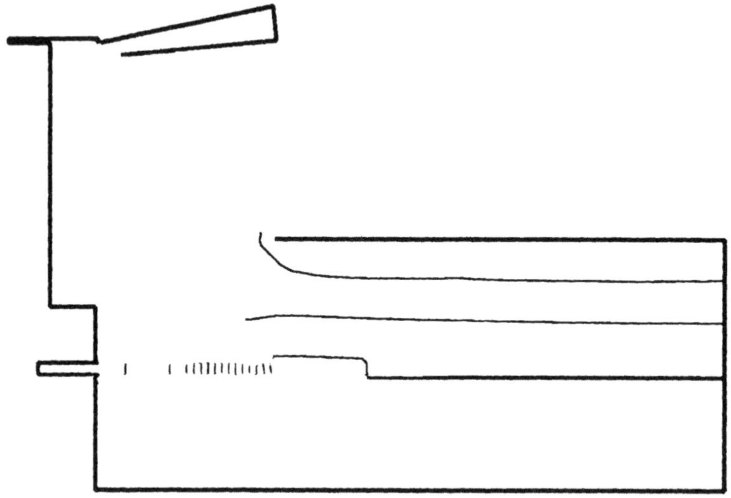

Figure 15 : isolines for electric potential for intensity=128000 A.

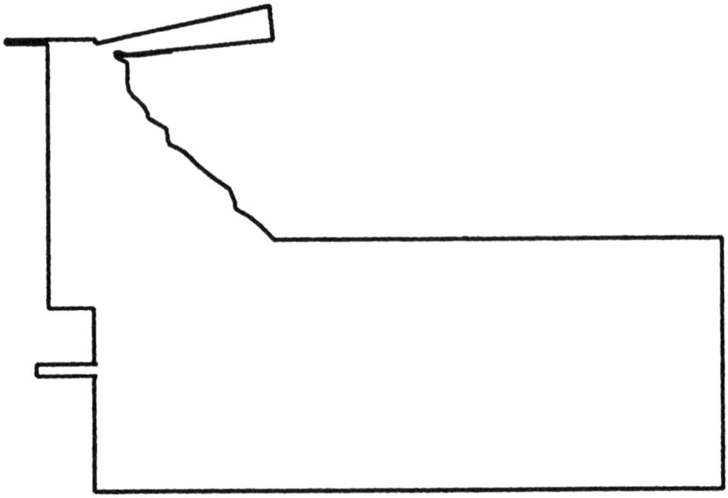

Figure 16 : free boundary for intensity=128000 A.

and the test of convergence

$$max_p|r^n(p) - r^{n+1}(p)| < 10^{-4}$$

the number of iterations was 8 .

Next we present some results for an aluminium electrolytic cell. They correspond to the full thermoelectric problem given in section 2 and real data of operation. We have used a finite element mesh with 1223 nodes. Figures (11) to (16) represent isolines for temperature, and electric potential, and the free boundary.

In all tests eutectic temperature of the bath and operation temperature were

$$T_{eu} = 954°C, \quad T_{op} = 966°C.$$

We show the obtained results for two different intensities namely 108000 and 128000 amperes.

Finite element mesh and figures have been made by using Modulef Library.

CONCLUSIONS

In this paper we have considered a mathematical model for numerical simulation of the thermoelectric behaviour of an aluminium reduction cell. A free boundary problem was obtained for which a weak formulation was given in a fixed domain. In addition to some theoretical results we proposed a finite element method associated to a fixed mesh for numerical solution. Numerical results both for a test having a solution in closed form and for a real reduction cell show the good behaviour of this method.

ACKNOWLEDGEMENTS

This work has been supported by INESPAL (La Coruña, Spain) under contract.

Authors are indebted to Mr. J.L. López Sangil, G. Tirapu, A. López Alba and C. Palacios from INESPAL, and Mr. R. Leblanc and M. Dupuy from ALCAN (Canada) for many useful discussions.

REFERENCES

1. Grjotheim K. and Kvande H.(Ed.) Understanding the Hall-Héroult Process for Production of Aluminium, Aluminium Verlag, Dusseldorf, 1986.

2. Grjotheim K., Krohn C., Malianovsky M., Matiasovsky, K. and Thonstad, J. Aluminium Electrolysis, Aluminium Verlag, Dusseldorf, 1982.

3. Bruggemen N.J. and Danka, D.J. Two-dimensional thermal modelling of the Hall-Héroult cell, Ligth Metals, pp. 203-209, 1990.

4. Arita, Y., Urata, N. and Ikeuchi, H. Estimation of frozen bath shape in aluminium reduction cell by computer simulation, Ligth Metals, pp. 59-72, 1978.

5. Utne, P. Freeze profile in side-break cells: calculations and measurements, Ligth Metals, pp.359-371, 1982

6. Sulmot, B. and Hudault, G. Application of a thermoelectric model to the investigation of reduction cell thermal equilibrium, Ligth Metals, pp. 73-86, 1978.

7. Saguez, C. Un probléme de Stefan avec source sur la frontiere libre, Rapport de Recherche n. 268, INRIA, Rocquencourt, 1977.

8. Fasano, A. and Primicerio, M. (Ed.) Free boundary problems: Theory and applications (Vol. I,II), Pitman, London, 1983.

9. Elliot, C. and Ockendon, J.R. Weak and variational methods for moving boundary problems, Pitman, London, 1982.

10. Bossavit, A. Damlamian, A. and. Frémond, M. (Ed.). Free boundary problems: Applications and Theory (Vol. III,IV). Pitman, London, 1985.

11. Hoffman, K.H. and Sprekels, J. (Ed.). Free boundary problems: Theory and Applications (Vol I, II), Pitman, London, 1990.

12. Wrobel, L.C., Brebbia, C.A. and Nowak, A.J. (Ed.). Advanced Computational Methods in Heat Transfer. Vol3, Computational Mechanics Publications, Southampton, 1990.

13. Brézis, H. Kinderlehrer, D. and Stampacchia, G. Sur une nouvelle formulation du probleme de l' écoulement a travers une dique, C.R. Acad. Sci. París. Ser. A.187, pp. 711-714, 1978.

14. Bayada, G. and Chambat, M. Existence and uniqueness for a Lubrication problem with nonregular conditions on the free boundary, Boll. U.M.I. (6) 3-B, pp. 543-557, 1984.

15. Bermúdez, A. and Moreno, C. Duality methods for solving variational inequalities, Comput. and Math. with Appl. 7, pp. 43-58, 1981.

16. Bermúdez, A. and Durany, J. Numerical Solution of cavitation problems in lubrication, Comput. Meth. Appl. Mech. Eng. 68, pp. 55-65, 1988.

An Alternative Approach for the Modelling and Simulation of Binary-Alloy Solidification using a Fixed-Grid Control-Volume Technique

M. Fabbri

Laboratório de Materiais e Sensores, Instituto Nacional de Pesquisas Espaciais, C.P. 515, 12225 São José dos Campos, Brazil

ABSTRACT

The quasi-equilibrium, time-dependent directional solidification of a binary alloy is numerically modelled by an implicit scheme, based on a control-volume discretization over a fixed grid. Features of the alloy's phase diagram are incorporated in detail, following the technique of Wilson, Alexiades and Solomon [15]. The method is flexible enough to accommodate the detailed set of boundary conditions and thermophysical properties that arise typically in crystal growth from the melt. Numerical results are shown that agrees well with known experimental facts about Bridgman growth of $(Pb, Sn)Te$.

INTRODUCTION

In the last decade, there has been an increasing interest in precise and/or reliable numerical simulation of binary alloy solidification processes,mainly for applications in the field of crystal growth. From the practical point of view, this is a natural consequence of some recent developments, namely, a high degree of sophistication in furnace design and control, the use of external electromagnetic fields to improve growth conditions, and the concrete possibility of growth in a microgravity environment.

A widespread technique for growing crystals from the melt is the Bridgman - Stockbarger (B.S.) configuration with an "adiabatic" zone (for a brief description of a typical assembly see, e.g., ref. [1]). B.S. growth is usually employed whenever one needs to grow not too-large binary semiconductor bulk crystals (crystal diameter less than 2 cm), with a small degree of radial segregation, provided that the phase diagram (P.D.) of the alloy is not very complicated. The simplicity of the B.S. technique also makes it very attractive for experiments aiming at the understanding of the complex interplay between convection, diffusion and heat transfer that takes place during crystal growth from the melt. The interpretation of typical B.S. experiments usually begins by mapping the concentration distribution and deffects of the as-grown crystals, through electron microprobe scanning and etching. Inhomogeneities on the concentration distribution, resulting from both axial and radial macrosegregation, are closely related to the solid-liquid (S-L) interface morphology and stability during growth, and these, in turn, depend strongly on the actual thermal gradients near the growing interface and on the mass transfer rates by diffusion and convection in the melt. Measurements of concentration profiles in directionally solidified $(Hg, Cd)Te$ alloys indicate that the S-L interface is typically neither an isotherm nor an isoconcentrate [2], which is also confirmed by microprobe measurements in quenched and etched $(Hg, Cd)Te$ [3] and $(Pb, Sn)Te$ [4]. This suggests that the relation between the radial compositional profile and the growth interface shape can be sensitive to details of the alloy's P.D.. On the other hand, the interface shape is strongly dependent on differences between the solid and liquid heat conductivities, as well as on the crucible (or ampoule) material [2,5]. Radial segregation and interface shape are linked through a strong coupling among the heat transfer to the cold and hot ends of the furnace [4], the amount of heat carried by the ampoule [6] (which causes the so-called interface effect [1]) and convection currents in the melt [4, 7].

Despite all of the complications mentioned above, crystal growth by the B.S. technique is essentially a quasi-equilibrium process, involving simple alloys with well-known P.D.s. Since constitutional supercooling is avoided as much as possible, one is concerned with a smooth and well defined S-L interface, which is uniquely determined by the actual values of concentration and temperature at the phase front. Numerical methods based on simple regularizations of the enthalpy-temperature-concentration relationship are not expected to yield reliable results in this case. This situation is in contrast with modelling of typical metallurgical processes, where the existence of fully developed mushy regions, or the lack of a precise knowledge about the alloy's P.D. justify a number of semiempirical approaches [8].

The first almost complete numerical simulation of the complex convective - diffusive moving boundary problem that arises in a typical B.S. growth was reported by Chang and Brown [9], who performed a steady-state calculation with a finite - element body - fitted coordinate system. Apart from the difficulty of extending their approach to incorporate all the features of the alloy's P.D., it becomes computational prohibitive if one is also interested in transients.

We describe in this work a fixed-grid numerical method based on a control-volume approach, which in principle is able to handle a host of complicated material properties and boundary conditions. The time integration is made implicit, which is known to be almost essential to account efficiently for the small Lewis number typical of semiconductor melts [10]. The details of the alloy's P.D. are included by a suitable equation of state, valid for any region of the Temperature (T)-Concentration (C) plane, which is obtained along the method developed by Solomon , Alexiades and Wilson [11] (hereafter referred briefly as SAW). In the original SAW method, time integration was made explicit. We shall see that a totally implicit form requires that the T, C fields are obtained in an autoconsistent, iterative manner for each time step. Also, time integration is done here in such a way that any cell retains its phase state during a time step; the phase state of a control volume is allowed to change only at the end of the time interval adopted for integration. This methodology is necessary because of the large discontinuities in the specific enthalpies and specific heat of a control-volume when it enters or leaves a region of phase change.

The use of a fixed-grid method implies that the phase-change front cannot be precisely located within a control volume, and that some kind of "mean field" prescription is necessary to model the physical properties of a cell undergoing phase change. Previous investigations with solidification of pure substances [12] indicate that these approximations do not preclude the method from yielding very good results concerning the velocity and shapes of S-L interfaces.

It is worth mentioning that Voller [13] has developed a fixed-grid enthalpy method, essentially an implicit-iterative scheme, where the latent heat contribution is isolated as a heat source in the energy equation. Voller's method seems to be quite efficient for the simulation of convective-diffusive processes during binary alloy solidification [14]; nevertheless, it is implemented through the specification of a melting range ϵ. Inclusion of details of the alloy's P.D. in Voller's scheme seems to require first a good model for the dependence of ϵ, as well as of the latent heat of fusion, on T and C. A close comparison between our method and Voller's will be the subject of a forthcoming communication.

PHASE DIAGRAM AND THE EQUATION OF STATE

We consider an alloy with a simple lens-type P.D.; the basic notation is shown in Fig. 1.

Following SAW [11], we write the internal energy of the alloys as

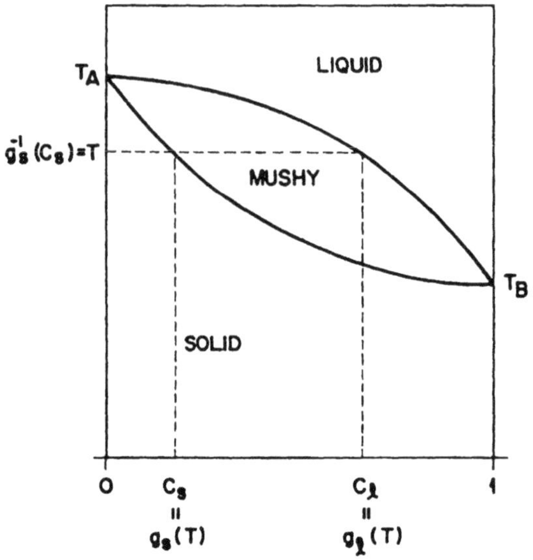

Fig. 1 The alloy A-B is supposed to display a lens-type phase diagram as shown.
A is taken to be the component with the highest melting point.

$$e = e(T, C) \quad , \tag{1}$$

which is assumed to be valid for all regions of the T-C plane (solid, liquid or "mushy"), through the prescription

$$e = \lambda e^\ell + (1 - \lambda) e^s \tag{2}$$

λ is the liquid fraction, defined by

$$\lambda = \begin{cases} 1 & \textit{in the liquid} \\ \dfrac{C - g_s(T)}{g_\ell(T) - g_s(T)} & \textit{in the "mushy" region} \\ 0 & \textit{in the solid,} \end{cases} \tag{3}$$

and

$$\begin{aligned} e^{\ell} &= e^{liq}(T,C) \ \ if \ \ \lambda = 1 \\ & e^{liq}(T,g_{\ell}) \ \ if \ \ 0 < \lambda < 1 \\ & 0 \ if \ \lambda = 0 \ ; \end{aligned} \tag{4.a}$$

$$\begin{aligned} e^{s} &= e^{sol}(T,C) \ \ if \ \ \lambda = 0 \\ & e^{sol}(T,g_{s}) \ \ if \ \ 0 < \lambda < 1 \\ & 0 \ \ if \ \ \lambda = 1. \end{aligned} \tag{4.b}$$

The explicit form of the equation of state is obtaineu by integration of the Gibbs relation:

$$de = cdT + hdC \tag{5}$$

For instance, the internal energy of any point in the liquid region is obtained by

$$e^{liq}(T,C) = L_A + \int_0^C h^{liq}(T_A,\xi)d\xi + \int_{T_A}^T c^{liq}(\tau,C)d\tau \quad , \tag{6}$$

where we have defined L_A as the heat of fusion of pure A.

Now, we assume the simplest possible behaviour of the thermophysical properties h (specific partial enthalpy) and c (specific heat). We suppose that c^{liq} and c^{sol} are supplied as experimental data for the pure components A and B, and that the dependence of these on temperature can be neglected. To a first approximation, we assume that

- c is a function of C only
- h is a function of T only

Then eq. (6) yields:

$$e^{liq}(T,C) = L_A + h^{liq}(T_A)C + c^{liq}(C)(T-T_A) \tag{7}$$

Proceeding along the same lines for the solid region, we get

$$e^{sol}(T,C) = c^{sol}(0)(T-T_A) + h^{sol}(T)C \tag{8}$$

Thermodynamic compatibility must now be imposed between eqs. (7) and (8), and it is easy to deduce that sufficient conditions are:

- h^{liq} and h^{sol} are linear functions of T
- c^{liq} and c^{sol} are linear functions of C

We therefore write

$$c^{liq,sol}(C) \;=\; [c^{liq,sol}(1) \;-\; c^{liq,sol}(0)]C \;+\; c^{liq,sol}(0) \tag{9}$$

$$h^{liq,sol}(T) \;=\; h^{liq,sol}(T_{A,B}) + [c^{liq,sol}(1) - c^{liq,sol}(0)](T - T_{A,B}) \tag{10}$$

One additional relation comes from the condition

$$e^{liq}(T_A^+,0) - e^{sol}(T_A^-,0) \;=\; L_A \;, \tag{11}$$

which imposes that

$$h^{liq}(T_A) - h^{sol}(T_B) \;=\; L_B - L_A + (T_B - T_A)[c^{sol}(0) - c^{liq}(1)] \;. \tag{12}$$

Eqs. (9), (10), (11) and (12) define the equation of state, to a first approximation, in terms of the known experimental values c^{liq}, c^{sol} and L for the pure components A and B.

BASIC EQUATIONS

These are exactly the same equations as in the SAW method [15], with some minor changes in the calculation of the interface values of the transport coefficients. Note that a control-volume approach permeates all the basic relations.

Main Assumptions

We disregard differences between solid and liquid mass densities, which are assumed constant throughout. Diffusion in the solid is neglected, as well as thermal-diffusion cross effects (Soret, Dufour). Convection in the liquid is also not considered here. All these effects are likely to be of importance during crystal growth, and can be included in the model without essential modifications. Cross-effects and diffusion in the solid were already considered in the original SAW scheme [15]. Voller and co-workers [14] have shown a couple of useful schemes to include convective effects and no-slip boundary conditions at the S-L interface, that could be employed here without modification.

Continuity of Mass

The continuity equation is written, as usual,

$$C_t = -\nabla \cdot \vec{J} \ , \tag{13}$$

where

$$C = \lambda C_{liq} + (1-\lambda)C_{sol} \tag{14}$$

$$C_{liq} = \begin{cases} 0 \ \ if \ \ \lambda = 0 \\ 1 \ \ if \ \ \lambda = 1 \\ g_\ell(T) \ \ if \ \ 0 < \lambda < 1 \ ; \end{cases} \tag{15.a}$$

$$C_{sol} = \begin{cases} 0 \ \ if \ \ \lambda = 1 \\ 1 \ \ if \ \ \lambda = 0 \\ g_s(T) \ \ if \ \ 0 < \lambda < 1 \ . \end{cases} \tag{15.b}$$

Mass flux between cells

These fluxes are given by a slight modification of Fick's law. The flux at a given interface between a cell around P and one of its neighbours is written as:

$$\vec{J}_{interf} = min(\lambda_p, \lambda_{neigh})[-D_{interface}(\nabla C_{liq})] \ , \tag{16}$$

where $min(\lambda_p, \lambda_{neigh})$ is a measure of the liquid-liquid contact and

$$D_{interf} = \frac{2D_p D_{neigh}}{D_p + D_{neigh}} \tag{17}$$

$$D_p = \begin{cases} D_{liq} \ \ if \ \ \lambda \neq 0 \ \ (the\ usual\ mutual\ diffusion\ coefficient\ in\ the\ liquid) \\ 0 \ \ if \ \ \lambda = 0 \end{cases} \tag{18}$$

Energy Conservation

The usual energy equation, with the help of the Gibbs relation, eq. (2), yields [15]:

$$\rho c T_t \; = \; -\nabla \cdot \vec{Q} - \rho \nabla \cdot (h\vec{J}) + \rho h \nabla \cdot \vec{J} \; , \tag{19}$$

with

$$c \; = \; \lambda c^{liq} + (1-\lambda) c^{sol} + L \frac{\partial \lambda}{\partial T} \tag{20}$$

$$h \; = \; \lambda h^{liq} \; + \; (1 \; - \; \lambda) h^{sol} \; + \; L \frac{\partial \lambda}{\partial C} \tag{21}$$

$$L \; = \; L(T) \; = \; e^{\ell} \; - \; e^{s} \tag{22}$$

Conductive Heat Flux Between Cells

These obey the usual Fourier law with the harmonic mean prescription for the interface conductivities:

$$\vec{Q}_{interface} \; = \; -k_{interface} \nabla T \tag{23}$$

$$k_{interface} \; = \; \frac{2 k_p k_{neigh}}{k_p \; + \; k_{neigh}} \tag{24}$$

We adopt for $k_p(C,T)$ a simple linear model which interpolates between the experimental values k^{liq} and k^{sol} for pure A and B, and extend it through the "mushy" region:

$$\begin{aligned} k_p(C,T) \; &= [k^{liq}(1) - k^{liq}(0)]C + k^{liq}(0) \quad \textit{in the liquid} \\ & [k^{sol}(1) - k^{sol}(0)]C + k^{sol}(0) \quad \textit{in the solid} \\ & \frac{k(g_\ell) - k(g_s)}{g_\ell - g_s}(C - g_s) + k(g_s) \quad \textit{in the ``mushy''} \end{aligned} \tag{25}$$

Interfacial value of the partial specific enthalpy

When required, we simple put

$$h^{interface} = (h_p + h_{neigh})/2 \qquad (26)$$

Note that we have assumed regular meshes when writting eqs. (17), (24) and (26).

CONTROL-VOLUME DISCRETIZATION

The mass and energy equations are now discretized by control volumes, for a cartesian regular grid covering the X-Y plane. We employ the usual notation of Patankar [16], and time advancing is made totally implicit. The result is:

$$\frac{C_P - C_P^0}{\Delta t}\Delta x \Delta y = \sum_{J=E,W,N,S} a_J min(\lambda_P, \lambda_J) D_j (C_{\ell_J} - C_{\ell_P}) \qquad (27)$$

$$\begin{aligned} c_P \frac{T_P - T_P^0}{\Delta t}\Delta x \Delta y = & \sum_{J=E,W,N,S} a_J \{ \frac{k_j}{\rho}(T_J - T_p) + \\ & + \frac{h_J - h_P}{2} min(\lambda_P, \lambda_J) D_j (C_{\ell_J} - C_{\ell_P}) \} + \\ & + \frac{\Delta x \Delta y}{\rho}(S_P^C + S_P^P T_P) \ , \end{aligned} \qquad (28)$$

where

$$a_E = a_W = \frac{\Delta y}{\Delta x} \ , \quad a_N = a_S = \frac{\Delta x}{\Delta y} \quad and \ C_{\ell_I} = C_{liq} \ at \ I-th \ cell.$$

A linearized source term was introduced in the energy equation (of the form $S^C + S^P T$), in order to handle easily the required time-dependent boundary conditions. The boundary condition for the C-equation is a zero-mass flux through the outer boundary, and is automatically included in eq.(27). In the numerical experiments that follow, the domain is enclosed by rigid walls of given thickness, which do not participate in mass transfer. The wall properties are introduced in eq.(28) by suitable values of k and c.

We note that, even in this simple model that neglects cross-effects, eq.(27) couples with the T-field whenever there is a term referring to a "mushy" cell.

SOLUTION METHODOLOGY

Interface Updating

The equation of state $e(T, C)$ is not differenciable over the liquidus and solidus curves; both specific heat and partial enthalpies display large discontinuities (caused by the terms $L\partial\lambda/\partial T$ and $L\partial\lambda/\partial C$) when a cell enters or leaves the phase-change region. A special integration scheme was therefore adopted, in which each cell retains its phase state during a time step; any cell is allowed to enter or leave the phase-change region only at the end of a time interval, and when the error involved is estimated to be within a prescribed tolerance. In this way, we can use a convenient, somewhat large value for Δt at a normal step of integration, and reduce it only when phase state updating is required at some cell. The resulting scheme, which we named the Interface Updating Algorithm (IUA), is displayed in Fig. 2.

The reduction of Δt, when required, was done by a simple linear interpolation between old and new values of the internal energy e (T, C).

Error Criteria

Admissible errors are all based on the same parameter input, namely, the allowable relative error in temperature, ϵ_T. Estimates for the acceptable errors in the concentration field, ϵ_C, and in the internal energy, ϵ_e, are obtained through worst-case analysis of the input data for the physical properties and the phase diagram. ϵ_e is used as the maximum error allowed when a cell enters or leaves a region of phase change.

T-C Solver

Solution for the T-C fields is obtained alternating line-column sweeps by a simple Gauss-Sidel procedure. The coupling between the T and C equations was treated considering that, in directional solidification experiments, the phase change is essentially thermal driven. We adopted the scheme depicited in Fig. 3, which showed convergence with reasonable efficiency, provided Δt was not too large ("too large" here means that the moving furnace should not be displaced by more than a fraction of the cell size during Δt).

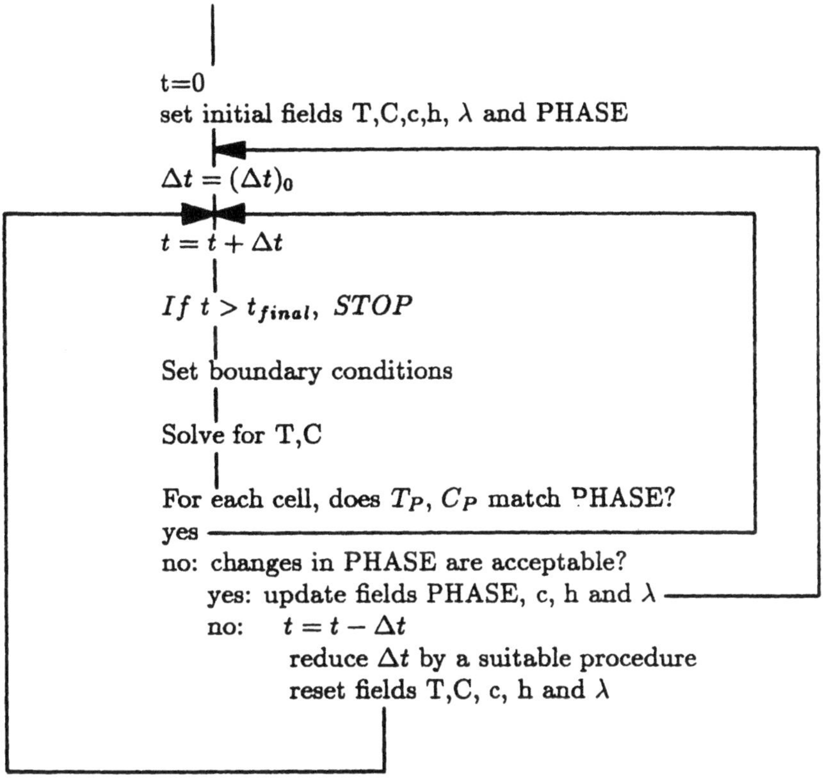

Fig. 2 Flux of the IUA calculations. The field PHASE indicates the phase state of each cell (solid, liquid or "mushy").

NUMERICAL RESULTS

Directional Solidification in Small-bore Tubes

Concentration Profiles Kinoshita and Sugii [17] have performed convection - free Bridgman growth of $(Pb, Sn)Te$ crystals in 2mm-bore tubes, for various values of the furnace translation speed R. The measured axial concentration of $SnTe$, at $R = 1mm/h$ and $10mm/h$, is displayed in Fig. 4, along with the numerical results we obtained employing the data shown in Table 1. The agreement seems very fair, given that the only adjusting constant was the mutual diffusion coefficient (D) of $SnTe - PbTe$ in the melt.

Mushy-zone Length Bridgman growth of binary crystals must be performed at a translation velocity below the critical value $Rcrit$ for constitutional supercooling. Most experiments are dimensioned by a direct application of the Tiller equation[18], which gives:

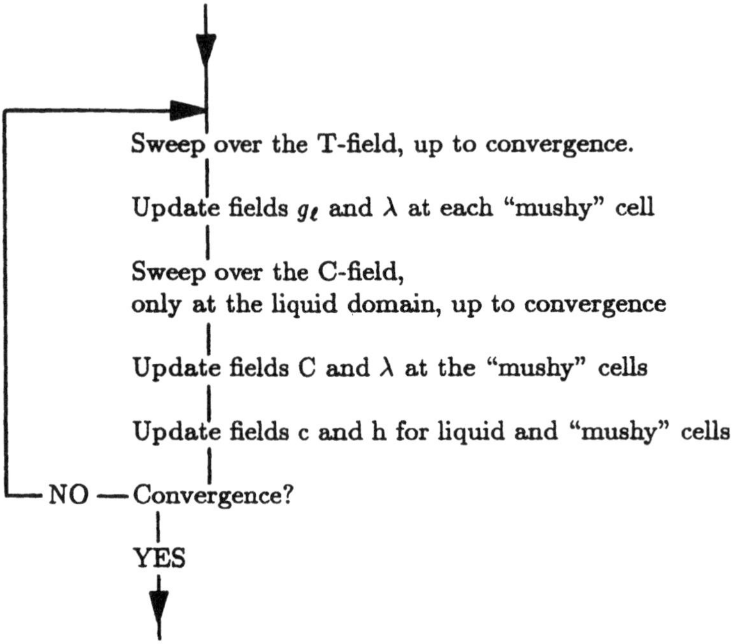

Fig. 3 Scheme of the T-C Solver.

$$Rcrit = \frac{GDK}{mC_0(1-K)} , \tag{29}$$

where

G is the temperature gradient around T_M, at the liquid side.
T_M is the melting temperature at $C = C_0$.
C_0 is the initial concentration in the melt.
K is the distribution coefficient at $C = C_0$
m is the slope of the liquidus line at $C = C_0$.

For $R > R_{crit}$, the mushy zone lenght ℓ_m is given by [18]:

$$1 - exp(-R\ell_m/D) = R_{crit} \cdot \ell_m/D \tag{30}$$

Due to the presence of the ampoule, the actual value of G at the interface is less than the thermal gradient of the unloaded furnace at T_M. In Fig. 5 we plotted $\ell_m \times G$, both as predicted by eq. (30) and as obtained from the numerical simulations, for $R = 10mm/h$. The results are consistent with the experimental findings of Kinoshita and Sugii[17], in that crystals grown at $R = 10mm/h$ and $G = 40^oC/cm$ showed a cellular substructure. In fact, the numerical results depicited in Fig. 5 indicates that it would be necessary a value of G greater than $70^oC/cm$ to avoid constitutional

TABLE 1 Geometry and Thermophysical Properties employed in the numerical simulations (c.g.s. units);all runs were made with an initial homogeneous melt at concentration $C_0 = 0.2$.

	PbTe Solid	PbTe Liquid	SnTe Solid	SnTe Liquid
k	0.019	0.022	0.011	0.017
c	0.27	0.27	0.63	0.63
L	340	-	280	-
ρ	8.2	8.2	6.4	6.4
D	-	-	-	10^{-4} (in liquid PbTe)

	Ampoule (1 mm thick)			Furnace				
Fig.	Length (mm)	Bore (mm)	k	T_{cold}	T_{hot}	ℓ(mm)	R (mm/h)	mesh
4a	50	2	0.028	600	1000	100	1	50×2
4b	50	2	0.028	600	1000	100	10	50×2
5	50	2	0.028	600	1000	various	10	50×2
6a,7a	50	10	0.028	800	1200	20	10	50×6
6b,7b	50	10	0.028	520	920	20	10	50×6

supercooling at $R = 10mm/h$ (note that these numerical simulations could not be used to predict values for ℓ_m less than the size of a control volume).

Bidimensional Profiles

Radial segregation is very sensitive to convective effects; therefore, comparison with experiments are of limited value by now. Neverthless, some

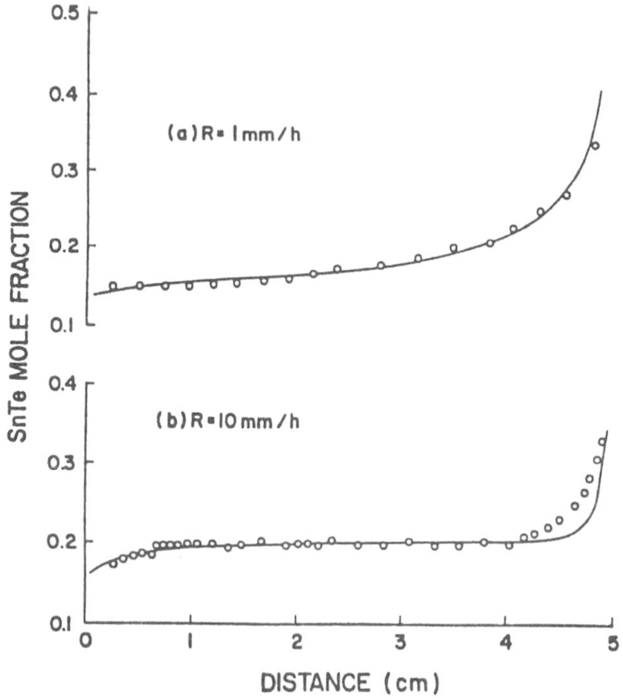

Fig. 4 Axial concentration profiles obtained in small-bore tube experiments as compared with the numerical results. Small circles are the measured values.

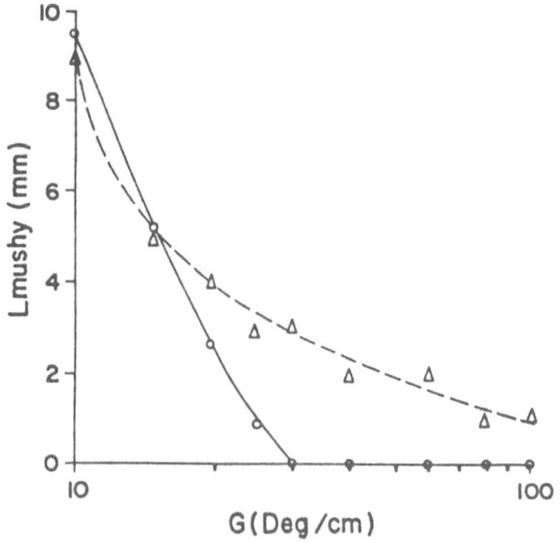

Fig. 5 The mushy-zone length as a function of the applied furnace temperature gradient. (o) predicted by a simplified application of the Tiller model, eq. (30). (Δ) results of numerical simulations, where G is imposed at the ampoule walls.

basic trends concerning sensitiveness on the furnace profile and on the ampoule material were investigated. As an illustration, we show in Figs. 6 and 7 some numerical results aimed to study the influence of the adimensional interface temperature $\phi = (T_M - T_{cold})/(T_{hot} - T_{cold})$ on the interface shape and radial segregation profiles. The $S-L$ interface at steady-state changes from concave to convex (toward the solid) as ϕ varies from 0 to 1 (Figs. 6a and 6b), as expected [4,6]; during transients, the interface shape is dictated mainly by the boundary conditions imposed at the left and right ampoule walls, and bears almost no resemblance to the radial concentration profiles (Figs. 7a and 7b). It should be noted that we are neglecting nucleation effects, which can substantially dominate the initial stages of growth.

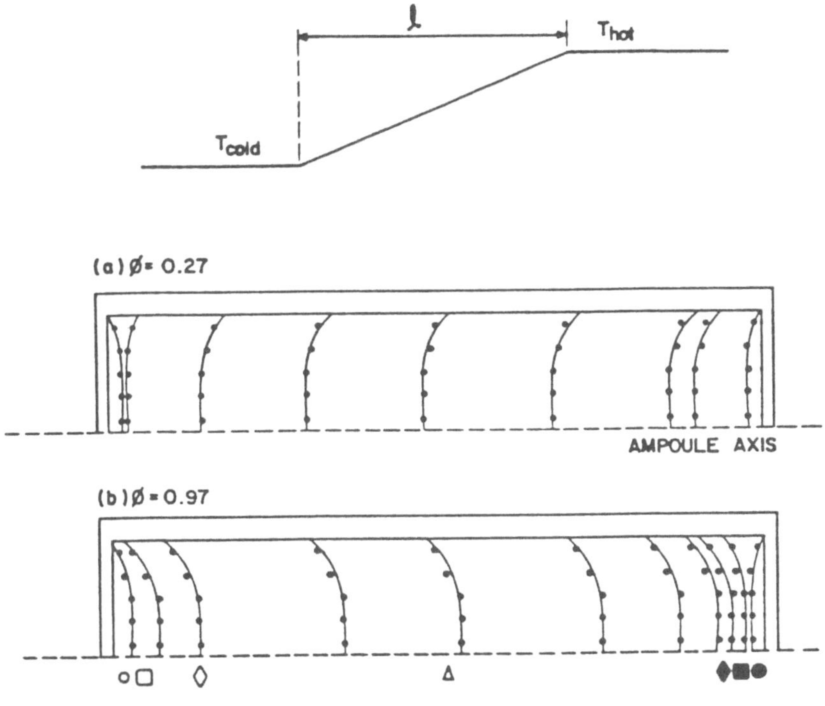

Fig. 6 Interface shape progression during directional solidification for two extremal values of the admensional interface temperature ϕ. The furnace profile moves from left to right.

Machine Timming All calculations were done on a simple PC-XT 8MHz microcomputer, with Turbo-Pascal and a 8087 coprocessor. No efforts were made to optimize the program. A typical complete run, such as the ones corresponding to Figs. 6 and 7, took nearly 20 hours of processing.

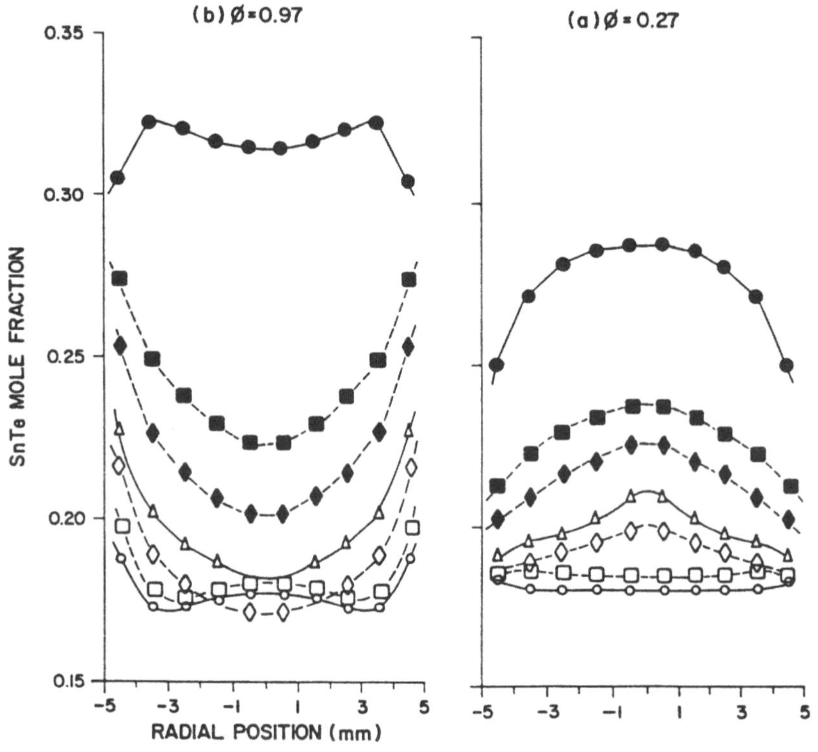

Fig. 7 Radial concentration profiles corresponding to Figs. (6a) and (6b). The values of concentration represent mean values over a 1mm - thick waffer, cut at the positions indicated in Fig. (6b).

DISCUSSION

The method here investigated seems to have the required flexibility to handle in detail the binary alloy solidification problem, when dealing with directional solidification processes. Its reliability lies both on a physical sounding basis for the modelling of the binary alloy phase change, and on an implicit control-volume implementation, which can handle almost any sort of subtleties in the boundary conditions and material properties. As mentioned before, convection in the melt and thermo-diffusion cross effects can be incorporated, without essential modifications. Nevertheless, some basic questions pertaining to free-boundary solidification problems remain to be addressed, particularly with respect to the interface morphology. It would be interesting to investigate if the use of a fixed grid is adequate to describe the richness of possibilities displayed in the $S - L$ interface geometry, specially when surface tension plays a significant role. Furthermore, the inclusion of non-equilibrium effects would probable require some reformulation in the basic control-volume mean-field approach.

It should be noted that the mean-field approximation for the "mushy" cell properties is subjected to some criticism, for it pressuposes an homogeneous C – field to prevail both at the solid (C_{sol}) and liquid (C_{liq}) components of a cell undergoing phase change; this seems to be reasonable only when the cell size is much less than the diffusion boundary layers in the liquid and solid phases. Application of the method to rapid freezing of alloys and to supercooled melts would need further refinements.

REFERENCES

1. Jasinski, T. and Witt, A.F. On Control of the Crystal-Melt Interface shape During Growth in a Vertical Bridgman Configuration, J. Crystal Growth, Vol. 71, p.p. 295-304, 1985.

2. Lehoczky, S.L. and Szofran, F.R. Directional Solidification and Characterization of $Hg_{1-x}Cd_xTe$ Alloys, in Materials Processing in the Reduced Gravity Environment of Space (Ed.Rindone, G.E.), pp. 409-420, North-Holland, Amsterdam, 1982.

3. Capper, P. et ali Quenching Studies in Bridgman-Grown $CdHgTe$, J. Crystal Growth, Vol. 63, pp. 154-164, 1983.

4. Huang, Y.; Debram, W.J. and Fripp, A.L. Interface Shapes During Vertical Bridgman Growth of (Pb, Sn)Te Crystals, J. Crystal Growth, Vol. 104, pp.315-326, 1990.

5. Jasinski, T.; Witt, A.F. and Rohsenow, W.M. Heat Transfer Analysis of the Bridgman - Stockbarger Configuration for Crystal Growth II. Analytical Treatment of Radial Temperature Variations, J. Crystal Growth, Vol. 67, pp.173-184, 1984.

6. Naumann, R.J. and Lehoczky, S.L. Effect of Variable Thermal Conductivity on Isotherms in Bridgmann Growth, J. Crystal Growth, Vol. 61, pp.707-710, 1983.

7. Motakef, S. Interference of Buoyancy-Induced Convection with Segregation During Directional Solidification: Scaling Laws, J. Crystal Growth, Vol. 102, pp. 197-213, 1990.

8. Szekely, J. and Themlis, N.J. Rate Phenomena in Process Metallurgy, Chapter 10, Interscience, New York, 1970.

9. Chang, C.J. and Brown, R.A. Radial Segregation Induced by Natural Convection and Melt/Solid Interface Shape in Vertical Bridgman Growth, J. Crystal Growth, Vol. 63, pp.343-364, 1983.

10. Derby, J.J. and Brown R.A. A Fully Implicit Method for Simulation of the One - Dimensional Solidification of a Binary Alloy, Chem. Eng. Sci., Vol. 41, pp.37-46, 1986.

11. Solomon, A.D.; Alexiades, V. and Wilson, D.G. A Numerical Simulation of a Binary Alloy Solidification Process, SIAM J. Sci. Stat. Comput., Vol 6(4), pp.911-922, 1985.

12. Fabbri, M. Fixed-Grid Simulation of Phase Change Problems: The Mushy-Cell Approach, pp.393-398, Proceedings of the III Brazilian Thermal Science Meeting, Itapema, SC, Brazil, 1990.

13. Voller, R. Implicit Finite-difference Solutions of the Enthalpy Formulation of Stefan Problems, IMA J. Num. Analysis, Vol.5, pp.201-214, 1985.

14. Voller, R.; Cross, M. and Markatos, N.C. An Enthalphy Method for Convection-Diffusion Phase Change, Int. J. Num. Meth. Eng., Vol. 24, pp.271- 284, 1987.

15. Wilson, D.G.; Alexiades, V. and Solomon, A.D. The 1984 Model of Binary Alloy Solidification, in Free Boundary Problems: Applications and Theory (Eds. Bossavit, A.; Damlamian, A. and Fremond, M.), Vol. III, pp.208-221, Pitman, 1985.

16. Patankar, S.V. Numerical Heat Transfer and Fluid Flow, Hemisphere, McGraw Hill, N.Y., 1980.

17. Kinoshita, K. and Sugii, K. PbTe-SnTe Mutual Diffusion Coefficient at Just Above the $Pb_{0.8}Sn_{0.2}Te$ Solidus Temperature, J. Crystal Growth, Vol. 67, pp. 375-379, 1984.

18. Tiller, W.A. et ali, The Redistribution of Solute Atoms During the Solidification of Metals, Acta Metall., Vol. 1, pp.428-437, 1953.

Mauricio Fabbri
Laboratory for Materials and Sensors
National Space Research Institute
C.P. 515, 12225 São José dos Campos, S.P.
Brazil

Melting of Anisotropic Crystals by the Boundary-Domain Integral Method

B. Sarler, A. Alujevic

Jozef Stefan Institute, University of Ljubljana, Yugoslavia

ABSTRACT

This paper presents the enthalpy based boundary-domain element method to cope with multidimensional melting or solidification of a pure substance (diffusive Stefan problem). Convective heat transport in the liquid phase is neglected and no mass transfer across interphase boundary is assumed. Material properties of solid and liquid phase are constant but could differ. Thermal conductivity of the solid phase is assumed anisotropic. All types of technically relevant boundary conditions are introduced. The tracking of a moving boundary is avoided by using the enthalpy transformation which treats each phase alternately. The resulting decoupled all-solid and all-liquid fixed domain diffusion problems are directly inversely formulated by the use of Green's function of the Fourier equation and solved by the introduction of constant boundary element and domain cell geometry splines, and constant time splines. As a case study, the effects of anisotropy are investigated by melting and solidification of a square with endothermic and exothermic heat sources and different orientations of crystalographic axes.

INTRODUCTION

The computational modelling of solid-liquid phase change has become a highly popular subject due to its pronounced influence in the development of the advanced technologies. An overview on related interdisciplinary research which involves theoretical, experimental, and computational aspects of phase transformation kinetics, solid mechanics, and transport phenomena could be found in the latest [1,2] detailed survey articles. An extensive collection of up to date references of the discipline, including principal key-words is available in [3].

The physical properties of materials are in general dependent upon the direction. This is due to anisotropy of crystals on microscopic scale, due to prefered collective direction of anisotropic crystals in polycristalline objects, or due to orientation of different constituents on macroscopic scale. As such, numerous applications necessitate the inclusion of anisotropic thermophysical properties when modelling melting or solidification. Despite that only two papers are known to the present authors on the subject:

In 1985 Gau and Viskanta published results [4] of a quasi-one-dimensional phase change experimental study with material (gallium) which has anisotropic thermal conductivity. For the conditions considered, the anisotropy had no influence during melting from above but the solidification from below was approximately 15% faster as predicted by Stefan's solution tuned to finite geometry.

Few years later Weaver and Viskanta [5] for the first time numerically investigate the effects of Stefan number, anisotropic thermal conductivity, aspect ratio, and initial liquid superheat on heat conduction of pure linear phase change material confined to a rectangular domain. The resulting anisotropic Stefan interphase conditions and energy equations for solid and liquid phase were solved by coordinate transformation, fully implicite finite difference scheme and resultant front tracking. They found out that anisotropy skews the solid-liquid interphase in contrast to a straight line in isotrophic case.

The discussed [4,5] anisotropic phenomena are, due to the relatively small amount of work done to date not yet sufficiently understood and additional studies are needed.

The discrete approximative methods based on front tracking (or two domain) method strategy are recognized as more accurate but less general as the fixed (or one) domain ones and thus hard to be used in a heat diffusion code able to support a broad spectra of different phase change situations. The extension of the fixed-domain principle to situations where the solid phase exhibits anisotropic behaviour was the principal incitement for present research. For this purpose recently developed [6] alternating phase truncation boundary element method and Chang's et al. [7] excellently detailed and general Green's function treatment of steady and transient heat conduction in anisotropic media have been used as a basis. The alternating phase truncation approach currently presents the only straightforward boundary element method alternative to cope with common physical situation where the thermal properties of phases differe.

The status of other boundary element applications associated with the multidimensional domain type phase change heat transfer modelling is as follows: Hong et al. [8] estimated the effect of latent heat on a temperature recovery basis, which is suitable not only for congruently melting materials, but also for alloys which solidify over a wide range of temperatures. The approximation was checked by comparison with the finite difference results for equal and constant thermal properties of the solid and liquid phase only. Their method could be easily implemented into any boundary element method for heat diffusion. In 1987 Tiba [9] formulated the multi-dimensional diffusive Stefan problem on the basis of the direct inverse formulation by using Green's functions of the Laplace equation in connection with incomplete enthalpy formulation. Consequently a non-linear algebraic equation system result, which has to be solved iteratively during each timestep. It also demonstrates the use of the method for different thermal diffusivities of phases but the extension of the method to cope with different thermal conductivities seems rather complicated. Due to the nature of the Green's function short timesteps have to be used.

The research on boundary element method [10] for heat transfer problems still lags theory and practical applications of finite difference and finite element method by approximately a decade. The present study is somehow an exception, since it probably appears to present the first domain-type attempt to solve an anisotropic melting or solidification problem.

FORMULATION

GOVERNING EQUATIONS

The melting or solidification problem for a pure substance, without convection in liquid phase, without mass transfer across interphase boundary, and linear material behaviour could be mathematically formulated as follows:

Material properties and geometry. Consider a connected fixed domain Ω with boundary Γ occupied by phase change material, represented by solid $\underline{K_S}$, $æ_S$ and liquid $\underline{K_L}$, $æ_L$ phase thermal conductivity and diffusivity, volumetric latent heat of phase change H^*, and melting temperature T^*. Sub-domain Ω_S of Ω is occupied by solid and Ω_L by liquid phases. The part of Γ which bounds Ω_S is denoted as Γ_S and similarly, Γ_L is the part of Γ which bounds Ω_L:

$$\Omega = \Omega_S + \Omega_L, \qquad \Gamma = \Gamma_S + \Gamma_L \tag{1}$$

The boundary between phases is denoted as Γ_{SL}. Sub-domain Ω_n, sub-boundary Γ_n; $n = S, L$, and interphase boundary Γ_{SL} in general need not to be connected.

Governing equation. The thermal field with temperature $T(p,\tau)$ and heat flux $F(p,\tau) = -\underline{K}(p)\cdot grad[T(p,\tau)]$ on Ω at point P and time τ is governed by the following non-homogenous diffusion equation with heat sources $A(p,\tau)$:

$$[K(p)/æ(p)]\cdot T(p,\tau)_{,\tau} = div\{\underline{K}(p)\cdot grad[T(p,\tau)]\} + A(p,\tau) \tag{2}$$

$$\underline{K}(p) = \left[\begin{array}{ll} \underline{K_S}; & P \in \Omega_S + \Gamma_S, \\ \underline{K_L}; & P \in \Omega_L + \Gamma_L \end{array}\right. \qquad æ(p) = \left[\begin{array}{ll} æ_S; & P \in \Omega_S + \Gamma_S \\ æ_L; & P \in \Omega_L + \Gamma_L \end{array}\right. \tag{3}$$

Since the solid phase is anisotropic and the liquid phase isotropic the form of conductivity tensors $\underline{K_S}$ and $\underline{K_L}$ and their inverses $\underline{K^S}$ and $\underline{K^L}$ in Cartesian coordinates are:

$$\underline{K_S} \equiv |K_{Sij}|, \quad \underline{K_L} \equiv |\delta_{ij}|, \quad K_{nij}\cdot K^{nij} = \delta_i{}^j;$$

$$\delta \equiv \text{Kronecker function} \tag{4}$$

$$K_n = det|K_{nij}|^{1/d}; \; d \equiv \text{dimensionality of the problem} \tag{5}$$

The second order tensors $\underline{K_S}$ and $\underline{K_L}$ should be symmetric and positive definite according to irreversible thermodynamics.

We seek the thermal field at time $\tau = t$ subject to the following initial conditions at $\tau \equiv 0$ and boundary conditions during time $0 \leq \tau \leq t$:

<u>Initial conditions.</u> Initial temperature $T(\mathbf{p},0)$ and phase $\Pi(\mathbf{p},0)$ on Ω is defined through known functions $T_0(\mathbf{p},0)$ and $\Pi_0(\mathbf{p},0)$ respectively:

$$T(\mathbf{p},0) = T_0(\mathbf{p},0), \qquad \Pi(\mathbf{p},0) = \Pi_0(\mathbf{p},0); \qquad P \in \Omega + \Gamma \tag{6}$$

<u>Boundary conditions.</u> The boundary Γ is, independent from Γ_S and Γ_L, divided into parts $\underline{\Gamma}^d$, $\underline{\Gamma}^n$, and $\underline{\Gamma}^r$:

$$\Gamma = \underline{\Gamma}^d + \underline{\Gamma}^n + \underline{\Gamma}^r \tag{7}$$

with Dirichlet, Neumann, and Robin boundary conditions respectively. On $\underline{\Gamma}^d$ the temperature is defined through function $T_0(\mathbf{p},\tau)$, on $\underline{\Gamma}^n$ the heat flow $F(\mathbf{p},\tau) = \mathbf{F}(\mathbf{p},\tau) \cdot \mathbf{n}(\mathbf{p})$, ($\mathbf{n}(\mathbf{p})$ presents the outward normal on Γ at point $P \in \Gamma$), is defined through function $F_0(\mathbf{p},\tau)$, and on $\underline{\Gamma}^r$ temperature or heat flow is defined with reference to known temperature $T_r(\mathbf{p},\tau)$ and heat transfer coefficient $h(\mathbf{p},\tau)$:

$$T(\mathbf{p},\tau) = T_0(\mathbf{p},\tau); \qquad P \in \underline{\Gamma}^d \tag{8}$$
$$F(\mathbf{p},\tau) = F_0(\mathbf{p},\tau); \qquad P \in \underline{\Gamma}^n \tag{9}$$
$$F(\mathbf{p},\tau) = h(\mathbf{p},\tau) \cdot [T(\mathbf{p},\tau) - T_r(\mathbf{p},\tau)]; \qquad P \in \underline{\Gamma}^r \tag{10}$$

<u>Stefan interphase conditions.</u> Temperature T_S on solid and T_L on liquid side (no subcooling or overheating is assumed) of interphase boundary Γ_{SL} is equal to phase change temperature and, since no mass transfer across Γ_{SL} is assumed, the velocity $V(\mathbf{p},\tau)$ of point P on Γ_{SL} is proportional to the discontinuity of solid F_S and liquid F_L side heat flux on it:

$$T_S(\mathbf{p},\tau) = T^{\bullet} = T_L(\mathbf{p},\tau); \qquad P \in \Gamma_{SL} \tag{11}$$
$$H^{\bullet} \cdot \mathbf{V}(\mathbf{p},\tau) = F_S(\mathbf{p},\tau) - F_L(\mathbf{p},\tau); \qquad P \in \Gamma_{SL} \tag{12}$$

<u>ENTHALPY FORMULATION OF DIFFUSION</u>

In order to solve the previously defined problem, the enthalpy approach [11] is chosen, which is one of the alternatives leading to the fixed domain method. With this concept we first introduce an enthalpy function which presents the sum of sensible and the latent heat required for a phase change and relate it with temperature and phase.

<u>Relations between enthalpy, temperature, and phase.</u> For the postulated pure phase change material the enthalpy H is defined through temperature and phase:

$$H(T,\Pi) = \begin{cases} (K_S/\text{æ}_S) \cdot (T - T^0); & T < T^{\bullet} \\ H_S; & T = T^{\bullet},\ \Pi = S \\ H_L; & T = T^{\bullet},\ \Pi = L \\ H_L + (K_L/\text{æ}_L) \cdot (T - T^{\bullet}); & T > T^{\bullet} \end{cases} \tag{13}$$

and temperature through enthalpy by the following relationship:

$$T(H) = \begin{cases} T^0 + (æ_S/K_S)\cdot H; & H < H_S \\ T^*; & H_S \le H \le H_L \\ T^* + (æ_L/K_L)\cdot(H - H_L); & H > H_L \end{cases} \qquad (14)$$

In the above two equations T^0 represents enthalpy reference temperature which is lower than T^*, H_S is maximum enthalpy of solid phase, and H_L is minimum enthalpy of liquid phase:

$$H_S = (æ_S/K_S)\cdot(T^* - T^0), \qquad H_L = H_S + H^* \qquad (15)$$

Phase is for $T(p,\tau) \neq T^*$ function of temperature only, and for the case $T = T^*$ additionally function of phase at some previous time:

$$\Pi(T,\Pi_0) = \begin{cases} S; & T < T^* \\ \Pi_0; & T = T^*, \\ L; & T > T^* \end{cases} \qquad \Pi_0 \equiv \Pi(p,\tau-\delta\tau);\ \delta\tau > 0 \qquad (16)$$

Now we rewrite the equation *(3)* by taking enthalpy as an independent variable:

$$[K(p)/æ(p)]\cdot T(H(p,\tau)),_\tau = div\{\underline{K}(p)\cdot grad[T(H(p,\tau))]\} + A(p,\tau) \qquad (17)$$

This equation already implicitly includes Stefan boundary conditions *(11,12)* in points where:

$$H(p,\tau) = \tfrac{1}{2}\cdot(H_S + H_L) \qquad (18)$$

BOUNDARY ELEMENT FORMULATION FOR ONE-PHASE PROBLEMS

We first assume one-phase diffusive Stefan problems which occur by freezing the liquid initially at fussion temperature (heat diffusion takes place in the solid phase only) or melting the solid initially at fussion temperature (heat diffusion takes place in the liquid phase only). This means that we deal either with solid or either with liquid thermal properties and equation *(17)* could be easily transformed to standard direct inverse form by weighting it (depending in which phase diffusion takes place) with solid T^{S*} or liquid T^{L*} Green's function of the Fourier equation during time $0 \le \tau \le t$ on the whole fixed domain Ω. This gives us, after some manipulations involving the Green's second identity, the desired boundary element method weighted residual scheme:

$$\frac{K_\Pi}{æ_\Pi}\cdot\int_0^t\int_\Omega [A(p,\tau)\cdot T^{\Pi*}(p,t;s,\tau)]\cdot d\Omega\cdot d\tau + \int_0^t\int_\Gamma [H(p,\tau)\cdot F^{\Pi*}(p,t;s,\tau) - Q(p,\tau)\cdot T^{\Pi*}(p,t;s,\tau)]\cdot d\Gamma\cdot d\tau =$$
$$= \frac{K_\Pi}{æ_\Pi}\cdot c_s\cdot H(s,t) - \frac{K_\Pi}{æ_\Pi}\cdot\int_\Omega [H(p,0)\cdot T^{\Pi*}(p,t,s,0)]\cdot d\Omega \qquad (19)$$

p and s are the position vectors of field point P,

and source point S. $F^{\eta *}$ is fundamental flux of the Fourier equation, defined by:

$$T^{\eta *} = a^{d/2}\cdot \exp[-a\cdot K_\eta \cdot K_\eta{}^{ij}\cdot(p_i-s_i)\cdot(p_j-s_j)];\quad a=1/[4\cdot \ae_\eta\cdot(t-\tau)]$$
$$F^{\eta *}(\mathbf{p},t;\mathbf{s},\tau) = -\ \underline{K_\eta}\cdot \mathrm{grad}[T^{\eta *}(\mathbf{p},t;\mathbf{s},\tau)] \tag{20}$$

p_i presents i-th Cartesian coordinate of point P. The introduction of enthalpy flux $\mathbf{Q}$ and enthalpy flow Q is a special consequence of the inverse formulation:

$$\mathbf{Q}(\mathbf{p},\tau) = -\ \underline{K}(\mathbf{p})\cdot \mathrm{grad}[\ H(\mathbf{p},\tau)\ ] \tag{21}$$
$$Q(\mathbf{p},\tau) = \mathbf{Q}(\mathbf{p},\tau)\cdot \mathbf{n}(\mathbf{p}) \tag{22}$$

Enthalpy flow and heat flow are related through:

$$Q(F,\eta) = \left[\begin{array}{l} (K_S/\ae_S)\cdot F;\ \eta = S,\\ (K_L/\ae_L)\cdot F;\ \eta = L \end{array}\right. \qquad F(Q,\eta) = \left[\begin{array}{l} (\ae_S/K_S)\cdot Q;\ \eta = S\\ (\ae_L/K_L)\cdot Q;\ \eta = L \end{array}\right. \tag{23}$$

The integral equation *(19)*, which includes Stefan interphase conditions, is applicable for one-phase problems (η represents the phase in which heat diffusion takes place), and for particular two-phase cases with $K_\eta \equiv K_S = K_L$ and $\ae_\eta \equiv \ae_S = \ae_L$.

<u>Solution procedure.</u> The one-phase problems and particular two-phase problems are solved by the classical enthalpy method:

- Initial and boundary conditions are transformed from temperature into enthalpy system:

$$H(\mathbf{p},0) = H[T_0(\mathbf{p},0),\eta_0(\mathbf{p},0)];\qquad P\in\Omega \tag{24}$$
$$H(\mathbf{p},\tau) = H[T_0(\mathbf{p},\tau),\eta(\mathbf{p},\tau)];\qquad P\in\Gamma^d \tag{25}$$
$$Q(\mathbf{p},\tau) = Q[F_0(\mathbf{p},\tau),\eta(\mathbf{p},\tau)];\qquad P\in\Gamma^n \tag{26}$$
$$Q(\mathbf{p},\tau) = Q\Big[h(\mathbf{p},\tau)\cdot\{T[H(\mathbf{p},\tau)]-T_r(\mathbf{p},\tau)\},\eta(\mathbf{p},\tau)\Big];\quad P\in\Gamma^r \tag{27}$$

- Integral equation *(19)* is solved by the boundary element method.
- Calculated thermal field variables are transformed back from enthalpy into temperature system:

$$\eta(\mathbf{p},t) = \eta[H(\mathbf{p},t),\eta(\mathbf{p},0)];\qquad P\in\Omega+\Gamma \tag{28}$$
$$T(\mathbf{p},t) = T[H(\mathbf{p},t)];\qquad P\in\Omega+\Gamma^n+\Gamma^r \tag{29}$$
$$F(\mathbf{p},t) = Q[F(\mathbf{p},t),\eta(\mathbf{p},t)];\qquad P\in\Gamma^d+\Gamma^r \tag{30}$$

<u>Remarks on boundary element method solution procedure.</u> The discretization equation *(19)* is, for the sake of simplicity, and to show the bounding accuracy of the lowest possible order, geometrically approximated (two-dimensional planar symmetry) by subdivision of resulting domain into M_Ω polygons (domain cells Ω_M) and the resulting boundary with N_Γ straight lines (boundary elements Γ_N). The field on each of them is approximated by constant space and time splines and their coefficients. The spline coefficients are chosen to match the geometrically centered ($\mathbf{p}_M$ and $\mathbf{p}_N$ *are center* points of the domain cell elements and boundary elements respectively) start (initial conditions), mean (heat sources),

and end (boundary variables) time values in each of the elements:

$$\int_{\Omega} H(p,0)\cdot T^{n*}(p,t;s,0)\cdot d\Omega \approx \sum_{M=1}^{M_\Omega} H(p_M,0)\cdot\int_{\Omega_M} T^{n*}(p,t;s,0)\cdot d\Omega \tag{31}$$

$$\int_0^t\int_{\Omega} A\cdot T^{n*}\cdot d\Omega\cdot d\tau \approx \sum_{M=1}^{M_\Omega} A(p_M,t/2)\cdot\int_0^t\int_{\Omega_M} T^{n*}\cdot d\Omega\cdot d\tau \tag{32}$$

$$\int_0^t\int_{\Gamma} H\cdot F^{n*}\cdot d\Gamma\cdot d\tau \approx \sum_{N=1}^{N_\Gamma} H(p_N,t)\cdot\int_0^t\int_{\Gamma_N} F^{n*}\cdot d\Gamma\cdot d\tau \tag{33}$$

$$\int_0^t\int_{\Gamma} Q\cdot T^{n*}\cdot d\Gamma\cdot d\tau \approx \sum_{N=1}^{N_\Gamma} Q(p_N,t)\cdot\int_0^t\int_{\Gamma_N} T^{n*}\cdot d\Gamma\cdot d\tau \tag{34}$$

The non-singular ($P \neq S$) integrals on the right side of the equations *(32,33,34)* are calculated by six-point Gaussian quadrature on the boundary, and by three (triangles) or nine (quadrilaterals) point Hammer's roule in the domain. Singular integrals ($P = S$) are calculated analytically as far as possible. The system of algebraic equations is formed from the approximation of the discretization equations by writing them for source point S coinciding with the meshpoints of boundary elements ($s \equiv p_N;\ N = 1,..,N_\Gamma$). The unknowns on the boundary are calculated from this system and then, explicitly through *(19)*, the unknowns on the domain at each timestep are determined. For further information on this standard boundary element method manipulations see for example [10].

BOUNDARY ELEMENT FORMULATION FOR TWO-PHASE PROBLEMS

Next we assume two-phase diffusive Stefan problems with different material properties of the phases. In order to conserve the domain type character of the method we proceed by the alternating phase truncation enthalpy transformations:

- The governing equation, initial, and boundary conditions are formulated in enthalpy system.
- The calculation over a timestep is divided into two half-steps: (similar decomposition in phase sense as the alternating direction finite difference method in the geometrical sense) the all-liquid and the all-solid one:
  In the all-liquid half-step, enthalpy is locally added to convert the phase change material occupying $\Omega_S + \Gamma_S$ (Ω_S and Γ_S are based on initial conditions) to the liquid phase with enthalpy H_L. Heat sources and boundary conditions are at these portions of the domain and boundary modified to preserve steady conditions with $H = H_L$ there. After the calculation (with timestep *t*) of all-liquid intermediate thermal field the enthalpy that was locally added is now locally subtracted to yield an intermediate

thermal field (first enthalpy conservation step).
In the all-solid half-step, enthalpy is locally subtracted to convert the phase change material occupying $\Omega_L + \Gamma_L$ ($\Omega_L + \Gamma_L$ are based on intermediate thermal field) to the solid phase with enthalpy H_S. Heat sources and boundary conditions are at these portions of the domain and boundary modified to preserve steady conditions with $H = H_S$ there. After the calculation (with timestep t) of all-solid thermal field the enthalpy that was locally subtracted is now locally added to yield the final thermal field (second enthalpy conservation step).

- The calculated enthalpy formulated thermal field at time t is converted to the temperature system, and if needed, the new interphase boundary position is determined.

Enthalpy transformations of all-liquid half-step. The all-liquid calculation is prepared by the introduction of enthalpy field variables $H^L(p,\tau)$ and $Q^L(p,\tau)$ of the liquid half-step, function $H'^L(p)$, which locally measures the added enthalpy to the initially solid phase, and function $A^L(p,\tau)$ which represents the modified distribution of the heat sorces $A(p,\tau)$. The initial and boundary conditions for this half-step are manipulated as follows:

$$H^L(p,0) = \max\{\, H(p,0),\ H_L \,\},\qquad H'^L(p) = H^L(p,0) - H(p,0); \qquad P \in \Omega \tag{35}$$

$$H^L(p,t) = \max\{\, H[T_0(p,t),\Omega_0(p,0)],\ H_L \,\},\qquad H'^L(p) = H^L(p,t) - H[T_0(p,t),\Omega_0(p,0)]; \qquad P \in \Gamma^{\underline{d}} \tag{36}$$

$$H'^L(p) = \max\{\, H(p,0),\ H_L \,\} - H(p,0),$$
$$Q^L(p,t) = \begin{cases} 0[J\cdot W/(m^5\cdot {}^\circ K)]; & H(p,0) < H_L \\ Q[F_0(p,t),L]; & H(p,0) \geq H_L \end{cases} \qquad P \in \Gamma^{\underline{n}} \tag{37}$$

$$H'^L(p) = \max\{\, H(p,0),\ H_L \,\} - H(p,0),$$
$$Q^L(p,t) = \begin{cases} 0[J\cdot W/(m^5\cdot {}^\circ K)]; & H(p,0) < H_L \\ h(p,t)\cdot[H^L(p,t) - H_L] + \\ \quad + (K_L/\text{æ}_L)\cdot h(p,t)\cdot[T^\circ - T_r(p,t)]; & H(p,0) \geq H_L \end{cases} \qquad P \in \Gamma^{\underline{r}} \tag{38}$$

Modified distribution of the heat sources for the half-step is defined as:

$$A^L(p,\tau) = \begin{cases} 0[W/m^3]; & H(p,0) < \tfrac{1}{2}\cdot(H_S + H_L) \\ A(p,\tau); & H(p,0) \geq \tfrac{1}{2}\cdot(H_S + H_L) \end{cases} \qquad P \in \Omega \tag{39}$$

Then the equation *(19)* is written for $H = H^n$, $Q = Q^n$, $A = A^n$; $n = L$. After the calculation of the unknowns, the following quantities on the boundary:

$$Q^L(p,\tau) = Q\Big[h(p,\tau)\cdot\{T[H(p,\tau)] - T_r(p,\tau)\},L\Big]; \qquad P \in \Gamma^{\underline{r}} \tag{40}$$

$$F^L(p,t) = F[Q(p,t),L]; \qquad P \in \Gamma \text{ and } P \notin \Gamma^{\underline{n}} \tag{41}$$

and $H^L(p,t)$ (explicitly through *(19)* for $\Omega = L$) in the domain are additionally determined.

<u>First enthalpy conservation step.</u> Ater the entire first half-step enthalpy field $H^L(p,\tau)$; $P \in \Omega + \Gamma$ is obtained, the intermediate thermal field $H^*(p)$ is calculated through it and function $H'^L(p)$ by:

$$H^*(p) = H^L(p,t) - H'^L(p); \qquad P \in \Omega + \Gamma \qquad (42)$$

in order to subtract the initially added enthalpy.

<u>Enthalpy transformations of all-solid half-step.</u> The all-solid calculaton is prepared by the introduction of enthalpy field variables $H^S(p,\tau)$ and $Q^S(p,\tau)$ of the solid half-step, function $H'^S(p)$, which locally measures the subtracted enthalpy from the intermediately liquid phase, and function $A^S(p,\tau)$ which represents the modified distribution of the heat sorces $A(p,\tau)$. The initial and boundary conditions for this half-step are manipulated as:

$$\begin{aligned} H^S(p,0) &= \min\{ H^*(p),\ H_S \}, \\ H'^S(p) &= H^*(p) - H^S(p,0); \end{aligned} \qquad P \in \Omega \qquad (43)$$

$$\begin{aligned} H^S(p,t) &= \min\{ H[T_0(p,t),\Omega_0(p,0)],\ H_S \}, \\ H'^S(p) &= H[T_0(p,t),\Omega_0(p,0)] - H^S(p,t); \end{aligned} \qquad P \in \Gamma^d \qquad (44)$$

$$H'^S(p) = H^*(p) - \min\{ H^*(p),\ H_S \},$$

$$Q^S(p,t) = \begin{cases} Q[F_0(p,t),S]; & H^*(p) < H_S \\ 0[J\cdot W/(m^5\cdot {}^\circ K)]; & H^*(p) \ge H_S \end{cases} \qquad P \in \Gamma^n \qquad (45)$$

$$H'^S(p) = H^*(p) - \min\{ H^*(p),\ H_S \},$$

$$Q^S(p,t) = \begin{cases} h(p,t)\cdot H^S(p,t) + \\ + (K_S/æ_S)\cdot h(p,t)\cdot[T^0 - T_r(p,t)]; & H(p,0) \le H_S \\ 0[J\cdot W/(m^5\cdot {}^\circ K)]; & H(p,0) > H_S \end{cases} \qquad P \in \Gamma^r \qquad (46)$$

Modified distribution of the heat sources for the half-step is defined as:

$$A^S(p,\tau) = \begin{cases} A(p,\tau); & H(p,0) < \tfrac{1}{2}\cdot(H_S + H_L) \\ 0[W/m^3]; & H(p,0) \ge \tfrac{1}{2}\cdot(H_S + H_L) \end{cases} \qquad P \in \Omega \qquad (47)$$

Then the equation *(19)* is written for $H = H^\Omega$, $Q = Q^\Omega$, $A = A^\Omega$; $\Omega = S$. After the calculation of the unknowns the following quantities on the boundary:

$$Q^S(p,\tau) = Q\Big[h(p,\tau)\cdot\{T[H(p,\tau)] - T_r(p,\tau)\},S\Big]; \quad P \in \Gamma^r \qquad (48)$$

$$F^S(p,t) = F[Q(p,t),S]; \qquad P \in \Gamma \text{ and } P \notin \Gamma^n \qquad (49)$$

and $H^S(p,t)$ (*explicitly through (19)* for $\Omega = S$) in the domain are additionally determined.

Second enthalpy conservation step. After the entire second half-step enthalpy field $H^s(p,t)$; $p \in \Omega + \Gamma$ is obtained, the final timestep thermal field $H(p,t)$ is calculated through it and function $H'^s(p)$ by:

$$H(p,t) = H^s(p,t) + H'^s(p,t); \qquad p \in \Omega + \Gamma \qquad (50)$$

in order to add the initially subtracted enthalpy.

The enthalpy to temperature system conversion. Manipulations during timestep are finished with the calculation of the values of temperature system at time t on $\Omega + \Gamma$:

$$T(p,t) = T[H(p,t)]; \qquad p \in \Omega \qquad (51)$$
$$T(p,t) = T_0(p,t); \qquad p \in \Gamma^d \qquad (52)$$
$$T(p,t) = T[H(p,t)]; \qquad p \in \Gamma \text{ and } p \notin \Gamma^d \qquad (53)$$
$$F(p,t) = F_0(p,t); \qquad p \in \Gamma^n \qquad (54)$$
$$F(p,t) = F^s(p,t) + F^L(p,t); \qquad p \in \Gamma \text{ and } p \notin \Gamma^n \qquad (55)$$

The characteristic of this formulation is, that in case of constant, but in general different material properties of each phase, we solve two systems of linear equations with constant augmented matrix coefficients. Under the assumption, that manipulations between temperature and enthalpy system could be neglected with respect to time needed for set-up and inversion of these two systems, the solution of the diffusive Stefan problem requires approximately twice the computer time and memory as the ordinary diffusion one. (One hour CPU time on 0.13·MFLOPS MicroVAX-II computer with 64bit flotaing point arithmetics could handle 50 time-steps of a phase-change problem with internal heat sources, discretized into 80 constant boundary elements and 400 constant rectangular domain cells as presented in Figure 1). The interphase boundary Γ_{SL} position calculation, if not required as a part of the output information, need not to be performed at all.

NUMERICAL EXAMPLES

VERIFICATION OF THE METHOD

Convergence of the method has been extensively checked [6,12] by a number of test cases including temperature comparison with two closed form solutions (Dirichlet jump solidification of an infinite rectangular corner, Neumann melting of a square with space and time dependent heat sources), and interphase boundary position comparison (Dirichlet jump solidification in a square) with other numerical methods. The accuracy of the calculations of thermal field and interphase boundary position was in good agreement with closed form solutions and other published results. The formulation shows, as other enthalpy based methods, nonmonothone oscillations in case of coarse mesh and their disapperance with finer mesh.

The appropriate computer implementation of the anisotropic formulation has been verified by agreement with the non-melting results on circular geometry as defined in Figure 2 from [7]. Next the gallium solidification results from [4] have been recalculated. Uniform mesh with 80 equidistant boundary elements and 400 eqidistant constant square domain cells reproduce the Weaver and Viskanta's results in a square within 5% accuracy (estimated by visual comparison of the isotherms).

TEST EXAMPLES

For the six numerical examples explicitly shown in this paper, a square two-dimensional plane domain with Cartesian coordinates $-1[m] \leq p_1 \leq 1[m]$, $-1[m] \leq p_2 \leq 1[m]$ was selected. Thermal diffusivities of phases are $æ_S = æ_L = 1[s/m^2]$, melting point $T^{*} = 1.5[^{\circ}C]$, melting enthalpy $H^{*} = 1[J/m^3]$. Enthalpy reference temperature is $T^0 = -1[^{\circ}C]$. Liquid phase is isotropic with $K_L = 1[W/(m \cdot K)]$. The first three examples have uniform initial conditions and uniform Dirichlet boundary conditions defined by: $T_0(\mathbf{p},0) = 0[^{\circ}C]$; $\mathbf{p} \in \Omega + \Gamma$, $T_0(\mathbf{p},\tau) = 0[^{\circ}C]$; $\mathbf{p} \in \Gamma$, $0[s] \leq \tau \leq \infty[s]$. The last three examples have uniform initial conditions and uniform Dirichlet boundary conditions defined by: $T_0(\mathbf{p},0) = 3[^{\circ}C]$; $\mathbf{p} \in \Omega + \Gamma$, $T_0(\mathbf{p},\tau) = 3[^{\circ}C]$; $\mathbf{p} \in \Gamma$, $0[s] \leq \tau \leq \infty[s]$. The first three examples have uniform exothermic heat sources $A(\mathbf{p},\tau) = 10[W/m^3]$; $P \in \Omega$, $0[s] \leq \tau \leq \infty[s]$. The last three examples have uniform endothermic heat sources (sinks) $A(\mathbf{p},\tau) = -10[W/m^3]$; $P \in \Omega$, $0[s] \leq \tau \leq \infty[s]$. The solid phase is isotropic with $K_S = 1[W/(m \cdot K)]$ for the first and fourth case, anisotropic with $K_{S11} = 2$, $K_{S22} = 0.5$, $K_{S12} = 0.0[W/(m \cdot K)]$ for the second and fifth case, and anisotropic with $K_{S11} = 1.625$, $K_{S22} = 0.875$, $K_{S12} = -0.6495190[W/(m \cdot K)]$ for the third and sixth case. Third and sixth case represent a rotation of the second and fifth case respectively crystalographic axes for an angle of $\pi/6$ in positive direction. The boundary of the square is discretized by 84 equally spaced constant boundary elements and 441 equally spaced quadratic domain cells. The timestep used is 0.01[s].

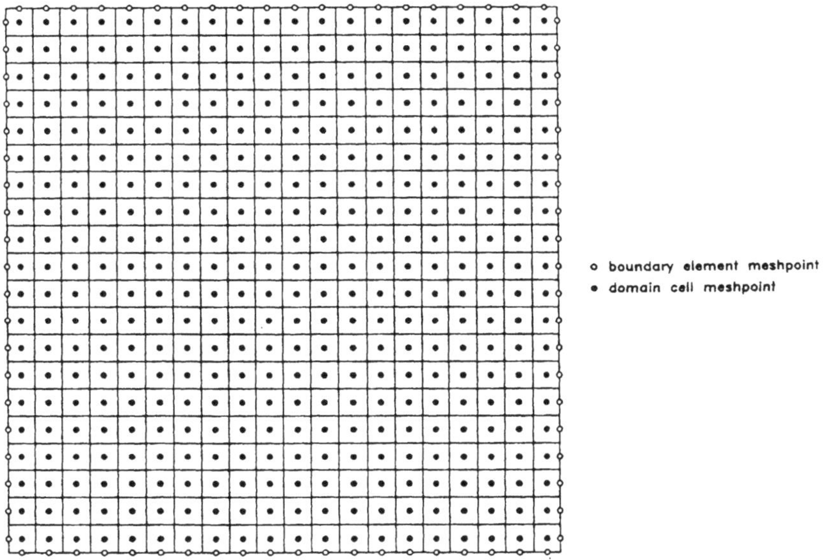

Figure 1: Scheme of the spatial discretization of a square.

τ	I		II		III		IV		V		VI	
	P1	P2	P1	P2	P1	P2	P1	P2	P1	P2	P1	P2
.0	0.00	0.00	0.00	0.00	0.00	0.00	3.00	3.00	3.00	3.00	3.00	3.00
.1	.973	.807	1.00	.823	1.03	.850	2.03	2.20	2.03	2.20	2.03	2.20
.2	1.50	1.23	1.50	1.22	1.50	1.30	1.50	1.78	1.50	1.78	1.50	1.78
.3	1.69	1.37	1.64	1.33	1.72	1.38	1.33	1.64	1.33	1.64	1.33	1.64
.4	2.07	1.40	1.94	1.35	2.02	1.39	.946	1.59	.946	1.59	.937	1.52
.5	2.30	1.45	2.11	1.36	2.20	1.40	.727	1.51	.718	1.52	.708	1.50
.6	2.45	1.50	2.24	1.37	2.32	1.40	.584	1.50	.547	1.50	.535	1.19
.7	2.59	1.68	2.31	1.37	2.42	1.50	.457	1.28	.443	1.28	.425	1.10
.8	2.67	1.73	2.37	1.38	2.47	1.50	.365	1.22	.361	1.24	.334	1.04
.9	2.72	1.75	2.41	1.50	2.50	1.58	.307	1.20	.304	1.21	.268	.989
1.	2.75	1.79	2.45	1.50	2.52	1.59	.278	1.19	.269	1.16	.229	.958

Table 1: Temperature [°C] at points *P1*: $p_1 = 0[m]$, $p_2 = 0[m]$ and *P2*: $p_1 = 0.47619[m]$, $p_2 = 0.47619[m]$, for the six calculated cases during time $\tau[s]$.

DISCUSSION OF THE RESULTS

Temperature histories at the center and on a point of the diagonal of the square for the six calculated cases are shown on Table 1. Figures from 2 to 7 illustrate the temperature distributions of the test cases at time *0.5[s]* where the melting or solidification is not finished yet. Figures 2, 3, and 4 show the melting and Figures 5, 6, and 7 the solidification process. Figures 2 and 5 are symmetric due to isotropic character of both phases. Figures 3 and 5 present the case where the othorombic crystallographic axes coincide with the coordinate axes. Since the thermal conductivity of the solid phase in p_1 direction is higher than in p_2 *direction, the solid phase in melting problem in* Figure 2 is elongated in p_1 and shortened in p_2 directions. Because of the same reason, the solid phase in solidification problem on Figure 6, is elongated in p_2 and shortened in p_1 directions. Figures 4 and 7 show the skewness of isotherms for melting and solidification cases where the crystal is not cut parallel to its symmetry axes. In the melting problem the skewness of the isotherms in solid phase is pronounced near the geometrycal borders of the square whereas in the case of solidification the skewness in liquid phase enhance with respect to the center of the square.

CONCLUSIONS

An analysis was made to solve a multidimensional general diffusive Stefan problem where the solid phase exhibits anisotropic behaviour. Due to the domain type of the approximation the method could serve as basis for the solution of a broad field of scientific and engineering applications. This method is currently in the stage of several physical and computational upgradings including binary-mixtures phase change formulation with convection in liquid phase [13], and use of dual reciprocity approach [14] which would enable reduction of the dimensionality also for such nonlinear parabolic type of problems.

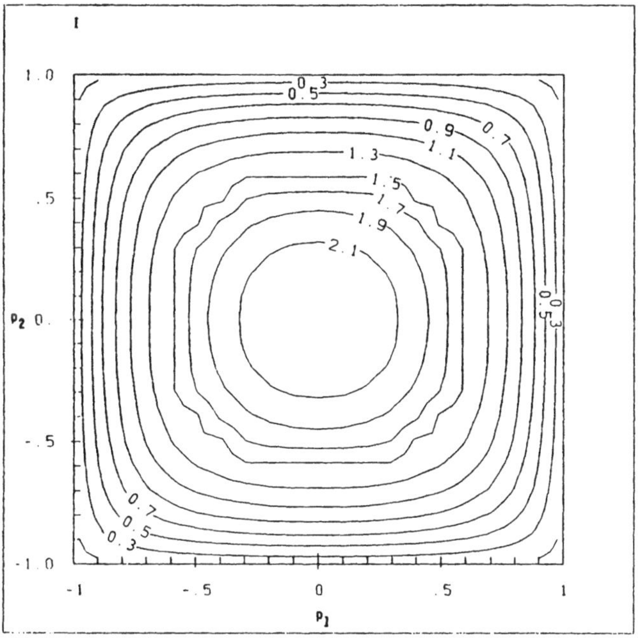

Figure 2: Isotherms of melting by exothermic heat sources at time τ = *0.5[s]*. Case I.

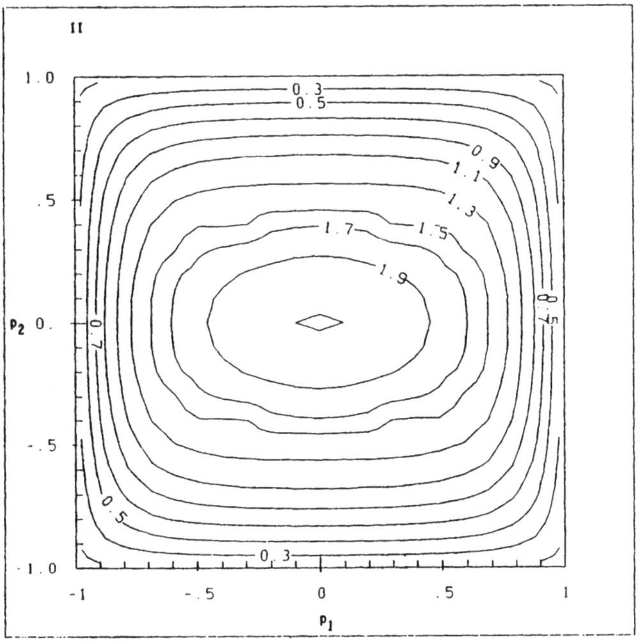

Figure 3: Isotherms of melting by exothermic heat sources at time τ = *0.5[s]*. Case II.

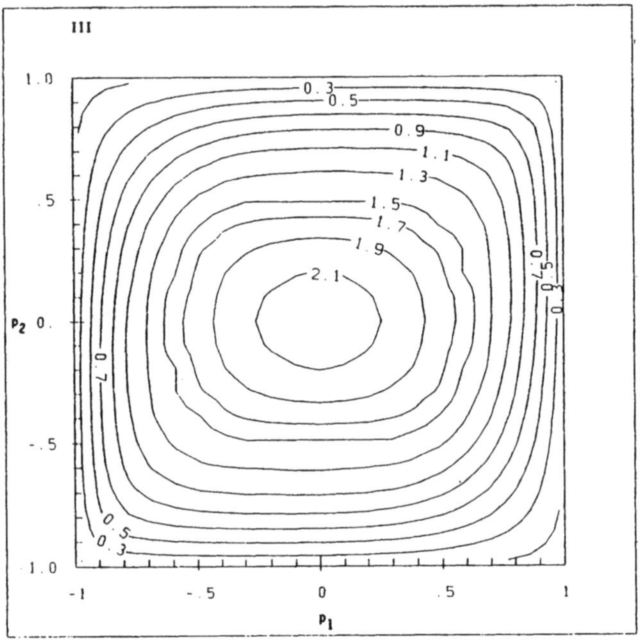

Figure 4: Isotherms of melting by exothermic heat sources at time τ = *0.5[s]*. Case III.

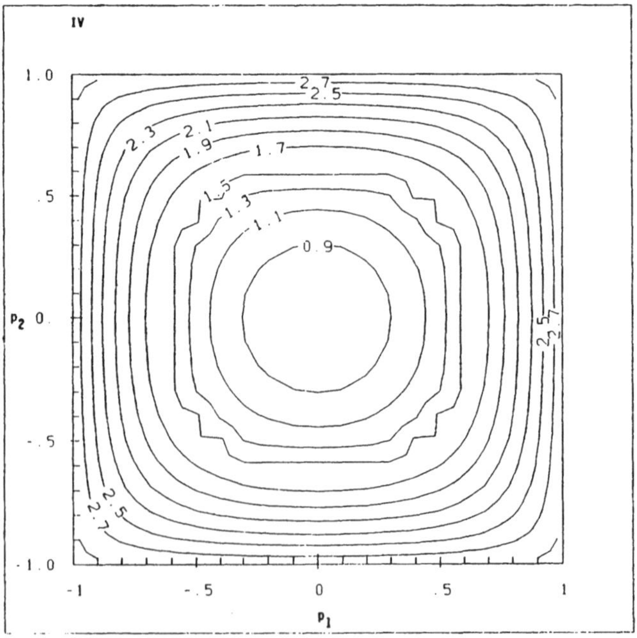

Figure 5: Isotherms of solidification by endothermic heat sources at time τ = *0.5[s]*. Case IV.

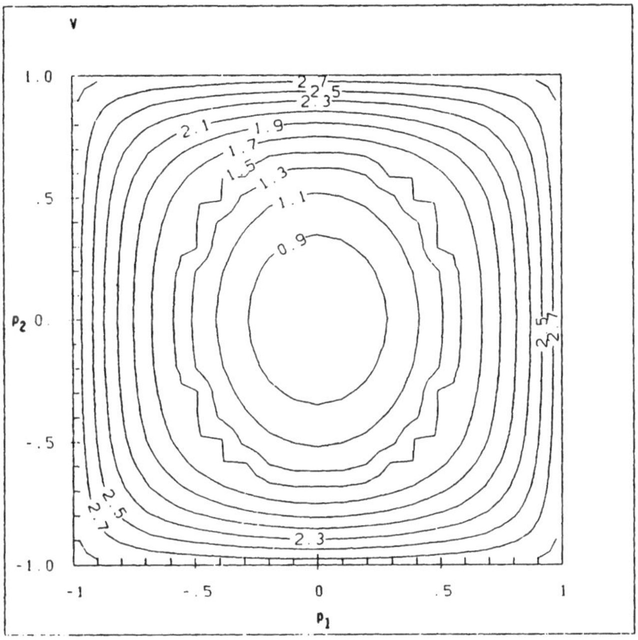

Figure 6: Isotherms of solidification by endothermic heat sources at time τ = *0.5[s]*. Case V.

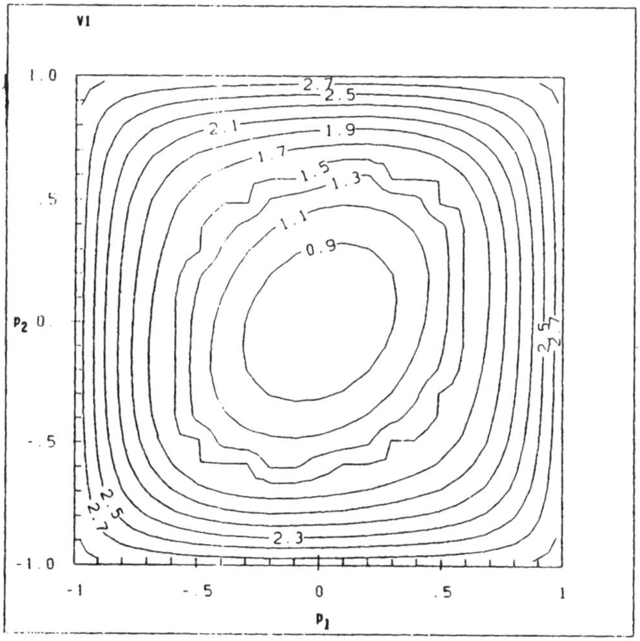

Figure 7: Isotherms of solidification by endothermic heat sources at time τ = *0.5[s]*. Case VI.

REFERENCES

[1] Yao, L.S. and Prusa, J. Melting and freezing, In: 'Advances in Heat Transfer, Vol.19', (Ed. Hartnett, J.P.), pp.1-95, 1989.
[2] Rappaz, M. Modelling of Microstructure Formation in Solidification Processes, Int.Mat.Reviews, Vol.34, pp.93-123, 1989.
[3] Sarler, B., Alujevic, A. and Malalan, J. Bibliography on Stefan problem, Technical Note IJS-DP-6077, Jozef Stefan Institute, Ljubljana, 1990.
[4] Gau, G. and Viskanta, R. Effect of Crystal Anisotropy on Heat Transfer During Melting and Solidification of a Metal, J.Heat Transfer, Vol.107, pp.706-708, 1985.
[5] Weaver, J.A. and Viskanta R. Effects of Anisotropic Heat Conduction on Solidification, Num.Heat Transfer, Part A, Vol.15, pp.181-195, 1989.
[6] Sarler, B., Alujevic, A., Kuhn, G. Solution of Multidimensional Stefan Problem by BEM, Z.angew.Math.Mech., Vol.71, pp.T611-T614, 1991.
[7] Chang, Y.P., Kang, C.S. and Chen, D.J. The Use of Fundamental Green's Functions for the Solution of Problems of Heat Conduction in Anisotropic Media, Int.J.Heat Mass Transfer, Vol.16, pp.1905-1918, 1973.
[8] Hong, C.P., Umeda, T. and Kimura, Y. Application of the Boundary Element Method in Two and Three Dimensional Unsteady Heat Transfer Involving Phase Change Solidification, In:' Boundary Elements, Proc.5th Int.Conf. on BEM, Hiroshima Japan', (Eds. Brebbia, C.A., Futagami, T. and Tanaka, M.), Springer-Verlag, Berlin and New York, 1983.
[9] Tiba, M. The Boundary Element Method in Two Phase Stefan's Problem, Engineering Analysis, Vol.4, pp.46-48, 1987.
[10] Wrobel, L.C. and Brebbia, C.A. (Eds.). Advances in Boundary Element Methods for Heat Transfer, Computational Mechanics Publications, Southampton, 1990.
[11] Voller, V.R., Swaminathan, C.R. and Thomas, B.G. Fixed Grid Techniques for Phase Change Problems: A Review, Int.J.Numer.Methods Eng., Vol.30, pp.875-898, 1990.
[12] Sarler, B., Alujevic, A. and Kuhn, G. Treatment of Multi-Dimensional Stefan Problems by the Alternating Phase Truncation Boundary Element Method, In: 'Boundary Elements in Machanical and Electrical Engineering', (Eds. Brebbia, C.A. and Chaudouet-Miranda, A.), Springer-Verlag, Berlin and New York, pp.371-391, 1990.
[13] Bennon, W.D. and Incropera, F.P. A Continuum Model for Momentum, Heat and Species Transport in Binary Solid-liquid Phase Change Systems, I. Model Formulation, Int.J.Heat Mass Transfer, Vol.30, (1987), pp.2161-2170.
[14] Partridge, P. W., Brebbia, C. A. and Wrobel, L.C. The Dual Reciprocity Boundary Element Method, Computational Mechanics Publications, Southampton, 1990.

An Efficient Finite Element Procedure for the Solution of Enthalpy Model for Phase Change Heat Conduction Problems

Z.-X. Gong (*), Y.-F. Zhang (*), A.S. Mujumdar (**)
(*) Tianjin Institute of Light Industry, Tianjin, P.R. China
(**) McGill University, Montreal, Quebec, Canada

ABSTRACT

An efficient finite element procedure incorporating a three -time-level scheme is proposed to solve the enthalpy model. This procedure does not need iteration. A newly-developed one-point quadrature program for the calculation of heat conductance matrices is used to further reduce the execution time.

INTRODUCTION

The finite-difference enthalpy model of Shamsundar and Sparrow [3] is one of the most widely used approaches for the numerical solution of phase change heat conduction problems. This approach uses the enthalpy as a dependent variable along with the temperature. Therefore, in the final discrete algebraic equation system (n equations contain 2n dependent variables), Gauss-Siedel iteration, or an equivalent nonsimultaneous iteration procedure has to be used to solve the algebraic equation system along with the equations of state. As such schemes provide slow convergence, especially due to the phase change nonlinearity, great number of iterations are required (particularly for two and three dimensional problems), resulting in very high computing time.

In this paper an efficient finite element procedure is proposed to solve multi-dimensional enthalpy model. By selecting the enthalpy rate as the to be solved variable and through analysis of the relationship between the enthalpy rate of and the temperature rate in phase change heat conduction process, it is found that Gauss elimination can be used to solve the algebraic equation system if a lumped capacity model is used and certain transformations of the matrix equation are made. A three-time-level scheme is also incorporated into the model. Thus, this procedure does not need iteration, which reduces greatly the execution time. To further reduce the execution time, a newly-developed one-point quadrature program [4] for the calculation of heat conductance matrices of bilinear four-node isoparametric elements is also used

MATHEMATICAL FORMULATION

According to Shamsundar and Sparrow [3], the governing equation for phase change heat conduction problems based on enthalpy can be described by

$$\frac{\partial(\rho H)}{\partial t} - \nabla\cdot(k\,\nabla T) = Q\ ,\ x \in \Omega \subset R^3,\ t \in [0,t] \tag{1}$$

in which H is the enthalpy per unit volume, Q the rate of internal heat generation, ρ the density, k the thermal conductivity, and

$$(\rho,\ k) = \begin{cases} \rho_s\ ,\ k_s & H < H_s^* \\ \rho_l\ ,\ k_l & H > H_l^* \end{cases} \tag{2}$$

Besides , there exist the following relationship between

enthalpy H and temperature T:

$$T - T_m = \begin{cases} (H-H_s^*)/c_s & H < H_s^* \\ 0 & H_s^* \le H \le H_l^* \\ (H-H_l^*)/c_l & H_l^* < H \end{cases} \quad (3)$$

where T_m is the melting temperature, c the specific heat, subscripts s and l indicate the solid and liquid phase, respectively, superscript $*$ represents the state of saturation.

FINITE ELEMENT DISCRETIZATION

The spacewise discretization of equation (1), subjected to the third kind of boundary condition can be accomplished using standard Galerkin's method [1].

Let the unknown function H and T be approximated, throughout the solution domain at any time t, by the relationship:

$$H = \sum_{i=1}^{n} N_i(x,y,z)\ H_i(t) \quad (4)$$

$$T = \sum_{i=1}^{n} N_i(x,y,z)\ T_i(t) \quad (5)$$

where N_i are the usual shape functions defined piecewise element by element, H_i and T_i being the nodal parameters. The simultaneous equations, allowing the solution for n values of H_i, are obtained typically for point i by equating to zero the integral over the domain Ω of the product of the weighing function N_i and the residual resulting from substitution into equation (1) of equation (4) and (5). After making use of Green's theorem, in order to avoid second derivatives in the integrals imposing unnecessary continuity conditions between elements, the n equations can be written down in a matrix form as:

$$[\,M\,]\,\{\,V_H\,\} + [\,K\,]\,\{\,T\,\} = \{\,F\,\} \quad (6)$$

Typical matrix elements are:

$$M_{ij} = \sum \int_{\Omega^e} \rho \, N_i \, N_j \, d\Omega \tag{7}$$

$$K_{ij} = \sum \int_{\Omega^e} k\left(\frac{\partial N_i}{\partial x}\frac{\partial N_j}{\partial x} + \frac{\partial N_i}{\partial y}\frac{\partial N_j}{\partial y} + \frac{\partial N_i}{\partial z}\frac{\partial N_j}{\partial z}\right) d\Omega$$

$$+ \sum \int_{\Gamma^e} h \, N_i \, N_j \, d\Gamma \tag{8}$$

$$F_i = \sum \int_{\Omega^e} Q \, N_i \, d\Omega + \sum \int_{\Gamma^e} h \, N_i \, T_f d\Gamma \tag{9}$$

In the above summations are taken over the contributions of each element, Ω^e is the element region and Γ^e refers only to elements with external boundaries on which the third kind of boundary condition is specified, h is the convection heat transfer coefficient.

It must be noted that the set of equations (6) is highly nonlinear due to the phase change nonlinearity.

TIME DISCRETIZATION

The discretization of the time derivative in equation (6) is most often achieved with a finite difference technique. Although many time-stepping schemes are available, the most popular are two-time-level methods in which iterations within each time step are required and therefore they are very time -consuming. To overcome this disadvantage a three-time-level scheme [5] is adopted. Therefore, equation (6) can be approximated as:

$$[M^*]\{V_H\} + [K^*]\{(0.5+\alpha)T^{n+1} + (0.5-2\alpha)T^n + \alpha T^{n-1}\}$$

$$= \{F^*\} \tag{10}$$

in which

$$\{V_H\} = \{H^{n+1} - H^n\}/\Delta t \tag{11}$$

and in correspondence with V_H

$$\{V_T\} = \{T^{n+1} - T^n\}/\Delta t \tag{12}$$

In equation (10) the superscript * designates that the matrices and vector are evaluated at time level t^*:

$$t^* = t_{n+\frac{1}{2}}$$

or

$$t^* = t_n$$

in correspondence with t^*

$$\{T^*\} = \{\frac{3}{2} T^n - \frac{1}{2} T^{n-1}\} \tag{13}$$

or

$$\{T^*\} = \{T^n\} \tag{14}$$

This three-time-level scheme was found, with $\alpha=\frac{1}{4}$, to work very satisfactory for a non-linear model equation [6] and to be stable, accurate and free from oscillations for solidification problems [7].

If the enthalpy rates are chosen as the to be solved variable and equation (12) is substituted into equation (10), it follows

$$[M^*]\{V_H\} + (0.5+\alpha)\Delta t[K^*]\{V_T\} = \{F^*\} - [K^*]\{(1-\alpha)T^n + \alpha T^{n-1}\} \tag{15}$$

Since there exists the following relationship between enthalpy and temperature for heat conduction problems without phase change,

$$H = \int C\, dT \tag{16}$$

we can get the following equation if a lumped heat capacity model [12] is adopted:

$$\{V_H\} = [C]\{V_T\} \tag{17}$$

in the equations above, [C] is the heat capacity matrix and it is diagonal, superscript e designates an element.

Equation (17) is not true if phase change takes place in the heat conduction process due to the fact that the temperature rates of the nodes in which phase change is taking place are equal to zero but the enthalpy rates of them are not. However, they can still keep true if the element matrix is modified as such that if phase change is taking place in nodal i, all the elements in row i and column i in the heat conductance matrix are taken to zero.

Since [C] is a diagonal matrix, equation (17) can be transformed into

$$\{V_T\} = [C]^{-1}\{V_H\} \tag{18}$$

Substituting equation (18) into equation (15), the following equation is obtained:

$$\Big[[M^*]+(0.5+\alpha)\Delta t[K^*][C]^{-1}\Big]\{V_H\}$$
$$=\{F^*\}-[K^*]\{(1-\alpha)\{T^n+\alpha T^{n-1}\} \tag{19}$$

It is convenient to solve equation (19) by Gauss elimination. After $\{V_H\}$ is obtained $\{H^{n+1}\}$ is updated by equation (11).

It is obvious that this procedure incorporating the three-time-level scheme does not need iteration and it is applicable to three dimensional problems.

ONE-POINT QUADRATURE

Because of the strong non-linearity of the problem, the element conductance matrices have to be calculated at each time step and the calculation occupies a large portion of the total work. Generally, the calculation of the element conductance matrices is performed by numerical quadrature. For the isoparametric bilinear quadrilateral elements adopted in this paper, four-point quadrature is often required. This is very time-consuming. In order to speed up the computation, a newly-developed one-point quadrature procedure is used here.

The major drawback of one-point quadrature is known as hourglass. Recent studies have indicated that the difficulty can be overcome by adding a stabilization matrix [4].

$$[K]^e = [K]^{e(1)} + [K]^e_{stab} \tag{20}$$

where $[K]^{e(1)}$ is the element matrix by one-point quadrature; the stabilization matrix is given by

$$[K]^e_{stab} = \bar{\varepsilon}\, A\, \{\gamma\}\{\gamma\}^T \tag{21}$$

where
$$\{\gamma\} = \frac{1}{A}\,[\{h\} - (\{h\}^T\{x\})\{b_1\} - (\{h\}^T\{y\})\{b_2\}]$$

$$\{b_1\}^T = \frac{1}{2A}\,[y_{24}, y_{31}, y_{42}, y_{13}]$$

$$\{b_2\}^T = \frac{1}{2A}\,[x_{42}, x_{13}, x_{24}, x_{31}]$$

$$x_{IJ} = x_I - x_J\ , \quad y_{IJ} = y_I - y_J$$

$$A = \frac{1}{2}\,(x_{31}y_{24} + x_{24}y_{31})$$

$$\{h\}^T = [1, -1, 1, -1]$$

$$\bar{\varepsilon} = \frac{A^2}{12}\, k\, (\{b_1\}^T\{b_1\} + \{b_2\}^T\{b_2\})$$

For non-linear heat conduction problems, almost two-thirds of the total solution time is saved when one-point quadrature is used and its accuracy is almost identical with that of four-point quadrature [8].

TEST EXAMPLES

To validate the present scheme, the finite element solutions obtained are testified against those in the literature. Both one- and two-dimensional solidification problems are examined. In all the example problems considered here, the thermal-physical properties are assumed to be constant in each phase but may differ between the solid and liquid phase.

Example 1. Solidification of an infinite slab of liquid

In the present context it is solved as a two dimensional problem and the results are compared with the analytical solution of Luikov [2] and with the numerical results of Comini [9].

The initial temperature is taken to be 283.15 K which is above freezing point(273.15 K) as previous investigations have shown that such conditions are more critical for a numerical calculation than those with the water initially at the freezing point [10]. The thermal-physical properties are as follows:

k_s=2.22 W/m.K, k_l= 0.556 W/m.K,

c_s=1762 J/m^3.K, c_l=4226 J/m^3.K,

ρ_s= ρ_l=1000 kg/m^3, λ =338000J/kg,

T_m=273.15 K .

The finite element mesh is shown as in Figure 1 in which L = 1.0m.

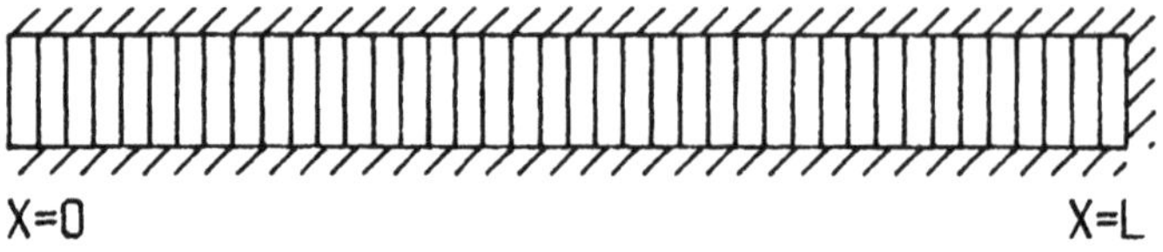

Figure 1 Finite Element Mesh

Figure 2 displays the the progress of the freezing front. As the slab involved is finite, comparisons with the theoretical infinite slab solution are terminated when temperature began to change appreciably at x=L. It can be seen from this figure that the results are in excellent agreement both with Luikov's analytical solution and with Comini's numerical results.

In this example several time step sizes are testified and it is found that relatively large time step can be used and the differences of the results between small and large time step sizes are small.

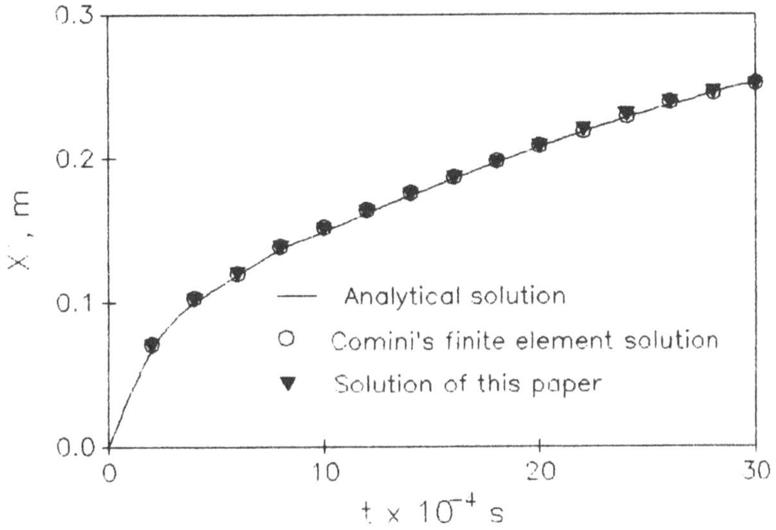

Figure 2 Comparison of Interface Position vs Time

Example 2. Solidification of a corner region

The corner region of a liquid body extending in the positive *x* and *y*-directions is frozen by bringing the surface temperature to -1.0 at time t=0. Thermal properties are $k_s = k_l = 1$, $c_s = c_l = 1.0$, $\rho = 1$, $\lambda = 1.5613$ in one case, 0.25 in the other.

The finite element mesh is shown as in Figure 3. Figure 4 is the comparisons between the present finite element results the analytical solution [11] of the freezing front history. It is found that the results are also in excellent agreement with the analytical solution.

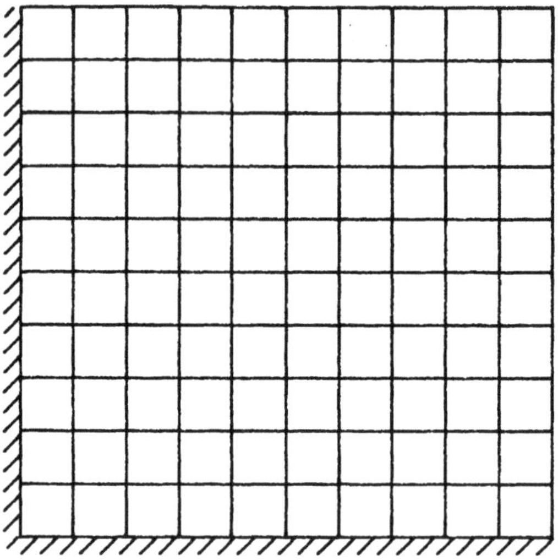

Figure 3 Finite Element Mesh

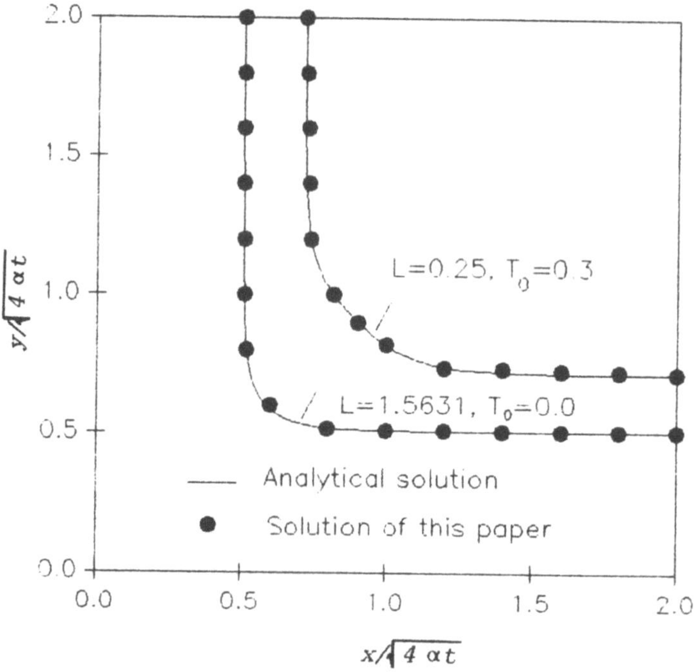

Figure 4 Comparison of Interface Position with Analytical Solution

CONCLUSION

An efficient finite element procedure incorporating a three-time-level scheme has been developed for solving multi-dimensional enthalpy model. This procedure does not need iterations and allows large time step. Numerical examples have demonstrated the accuracy obtainable.

REFERENCE

1. Zienkiewicz, O.C., The Finite Element Method, McGraw-Hill, New York, 1977.
2. Luikov, A.V. Analytical Heat Diffusion Theory, Academic Press, New York, 1968.
3. Shamsundar, N. and Sparrow, E.M., Analysis of Multidimensional Conduction Phase Change via the Enthalpy Model, J. Heat Transfer, Trans. ASME, Vol 97, pp. 333-340, 1975.
4. Liu, W.K., and Belytschko, T., Efficient Linear and Non-linear Heat Conduction with a Quadrilateral Element, Int. J. Num. Meths. Eng., Vol. 20, pp. 931-948, 1984.
5. Dalhuijsen, A.J., and Segal, A., Comparison of Finite Element Techniques for Solidification Problems, Int. J. Num. Meths. Eng., Vol 23, pp. 1807-1829, 1986.
6. Hogg, M., A Comparison of Two- and Three-Level Integration Schemes for Non-linear Heat Conduction, in R.W. Lewis, K. Mogan and O.C. Zienkiewicz(eds), *Numerical Methods in Heat Transfer,* Wiley, Chichester, pp. 75-90, 1981.
7. Thomas, B.G., Samarasekera, I.V. and Brimacombe, J. K., Comparison of Numerical Modeling Techniques for Complex Two-dimensional Transient Heat-Conduction Problems, Met. Trans. B, Vol 15, pp. 307-318, 1984.
8. Gong, Z.X. and Zhang, Y.F. Implementation and Accuracy of One-point Quadrature to Nonlinear Heat Conduction Problems

J. Tianjin Institute of Light Industry(in Chinese), Vol. 2, pp. 71-76, 1988.

9. Comini, G., Guidice, S.D., Lewis, R.W. and Zienkiewicz, O.C., Finite Element Solution of Non-linear Heat Conduction Problems with Reference to Phase Change, Int. J. Num. Meths. Eng., Vol. 8, pp. 613-624, 1974.
10. Bonacina, C., Comini, G., Fasano, A. and Primicerio, M., Numerical Solution of Phase Change Problems, Int. J. Heat Mass Transfer, Vol. 16, pp. 1825-1832, 1973.
11. Budhia, H. and Kreith, F., Heat Transfer with Melting or Freezing in a Wedge, Int. J. Heat Mass Transfer, Vol. 16, pp. 195-211, 1983.
12. Pham, Q.T., The Use of Lumped Capacitance in the Finite-Element Solution of Heat Conduction Problems with Phase Change, Int. J. Heat Mass Transfer, Vol. 29, pp. 285-291, 1986.

Cyclic Phase Change Heat Conduction in Thin Composite Slabs

Z.-X. Gong (*), Y.-F. Zhang (*), A.S. Mujumdar (**)
(*) Tianjin Institute of Light Industry, Tianjin, P.R. China
(**) McGill University, Montreal, Quebec, Canada

ABSTRACT

A one-dimensional conduction-controlled phase change model is derived for composite slabs to simulate the cyclic operation of phase change energy storage systems. A fixed-mesh finite element scheme is adopted to solve the resulting multiple moving boundary problem. Dynamic simulations are carried out. The cumulative energy transferred per cycle, the instantaneous heat flux, the time wise interface locations and the temperature profiles are presented for one mode of operation viz, equal thermal swings with equal melting and freezing periods.

INTRODUCTION

The latent heat storage system has attracted considerable attention, particularly for the utilization of solar energy for low temperature applications. The latent heat that accompanies some phase transformations offers a prospect of storage with high energy density and stabilization of the temperature , i.e. a small temperature variation from storage to retrieval. In a Phase Change Thermal Energy Storage (PCTES) system energy is stored during melting and recovered during freezing. An example of a PCTES system is the use of phase change solar walls in homes [2, 3].

Despite the significance of cyclic melting and freezing, it has received very little attention. A thorough search of the literature revealed little information about cyclic phase change heat transfer even for the case of a slab subjected to a time-dependent variation in temperature around the fusion temperature of the phase change material(PCM). With regard to periodic phase change heat transfer, only four relevant references (Bransier [4], Kalhori and Ramadhyani [5], and Jariwala et al. [6], Hasan [7]) were located.

Bransier [4] was the first to analyze cyclic melting /freezing of a PCM. His analysis was based on a one-dimensional conduction model. The temperature of the heat transfer fluid was assumed to vary sinusoidally about the fusion temperature. The numerical scheme he adopted was the apparent heat capacity method which is sensitive to the selection and implementation of the artificially extended temperature range for the distinct phase change temperature and did not allow him to use large time step. Kalhori and Ramadhyani [5], while studying experimentally the melting characteristics of finned and unfinned vertical cylinders embedded in a 99% pure n-ecosane PCM, carried out a single cyclic melting/freezing experiment. Jariwala et al.[6] carried out an experimental study of the cyclic thermal performance of a latent heat storage unit which consisted of a cylindrical helically coiled copper tube embedded in a cylindrical tank of a commercial paraffin wax and developed a quasi-steady one-dimensional model which was reported to be surprisingly accurate. Hasan [7] developed a one-dimensional conduction-controlled phase change model and a front-tracing, self-marching finite-difference scheme to solve the resulting multiple moving boundary problem. Experiments were also done by alternately melting and freezing n-octadecane in a thin rectangular cell. It was found there was excellent agreement between numerical and experimental results.

In this paper a one-dimensional conduction phase change model is derived for alternate melting and freezing of thin composite PCMs to simulate the cyclic operation of PCTES systems. One surface of the slab is subjected to successive periods of constant temperature above and below the fusion temperature of the PCMs while the other surface is adiabatic. A fixed-mesh finite element scheme is chosen and modified to solve the resulting multiple moving boundary problem. Dynamic simulations are made for cyclic melting and freezing of a composite slab and a single PCM slab until periodic steady state (PSS) is attained. Comparison of the cumulative energy transferred per cycle of the two slabs is carried out.

PHYSICAL MODEL AND BASIC FORMULATION

The physical system modeled is a large plane slab composed of two PCMs one surface exchanges heat while the opposite is insulated. The following assumptions are made:

(1) The PCMs are isotropic and homogeneous.
(2) Heat transfer in the PCMs is by one-dimensional conduction.

(3) The densities of the solid and liquid are equal.
(4) The solid/liquid or liquid/solid interface takes place at a discrete temperature.
(5) There is no heat resistance at the interface of the PCMs.

Initially the slab is a single phase liquid or solid with a uniform temperature of T_{in}. The transfer surface is exposed to a rectangular wave temperature variation around the fusion temperature of the PCMs. If the PCMs are initially liquid, the temperature variation at x=0 is

$$T_w = T_{in}, \qquad t \le 0 \tag{1}$$

$$T_w = T_{wf}, \quad \text{for} \quad nt_p < t < nt_p + t_f \qquad n=0,1,2,\ldots \tag{2}$$

$$T_w = T_{wm}, \quad \text{for} \quad nt_p + t_f < t < (n+1)t_p$$

where

$$t_p = t_m + t_f \tag{3}$$

and $T_{wf}(<T_m)$ is the surface temperature during freezing, $T_{wm}(>T_m)$ is the surface temperature during melting and t_p is the constant cycle time.

Although cyclic freezing and melting create multiple moving interfaces, the energy equation of a composite slab can be described as follows if a fixed mesh temperature based finite element method is used to solve this problem:

$$\rho c \frac{\partial T}{\partial t} = \frac{\partial}{\partial x}\left(k \frac{\partial T}{\partial x}\right) + Q \qquad \text{for} \quad 0 < x < L, \quad t > 0 \tag{4}$$

in which ρ, c, k are the density, specific heat and thermal conductivity respectively, Q is the heat source or sink.

The following are the initial, boundary and moving boundary conditions:

1. Initial Conditions:

$$\text{at } t=0, \qquad T(x,0) = T_{in} \qquad \text{for } 0 \le x \le L \tag{5}$$

$$\text{at } t=0, \qquad s_1(t) = 0, \qquad s_2(t) = fL \tag{6}$$

where s_1 and s_2 represents respectively the phase front in PCM 1 and PCM 2, f is a coefficient $(0 \le f \le 1)$.

2. Boundary Conditions:

$$\text{at } x = 0, \qquad T(0,t) = T_w = \text{constant}, \qquad t > 0 \tag{7}$$

$$\text{at } x = L, \qquad \frac{\partial T}{\partial x} = 0, \qquad t > 0 \qquad (8)$$

3. Moving Boundary Conditions:

$$\text{at } x = s_1(t)$$

$$T_s(x,t) = T_l(x,t) = T_{1m} \qquad (9)$$

$$k_{1s}\frac{\partial T_s}{\partial x} - k_{1l}\frac{\partial T_l}{\partial x} = \rho_1\lambda_1\frac{ds_1}{dt} \qquad t>0 \qquad (10)$$

$$\text{at } x = s_2(t)$$

$$T_s(x,t) = T_l(x,t) = T_{2m} \qquad (11)$$

$$k_{2s}\frac{\partial T_s}{\partial x} - k_{2l}\frac{\partial T_l}{\partial x} = \rho_2\lambda_2\frac{ds_2}{dt} \qquad t>0 \qquad (12)$$

where λ is the latent heat of fusion, k_s and k_l represent respectively the thermal conductivity of the solid and liquid phase, subscript 1 and 2 designate respectively PCM 1 and PCM 2.

FINITE ELEMENT MODEL

In order to convenience the future work which will extend the model to two-dimensional problems a fixed-mesh temperature based finite element method is used to solve the the mathematic model described above.

After the spacewise discretization of equation (4) is accomplished using Galerkin's method [1], we get the following matrix equation:

$$[\ C\ ]\{\ \dot{T}\ \} + [\ \ K\ ]\{\ T\ \} = \{\ F\ \} \qquad (13)$$

where [C] and [K] are heat-capacity and conductivity matrices, respectively, and the matrix elements are:

$$C_{IJ} = \sum \int_{\Omega^e} \rho c\, N_I \cdot N_J \, d\Omega \qquad (14)$$

$$K_{IJ} = \sum \int_{\Omega^e} k\, N_{I,i} \cdot N_{J,j} \, d\Omega \qquad (15)$$

$$F_I = \sum \int_{\Omega^e} Q\, N_I \, d\Omega \qquad (16)$$

In the above, summations are taken over the contributions of each element, Ω^e is the element region.

If the three-time-level scheme in reference [9] is used to

perform the discretization of the time derivative in equation (13) we get

$$[C^*]\{T^{n+1}-T^n\}/\Delta t + [K^*]\{(0.5+\alpha)T^{n+1}+(0.5-2\alpha)T^n+\alpha T^{n-1}\} = \{F^*\} \quad (17)$$

The superscript * designates that the matrices and vector are evaluated at time level t^*:

$$t^* = t_{n+\frac{1}{2}}$$

or

$$t = t_n$$

in correspondence with t^*

$$\{T^*\} = \{\frac{3}{2}T^n - \frac{1}{2}T^{n-1}\} \quad (18)$$

or

$$\{T^*\} = \{T^n\} \quad (19)$$

This three-time-level scheme was found, with $\alpha=\frac{1}{4}$, to work very satisfactory for a non-linear model equation and to be stable, accurate and free from oscillations for solidification problems [9].

With $\alpha=\frac{1}{4}$, Equation (17) can be written as:

$$([C^*] + \frac{3}{4}\Delta t[K^*])\{T^{n+1}\} = [C^*]\{T^n\} - \frac{1}{4}\Delta t[K^*]\{T^{n-1}\} + \Delta t\{F^*\} \quad (20)$$

Based on the three-level-enthalpy method in reference [8], the solution steps of equation (20) are as follows:

1. Calculation of the equivalent heat capacity

At the start of each time step, the enthalpy change ΔH_i^* at each node is estimated from the known temperature T_i^n of the node and those of its immediate neighbors at the present time level n

$$\Delta H_i^* = (-\sum_j K_{ij}^n T_j^n - F_i^n)\Delta t/M_i \quad (21)$$

in which M_i is the mass of node i. The estimated new nodal enthalpy is

$$H_i^* = H_i^n + \Delta H_i^* \quad (22)$$

Thus, the estimated new temperature is

$$T_i^* = f_T(H_i^*) \quad (23)$$

and the specific heat of the node is then defined as

$$c_i^n = \Delta H_i^* / (T_i^* - T_i^n) \tag{24}$$

2. Solution of the discrete matrix equation

Using the values of c_i^n from equation (24) to calculate the lumped C-matrix and then solving equation (20), we obtain $\{T^{n+1}\}$.

3. Correction of the temperature

After we get $\{T^{n+1}\}$, the following temperature correction step is applied:

$$T_i^{n+1}(\text{corrected}) = f_T(f_H(T_i^n) + c_i^n(T_i^{n+1} - T_i^n)) \tag{25}$$

It should be noted that the computer is unable to handle equation (24) when $T_i^* = T_i^n$ i.e. when the step change in the enthalpy-temperature curve is encountered. The following methods may be used to overcome this problem:

(i) Set $T_i^{n+1} = T_i^n$ and eliminate T_i^{n+1} from the set of equations, in a manner similar to the handling of the first kind of boundary condition.

(ii) Give the step change in enthalpy (or latent heat peak) a finite but arbitrary small width, to prevent c_i^n in equation (24) from becoming infinite.

The second approach is the most convenient and will be used in this paper.

ENERGY STORED AND RECOVERED

When the two PCMs are initially liquid with a uniform temperature of T_{in} the energy stored in the slab at t=0 is

$$Q_{in} = \rho_1 fL\ f_{1H}(T_{in}) + \rho_2(1-f)L\ f_{2H}(T_{in}) \tag{26}$$

The maximum amount of energy which can be stored or recovered (per cycle) of the composite slab is

$$Q_M = \rho_1 fL\ [f_{1H}(T_{wm}) - f_{1H}(T_{wf})] + \rho_2(1-f)L\ [f_{2H}(T_{wm}) - f_{2H}(T_{wf})] \tag{27}$$

At any time t the energy stored in the slab is

$$Q_T = \sum_{i=1}^{r} M_i f_{1H}(T_i) + \sum_{i=r}^{n} M_i f_{2H}(T_i) \tag{28}$$

ACCURACY TEST

In order to test the finite element procedure, a repetition of Hasan's work [7] on the cyclic melting and freezing of a n-octadecane slab is completed. The thickness of the slab is 2.67cm. The thermal-physical properties of the n-octadecane are

T_m=27.4°C, λ=243.5 kJ/kg, k_l=0.148W/m.°C, k_s=0.358W/m.°C ,
c_l=2.23kJ/kg.°C, c_s=2.15 kJ/kg, ρ=768kg/m^3.

Initially the PCM is solid at a uniform temperature of 27°C. One surface of the slab is subjected to a rectangular wave temperature variation $|T_w - T_m|$ =19°C around the fusion temperature with a period of 1.5hrs, while the other is adiabatic. This problem is calculated until PSS is achieved.

Figure 1 shows the progress of the interface vs dimensionless time. The interface positions are compared with Hasan's numerical and experimental results in this figure. It was found that the results of the finite element model are in excellent agreement with Hasan's both numerical and experimental ones.

It should be interpreted that Figure 1 and the following Figures are potted in dimensionless forms. The dimensionless parameters in the figures are as follows:

$\theta=(T-T_{1m})/\lambda_1/c_{1s}$, $\tau=\alpha_{1s}t/L^2$, $S=s/L$, $X=x/L$.

DYNAMIC SIMULATION

Example 1. Simulation of a composite PCM slab

The composite slab to be computed is composed of two PCMs, PCM 1 and PCM 2, as in Figure 2 in which L=0.5m and f=0.5. The thermo-physical properties of the two PCMs is as follows:

$\rho_1=\rho_2 100kg/m^3$, $c_{1s}=c_{1l}=c_{2s}=c_{2l}$=400J/kg.K,

$k_{1s}=k_{1l}=k_{2s}=k_{2l}$=1.0W/m.K, $\lambda_1=\lambda_2$=40000J/kg.

T_{1m}=0°C, T_{2m}=5°C

Initially the PCMs are assumed to be a single phase of liquid at a uniform temperature of 10°C. One side of the slab is subjected to a rectangular wave temperature variation around the fusion temperature of PCM 1, while the other is adiabatic. Transient computations are made until PSS is achieved for one mode of operation, namely, equal thermal swings with equal melting and freezing periods, i.e. $|T_{wf}| = T_{wm}$= 30°C and $t_f=t_m$ =10000s,

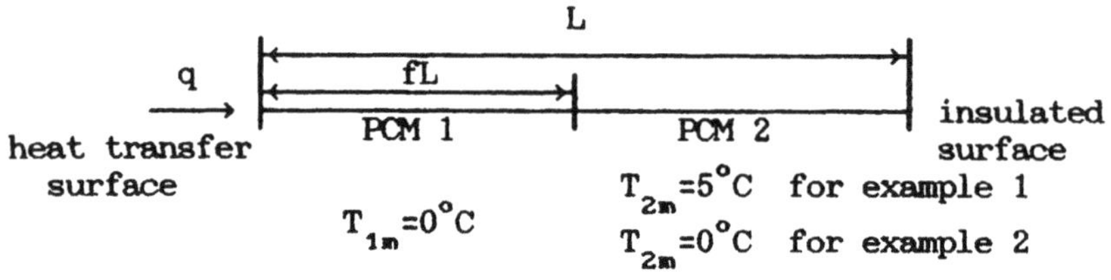

Figure 2 Description of the composite slab

Example 2. Simulation of a single PCM slab

This example is designed only for comparison with example 1. Therefore, the geometric and boundary conditions are the same as those in example 1 and all the thermo-physical properties are also the same as those in example 1 except for $T_{2m}=0°C$.

Dynamic simulations of the two examples are carried out until PSS is reached. The numerical results are shown in Figure 3 to figure 8

RESULTS AND DISCUSSION

Figure 3 shows the movement of the interfaces against dimensionless time during the initial transient cycles until periodic steady state (PSS) is reached for example 1. In this figure $S_i=0$ is the heat transfer surface, $S_i=1$ is the insulated boundary and $S_i=0.5$ is the interface of the two PCMs. During a freezing period if a solid/liquid front in PCM 1 starts from $S_i=0$ and gets to X=0.5, PCM 1 becomes solid, while if a solid/liquid interface in PCM 2 reaches the insulated boundary at $S_i=1$, PCM 2 is solid. Similarly, during melting if a liquid/solid front in PCM 1 arrives at X=0.5 PCM 1 is melted completely, and if a liquid/solid interface in PCM 2 reaches the insulated boundary, PCM 2 is liquid. During the first freezing period a solid/liquid interface in PCM 1 (curve 1F) originates from the heat transfer surface and advances freezing the liquid ahead until it reaches X=0.5 at $\tau \doteq 0.7$. In PCM 2 a solid/liquid interface (1F) starts to advance from X=0.5 at $\tau \doteq 0.18$. This solid/liquid front reaches $X \doteq 0.78$ at the end of the first freezing period($\tau_p=1$). With the initiation of the melting period (at $\tau=1$) a liquid/solid front (1M) in PCM 1 begins and moves forward rapidly, while just after the beginning of the melting period (at $\tau=1.085$) another liquid/solid interface in PCM 1 originates from X=0.5 and moves slowly backward. At $\tau \doteq 1.185$ the two liquid/solid front meet at $X \doteq 0.46$. The cause of the liquid/solid interface near X=0.5 in

PCM 1 is due to the inverse heat flow from PCM 2 to PCM 1 because at the end of the first freezing period the temperature near X=0.5 in PCM 1 is negative, while the temperature of a part of PCM 2 near X=1 is higher than 5.0°C since phase change has not taken place in this part during the first freezing period. In the melting period a liquid/solid front (1M) in PCM 2 begins at X=0.5 at $\tau\doteq1.42$ and reaches $X\doteq0.73$ at $\tau=2.0$. While at the beginning of the melting period the first solid/liquid front in PCM 2 keeps moving forward until $\tau\doteq1.40$ to reach $X\doteq0.85$. Afterwards the front nearly keeps still until $\tau\doteq2.92$. During the second freezing period a second solid/liquid (2F) front in PCM 1 originates and reaches X=0.5 at $\tau\doteq2.69$, while in PCM 2 the first liquid/solid front nearly keeps still and a second solid/liquid (2F) front originates and moves into the liquid from X=0.5 at $\tau\doteq2.15$ and at $\tau\doteq2.88$ it meets the first liquid/solid front at $X\doteq0.73$. The second melting period almost repeats the first one. With continued cyclic melting and freezing the first solid/liquid front in PCM 2 reaches the insulated boundary at $\tau\doteq5.0$. In PCM 2 for longer times only the PCM near X=0.5(S_{2i}< 0.70) melts and freezes. An important fact to note in Figure 3 is that, after the liquid PCM 2 is frozen to the insulated surface the movement of the solid/liquid and liquid/solid interfaces in the two PCMs is repetitive from cycle to cycle.

Figure 4 presents transient propagation of the interface until PSS is reached for example 2. Initial temperature of the slab is 10°C. during the first freezing period a solid/liquid front originates and moves forward from X=0 and reaches X=0.72 at $\tau=1.0$. With the initiation of the melting period a liquid/solid interface begins to diminish the thickness of the solid layer and ultimately melts the solid layer at X=0.75 at $\tau=2.0$ while the first solid/liquid front keeps almost still. During $1.0<\tau<2.0$ three layers exist: a liquid near the heat transfer surface, a diminishing solid layer and the original liquid layer between the solid layer and the insulated boundary. In the following cycles the variation of the solid/liquid interface during freezing and the liquid/solid interface during melting is almost the same as the first cycle.

Figure 5 displays the normalized heat stored or recovered, Q_T/Q_M, in each melting or freezing period from start-up until PSS is reached. The results are for the two examples. The open circle line (curve a) is for example 1 and the filled circle line (curve b) is for example 2. In this figure the discrete point values of Q_T/Q_M, each corresponding to either a freezing or a melting period, are joined by line segments to show trends. It can be found that oscillation dies gradually with continued cycling after the third freezing cycle and at the fifth freezing cycle PSS has been achieved in example 1. In example 2 Q_T/Q_M has almost no oscillation except that it has a small jump from $\tau=2$ to $\tau=3$ and at the third cycle PSS is achieved. An important observation in figure 5 is that Q_T/Q_M is higher in example 1 than in example 2, i.e. the cumulative energy stored and recovered per cycle is higher in the composite slab than in the single PCM slab under the same circumstances. However, an examination of Figure 3 and Figure 4 shows that at the end of each melting or freezing cycle the distance of the phase front reached is a little shorter in the composite slab than in the single PCM slab at PSS, e.g. the cumulative latent heat stored and recovered per cycle is less in the composite slab than in the single PCM slab. In this case, what causes the cumulative energy transferred per cycle is higher in the composite slab than in the single PCM slab? Explanation is as such that more sensible heat is transferred per cycle in the composite slab than in the single PCM slab.

Figure 6 displays the dimensionless instantaneous heat flux at the heat transfer surface for the two examples. The results are plotted in a two panel format one for the freezing period and the other for the melting period. The solid line (curve a) is for example 1 and the dashed line (curve b) for example 2. Note during freezing or melting the instantaneous heat flux falls rapidly near $\tau=0$ while the rate of decrease slows with time. This decline is caused by the increasing conduction resistance of the thickening freezing or melting layer. In more than 80% of a freezing or a melting period the heat flux of the composite slab is higher than that of the single PCM slab. This proves a fact that the energy stored and recovered per cycle is higher in the composite slab than in the single PCM slab. The change in the curvature of the flux curve during freezing (curve a) occurs when the advancing solid/liquid interface meets the pre-existing liquid/solid interface and the entire slab is solid phase. This occurs at $\tau=0.54$. Subsequently only sensible heat is removed. Curve b has no curvature change since only at the end of the cycle ($\tau=2.0$) can the liquid/solid interface meet the pre-existing solid/liquid interface.

Figure 7 gives transient temperatures corresponding to Figure 3 at the dimensionless distances of 0.2 (curve a), 0.5 (curve b) and 0.8 (curve c) from the heat transfer surface. X=0.2 is in PCM 1 and X=0.8 in PCM 2. X=0.5 is the interface position of the two PCMs. In this figure the maximum value of θ which a superheated melt can attain is 0.3 and the minimum value of a subcooled solid can attain is -0.3 since $|\theta_w|=0.3$. The temperature at X=0.2 achieves PSS after the first two cycles, i.e. in the third and higher cycles the temperature variations at X=0.2 are identical with maximum and minimum values of 0.223 and -0.238. Farther from the heat transfer surface it takes longer to reach PSS, e.g. four cycles are required at X=0.5. At the PSS the temperature at X=0.5 oscillates between θ =0.11 during melting and θ=-0.14 during freezing. At X=0.8 the temperature only gets to the fusion temperature of PCM 2 (5°C) during freezing since the liquid/solid interface reaches only to X≐0.7 at the end of each freezing period.

Figure 8 displays transient temperature profiles corresponding to Figure 4 at X=0.2 (curve a), X=0.5 (curve b), X=0.8 (curve c). At the third cycle (τ=6 to 8) the profiles attain PSS. This figure also shows that only after the first cycle the temperature variations at X=0.2 reach PSS. At PSS the maximum temperature at X=0.2 during melting is 0.214 and the minimum during freezing is -0.214. At X=0.5 temperature attains PSS at the second cycle and afterwards the temperature varies from θ=-0.091 during freezing to θ=0.091 during melting. At X=0.8 the temperature reaches only to the fusion temperature (0°C) during freezing since the solid/liquid interface penetrates only to X≐0.73 at the end of each melting period.

CONCLUSION

A one dimensional conduction phase change model is developed to simulate the cyclic melting and freezing of composite slabs. A fixed mesh finite element procedure is chosen and modified to solve the multiple moving boundary problem. Numerical example shows that the cumulative energy transferred per cycle is higher in the composite slab than in the single PCM slab under the same conditions.

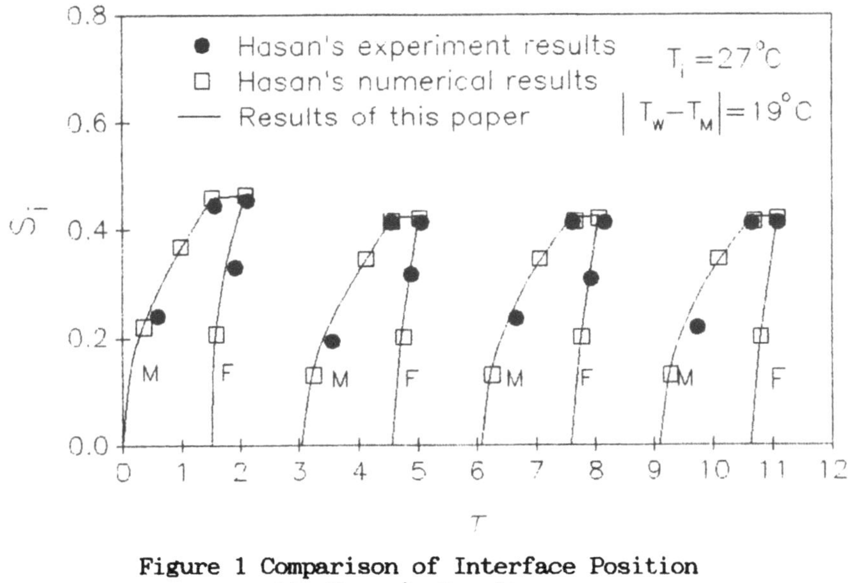

Figure 1 Comparison of Interface Position
with Hasan's Results

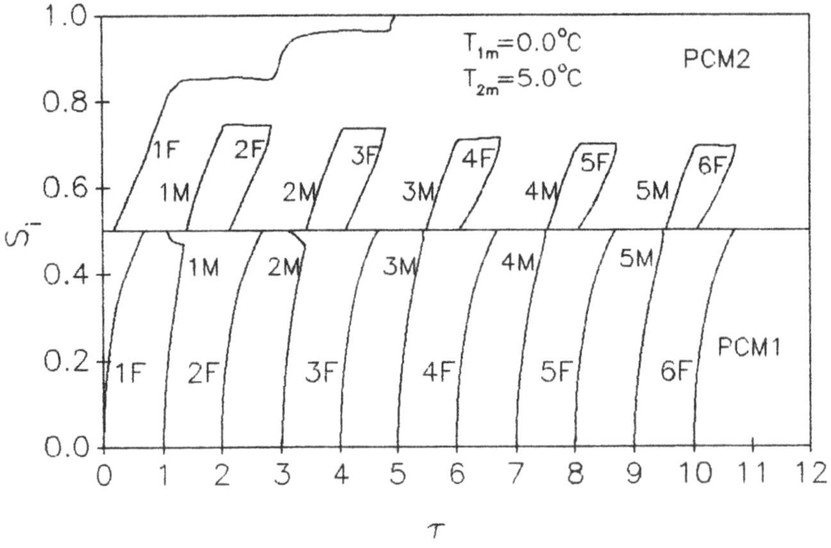

Figure 3 Interface position vs time during transients

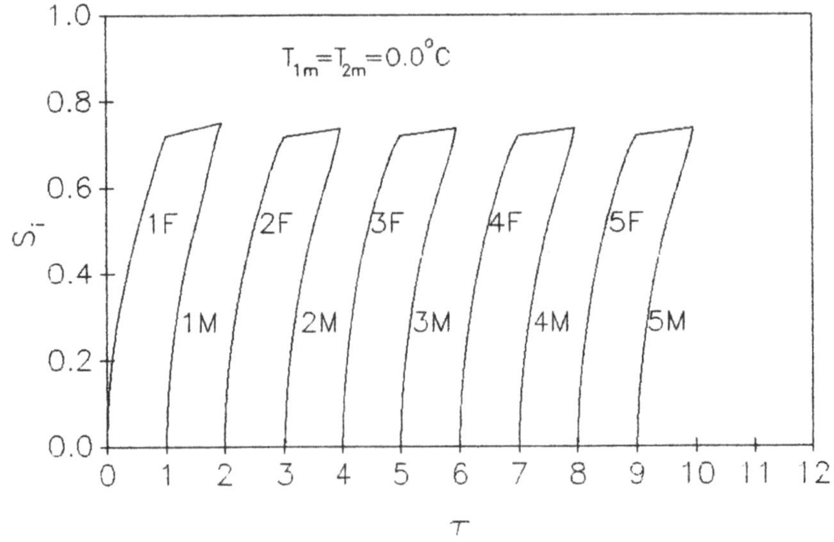

Figure 4 Interface position vs time during transients

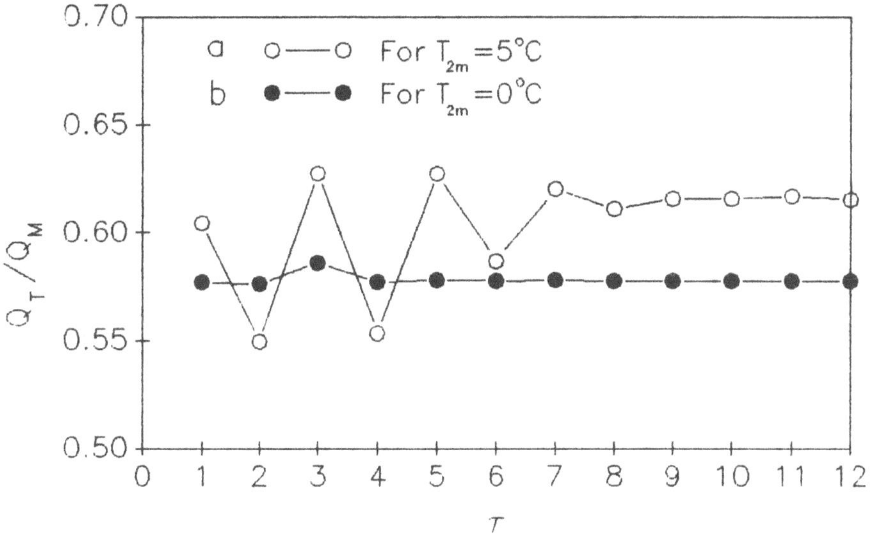

Figure 5 Dimensionless cumulative energy transferred vs melting/freezing period

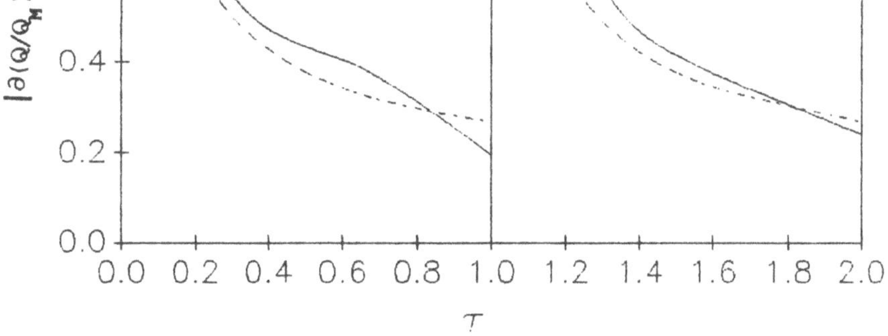

Figure 6 Dimensionless instantaneous heat flux at the heat transfer surface vs time

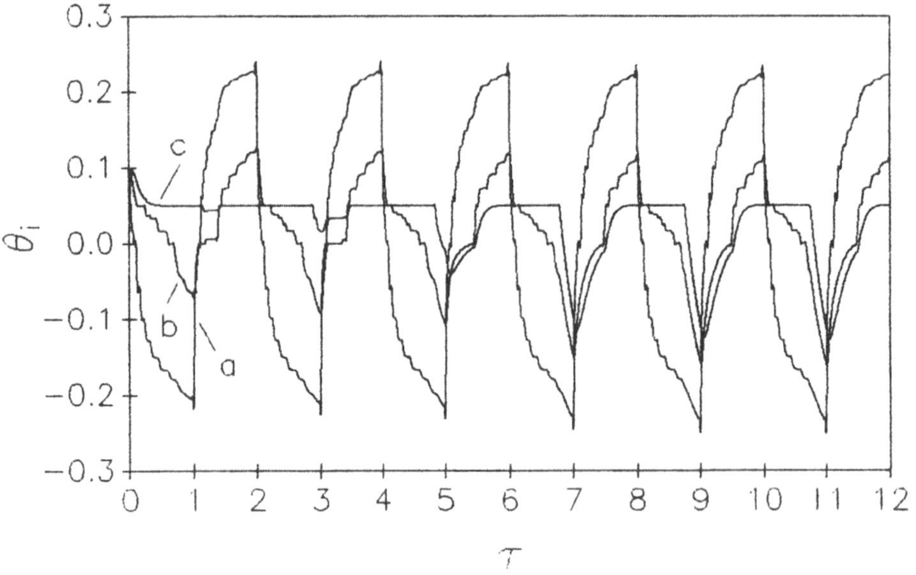

Figure 7 Transient temperature profiles at X=0.2(curve a), 0.5(curve b), 0.8(curve c) vs time

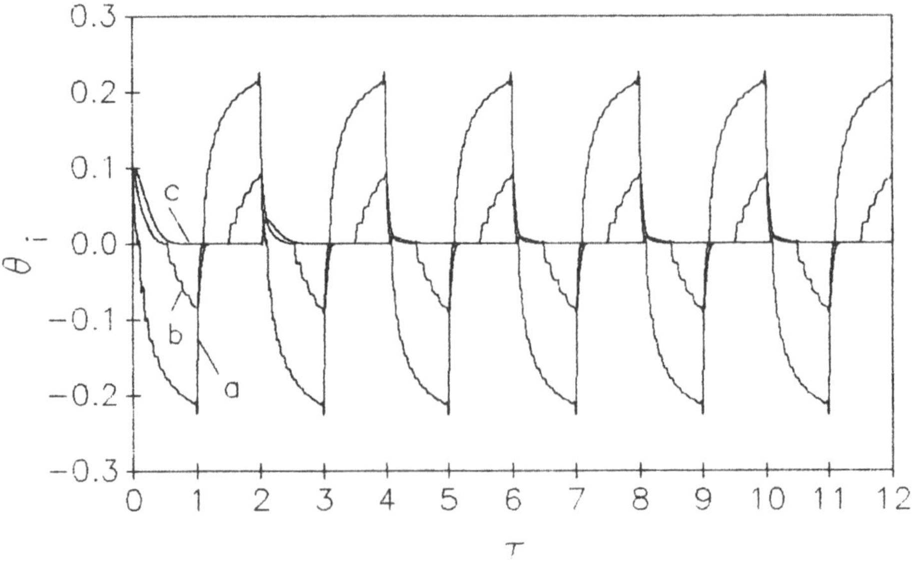

Figure 8 Transient temperature profiles at X=0.2(curve a), 0.5(curve b), 0.8(curve c) vs time

REFERENCE

1. Zienkiewicz, O.C., *The Finite Element Method,* McGraw-Hill, New York, 1977.
2. Guceri, S.I. and Faunce, S.F., Modeling of a Thermal Wall Panel Using Phase Change Materials, Energy, Vol. 4, pp. 695-699, 1979.
3. Bernard, C., Body, Y., Gobin, D. and Guerrier, B., Use of a Variable Parameter Test Cell for the Study of Latent Heat Solar Walls, Solar Energy, Vol. 29, No. 2, pp. 101-109, 1982.
4. Bransier, J., Storage Periodique par Chaleur Latente: Aspects Fonasmentaux \lies a la Cinetique des Transferts, Int. J. Heat Mass Transfer, Vol. 22, pp. 875-883, 1979.
5. Kalhori, B. and Ramadhyani, S., Studies on Heat Transfer from a Vertical Cylinder, With or Without Fins Embedded in a Solid Phase Change Medium, J. Heat Transfer, Vol. 107, pp. 44-51, 1985.
6. Jariwala, V., Mujumdar, A.S., and Weber, M.E., The Periodic Steady State for Cyclic Energy Storage in Paraffin Wax, Can. J. Chem. Eng., Vol. 65, pp. 899-906, 1987.
7. Hasan, M., Cyclic Phase Change: Energy Storage and Recovery, PhD thesis(Dept. Chem. Eng., McGill University, Canada), 1988.
8. Pham, Q.T., The Use of Lumped Capacitance in the Finite-Element Solution of Heat Conduction Problems with Phase Change, Int. J. Heat Mass Transfer, Vol. 29, No. 2, pp. 285-291, 1986.
9. Dalhuijsen, A.J. and Segal, A., Comparison of Finite Element Techniques for Solidification Problems, Int. J. Num. Meths. Eng., Vol. 23, 1087-1829, 1986.

Simulation of Natural Convection Melting of a Pure Metal from Below

M. Lacroix

Groupe de recherche en ingénierie des procédés et systèmes, Université du Québec à Chicoutimi, Québec, Canada, G7H 2B1

ABSTRACT

The role of natural convection during melting from below of a pure metal was studied numerically. A stream function-vorticity-temperature formulation is employed in conjuction with body fitted coordinates for mapping the irregular shape of the timewise changing solid-liquid interface. Numerical simulations are carried out for temperature differences of 1 to 40°C between the heated wall and the phase front. Results reveal complicated time-dependent flow structure in the melt evolving from multiple cells to a single cell. These cells are found to significantly perturb the heat fluxes and, as a result, distort the advance and the shape of the solid-liquid interface.

INTRODUCTION

Melting of metals is of particular interest in material processing, metallurgy and purification of metals. Due to the fact that metals have high thermal conductivity, it has traditionally been assumed, in analysing solid-liquid phase-change heat transfer problems, that conduction is the sole mechanism of heat transfer [1]. However, it has also been recognized that natural convection in the melt during melting of metals can affect the rate of melting and the advance and the shape of the solid-liquid interface.

During the past decade, the problem of natural convection dominated melting of a pure metal from an isothermal wall has received increasing attention. Webb et al [2] appear to be among the first authors to report on a numerical analysis of melting of a pure metal from an isothermal vertical wall. A control volume-based discretization scheme adapted for irregular geometries was employed. Brent et al. [3] have tackled the same problem using a fixed grid enthalpy-porosity approach. Lacroix [4] considered a similar problem using a moving grid technique. All these studies revealed the important role of natural convection on the melting of a high thermal conductivity, low Prandtl number fluid. The phase change boundary, the melting rate and heat transfer were greatly affected by buoyancy driven natural convection.

On the other hand and despite its many industrial applications, the problem of natural convection melting from below of a pure metal has received very little attention. Perhaps the paucity of attention directed to this problem is related to the difficulty in observing this phenomenon in the laboratory and the mathematical challenge of simulating it. Gau and Viskanta [5] have studied the problem experimentally and have proposed a semi-empirical one-dimensional model for the motion of the solid-liquid interface. However, no attempt was made to model the multidimensional melt flow. It is precisely the aim of this paper to report on a numerical study of solid-liquid phase change heat transfer in such unstable situations when natural convection is present in a pure metal melt.

It is clear that the present flow problem is of three dimensional nature. Another recent experimental study [6] has presented the results of the effect of melting of a solid phase change material (PCM) from below. The resulting thermally unstable flow field yielded Bénard-like convection cells. These cells were manifested in the phase front by hexagonal, hemispherically-capped domes where high temperature rising fluid increased the local melting rate. Despite these observations, it is felt that a two-dimensional analysis of this flow problem could still lead to improved understanding of it. It may be seen as a stepping stone towards a full three dimensional analysis. Besides, with the available computer capabilities, results from three-dimensional moving boundary simulations would likely be questionable as relatively coarse grids ought to be used in order to limit CPU times.

The computational methodology is first outlined and then applied to the problem of melting from below of a pure metal. Numerical simulations are carried out for temperature differences ranging from 1 to 40°C between the heated wall and the solid-liquid interface.

GOVERNING EQUATIONS

The PCM is contained in a two-dimensional rectangular cavity of height L and width H (Fig. 1).

The PCM is assumed to be initially at its fusion temperature T_f, eliminating the need for solution of the energy equation in the solid. The left ($x = 0$) and right ($x = H$) boundaries are adiabatic. The walls and the phase

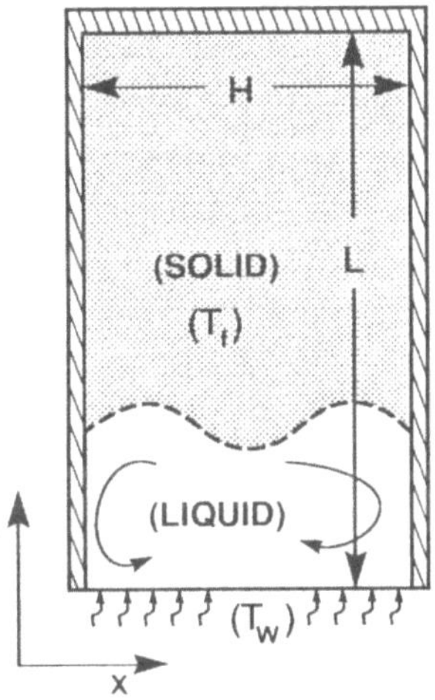

Fig. 1 Schematic of the physical system. The broken line represents the melting front.

front form hydrodynamically no-slip boundary conditions. At time t = 0, the temperature of the bottom horizontal wall (y = 0) is raised impulsively to a prescribed temperature above the fusion point $T_w > T_f$.

It is assumed that the thermophysical properties are constant except for the density in the buoyancy term. Density changes and viscous dissipation are neglected. The liquid is Newtonian and incompressible and the flow is laminar and two-dimensional.

To overcome the difficulties which arise due to the complex, timewise changing physical domain, i.e. the moving solid-liquid interface, the governing dimensionless conservation equations are readily cast in a generalized curvilinear grid (ξ, η). The advantage of this transformation is the fact that the boundary conditons in the transformed space are specified on straight boundaries and the computational grid, (ξ, η) is rectangular and uniformly spaced. Adopting this technique [4], the general transport equation has the form

$$\frac{\partial(\rho\phi)}{\partial t} + \frac{1}{J}\left[\frac{\partial x}{\partial \eta}\frac{\partial y}{\partial t} - \frac{\partial y}{\partial \eta}\frac{\partial x}{\partial t}\right]\frac{\partial(\rho\phi)}{\partial \xi} + \frac{1}{J}\left[\frac{\partial y}{\partial \xi}\frac{\partial x}{\partial t} - \frac{\partial x}{\partial \xi}\frac{\partial y}{\partial t}\right]\frac{\partial(\rho\phi)}{\partial \eta} + \frac{1}{J}$$
$$\frac{\partial}{\partial \xi}\left[\rho U\phi - \frac{\Gamma}{J}\left(\alpha\frac{\partial\phi}{\partial \xi} - \beta\frac{\partial\phi}{\partial \eta}\right)\right] + \frac{1}{J}\frac{\partial}{\partial \eta}\left[(\rho V\phi) - \frac{\Gamma}{J}\left(\gamma\frac{\partial\phi}{\partial \eta} - \beta\frac{\partial\phi}{\partial \xi}\right)\right] = S(\xi,\eta) \tag{1}$$

ϕ is the dependent variable, S(x,y) is a source term, Γ is an exchange coefficient and ρ is a constant. These parameters are defined in table 1.

Table 1 Variables and parameters in equation (1)

Function	ϕ	0	Γ	S(x,y)
Stream function	ψ	0	1	ω
Vorticity	ω	1	Pr	$\frac{Ra \cdot Pr \cdot}{J}\left(\frac{\partial y}{\partial \eta}\frac{\partial \theta}{\partial \xi} - \frac{\partial y}{\partial \xi}\frac{\partial \theta}{\partial \eta}\right)$
Temperature	θ	1	1	0

The geometric factors U, V. α, β, γ and the Jacobian J are defined as

$$U = \partial\psi/\partial\eta$$

$$V = -\partial\psi/\partial\xi$$

$$\alpha = \left(\frac{\partial x}{\partial \eta}\right)^2 + \left(\frac{\partial y}{\partial \eta}\right)^2 \tag{2}$$

$$\beta = \frac{\partial x}{\partial \xi}\frac{\partial x}{\partial \eta} + \frac{\partial y}{\partial \xi}\frac{\partial y}{\partial \eta}$$

$$\gamma = \left(\frac{\partial x}{\partial \xi}\right)^2 + \left(\frac{\partial y}{\partial \xi}\right)^2$$

$$J = \frac{\partial x}{\partial \xi}\frac{\partial y}{\partial \eta} - \frac{\partial x}{\partial \eta}\frac{\partial y}{\partial \xi}$$

The initial conditions (t = 0) are $\psi = \omega = \theta = 0$. For t > 0, the boundary conditions are given in table 2. Furthermore, since no heat conduction occurs in the solid phase, all heat transferred to the interface is utilized for melting. Then an energy balance for the interface yields the following dimensionless condition for the moving boundary

$$\frac{dy}{dt} = -\frac{Ste}{\tilde{\rho} J} \cdot \frac{\partial x}{\partial \xi} \cdot \frac{\partial \theta}{\partial \eta} \tag{3}$$

Ra, *Pr* and *Ste* represent the Rayleigh, Prandtl and Stefan numbers respectively. $\tilde{\rho}$ represents the ratio of the solid and the liquid densities.

Table 2 Boundary conditions for the problem

Boundary	Stream function ψ	Vorticity ω	Temperature θ
Bottom heated wall	$\psi = 0$	$\omega = -\frac{\gamma}{J^2} \cdot \frac{\partial^2 \psi}{\partial \eta^2}$	$\theta = 1$
Solid-liquid interface	$\psi = 0$	$\omega = -\frac{\gamma}{J^2} \cdot \frac{\partial^2 \psi}{\partial \eta^2}$	$\theta = 0$
Left wall	$\psi = 0$	$\omega = -\frac{\alpha}{J^2} \cdot \frac{\partial^2 \psi}{\partial \xi^2}$	$\frac{\partial \theta}{\partial \xi} = \frac{\beta}{\alpha} \frac{\partial \theta}{\partial \eta}$
Right wall	$\psi = 0$	$\omega = -\frac{\alpha}{J^2} \cdot \frac{\partial^2 \psi}{\partial \xi^2}$	$\frac{\partial \theta}{\partial \xi} = \frac{\beta}{\alpha} \frac{\partial \theta}{\partial \eta}$

The finite difference equations are obtained on integrating the general governing equation (1) over each of the control volume in the (ξ, η) plane. The linearized equations are then solved iteratively for θ, ω and ψ using an alternating line-by-line solver. A quasi-static approximation is adopted for the present study, i.e., the conservation equations (1) are solved and then the next position of the phase front is determined from the energy balance equation (3) using the current values of the temperature field.

The present mathematical model has been successfully applied to numerous phase change problems. It has also been validated with experimental data for natural convection dominated melting heat transfer from below of gallium [7]. Further details of the mathematical model and its various applications may also be found in references [4] and [7].

RESULTS AND DISCUSSION

The foregoing model was used for simulating convection dominated melting heat transfer from below of gallium. The experiments were conducted

Table 3 Test cases studied

Run	$T_W - T_f$	Pr	Ra	Ste
D01	1°C	0.0208	$7.60 \cdot 10^4$	0.0044
D05	5°C	0.02.08	$3.80 \cdot 10^5$	0.0248
D10	10°C	0.02.08	$7.60 \cdot 10^5$	0.0496
D20	20°C	0.0208	$1.52 \cdot 10^6$	0.0993
D40	40°C	0.0208	$3.04 \cdot 10^6$	0.1986

for a rectangular cell of height L = 8.89 cm and width H = 6.35 cm. Five cases were considered in which the tempeature difference between the heated wall and the solid-liquid interface $T_w - T_f$ ranged from 1 to 40°C. These cases are defined in terms of Rayleigh and Stefan numbers (table 3). Following a grid refinement study, all calculations were done with a grid size of 35 (x direction) x 25 (y direction) non uniformly distributed nodes. This makes it possible to concentrate several grid points in the critical regions near the heated wall and the solid-liquid interface where large temperature gradients prevail.

The total simulated melting time in all cases was 12 minutes with a constant time step of 0.5 s. The CPU time for the most demanding case (no. D40) was approximately 4 hours on a SUN4–260. As an example, figure 2 shows the time evolution of the velocity vector maps and isotherms for experiment D10. Increments between adjacent isotherms are constant. The isotherm along the bottom heated wall is $\theta = 1$. Along the top moving surface, the isotherm is $\theta = 0$. At early times, heat transfer in the melt zone is predominated by conduction. Heat fluxes are uniform and the solid-liquid interface remains flat. However this situation is potentially unstable as layers of cold and denser fluid adjacent to the solid-liquid interface lie above layers of hot and lighter fluid near the bottom heated surface. For as long as the temperature gradients remain perfectly vertical, the source term for the vorticity equation (Table 1) is null and the flow is stagnant. As melting progresses, the melt cavity expands and the internal Rayleigh number Ra_δ, based on the local melt gap width δ, increases. Eventually this number reaches a critical value for which the slightest disturbance i.e. numerical perturbation, triggers natural convection. In a previous study for the melting of a PCM in a cylindrical enclosure heated from below, Ra_δ was found to be 2197 [8]. In the present study, no attempt was made to determine the exact value of this number. It was observed, however, that convective motion was triggered for a critical Rayleigh number of this magnitude. As a result, two Bénard cells first appear, at both ends of the cavity. The liquid rises in the central region of the cavity and falls around its edges. Similar observations were

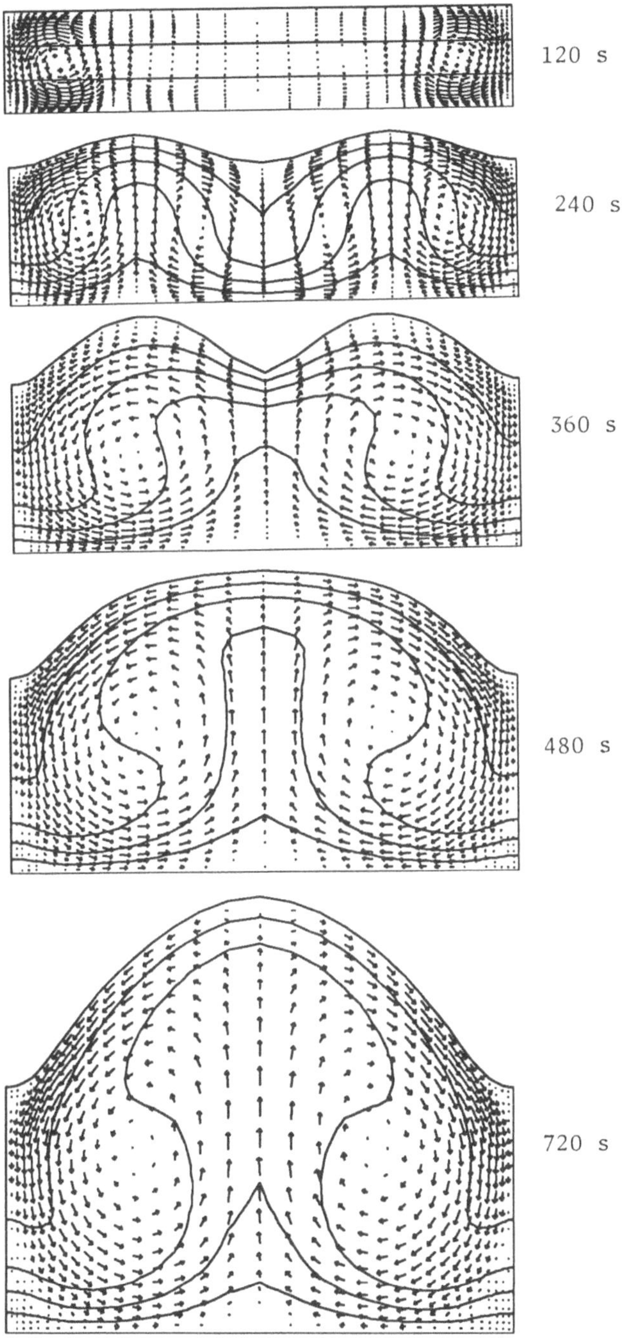

Fig. 2 Predicted velocity field and isotherms for case D10

quickly gives rise to two additional counterrotating convective cells in the central region of the cavity. As time passes, these cells grow in size and the intensity of the convective motion augments. Consequently, the shape of the solid-liquid interface becomes irregular. High melting rates are observed at locations where hot fluid coming up from the heated surface impinges on the phase-front while low melting rates occur at locations where cold fluid leaves the interface to flow down. The outside convective cells continue to grow in size at a faster rate than that for the inside cells and after a while ($\sim$ 360 s) the latter engulf the former yielding two large counterrotating cells. The size of these eddies still increases as melting proceeds and so does the intensity of the uprising fluid motion in the central region. As a result, the melting rate at the middle of the solid-liquid interface is gradually enhanced and the phase front advances more rapidly. The interface contour changes slowly from camel bump shape to a bell shape. For the present case, melting terminates with two symmetric counterrotating cells inside the melt. For cases D20 and D40 however, the size of the cavity becomes large enough so that eventually the two cells merge together to yield one large counterclockwise rotating cell. This merger destroys the symmetry of the flow and temperature fields and perturbs considerably the advance and the shape of the phase front.

The time history of the local Nusselt number profile at the solid-liquid interface is depicted in Fig. 3. The Nusselt number at the melting front is calculated from the normal temperature gradients $\nabla\theta \cdot \vec{n}$ along the interface contour. As time passes, the melt cavity expands and the thermal resistance across the liquid layer increases. This results in a general decrease in the magnitude of the local Nusselt number. The Nusselt number is seen to be uniform initially and deviates with the onset of buoyancy-induced fluid motion. The heat transfer is highest where warm fluid arriving from the heat source impinges on the melt front. For example, four recirculating cells are established in the melt after 240 s and two equal maxima for the Nusselt number are reached at x = 0.3 and x = 0.7. After 480 s, only two recirculation

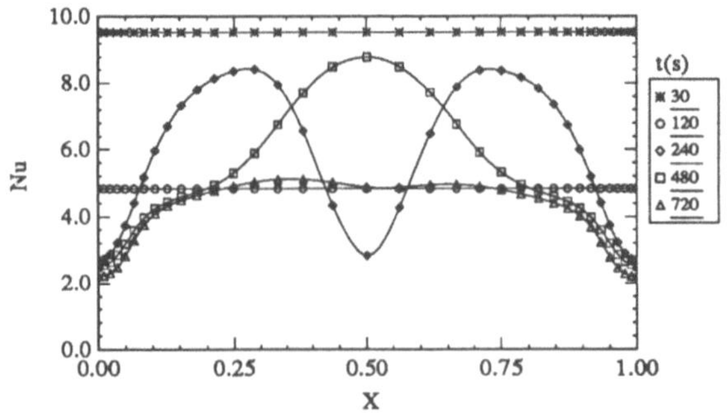

Fig. 3 Local Nusselt number at the solid-liquid interface for case D10

bubbles remain and a maximum value for the Nusselt number is observed at x = 0.5.

As expected, these trends in the local heat transfer rates are reflected in the average Nusselt number at the solid-liquid interface. This number is defined, in the computational space, as

$$\overline{Nu} = -\int_0^1 \frac{\sqrt{\gamma}}{J}\frac{\partial \theta}{\partial \eta} d\xi$$

and is calculated from the converged temperature field at selected time steps in the melting process. Figure 4 shows the calculated average Nusselt number at the phase front for all cases studied. The results display a rapid decrease in the

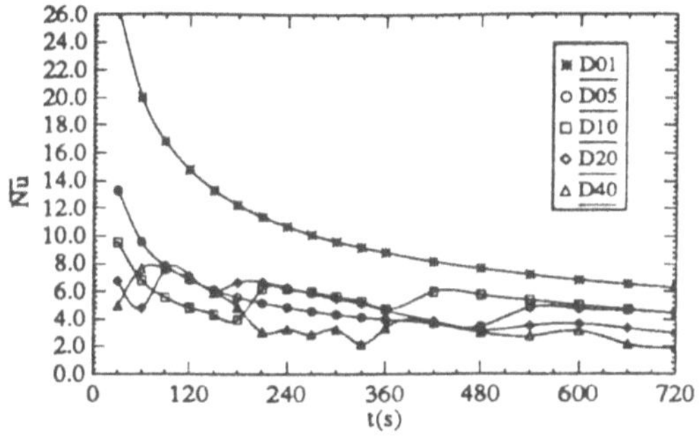

Fig. 4 Temporal variation of the average Nusselt number at the solid-liquid interface

heat transfer rate at the early stages of melting which is indicative of transient heat conduction. A higher Rayleigh number results in an earlier departure from the pure conduction behavior. As soon as natural convection sets in with the appearance of four recirculation cells heat transfer starts increasing. As the melting progresses, the melt cavity expands resulting in a decrease of the magnitude of the Nusselt number. When the four convective cells merge to yield two cells $\overline{Nu}$ increases again for a while and then decreases steadily with the cavity expansion until the time when the two large recirculation bubbles become one (cases D20 and D40). $\overline{Nu}$ experiences a new increase and then starts declining again.

The temporal variation of the molten volume fraction was also determined from a numerical integration of the melt cavity volume,

$$MVF = \frac{H}{L}\int_0^1 S(x,t)dx$$

The MVF results are plotted in Fig. 5. It is seen that the fraction of melted

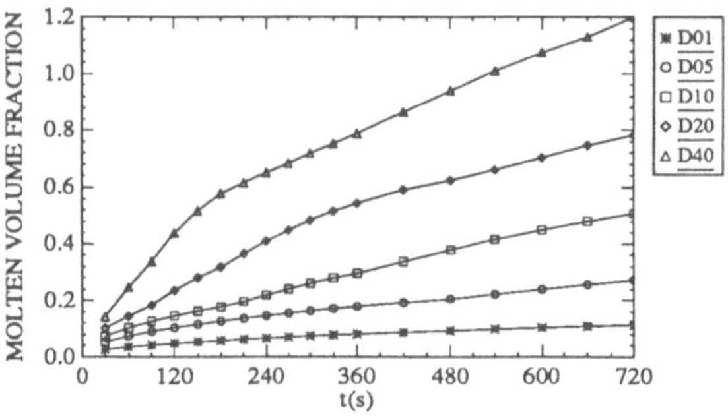

Fig. 5 Predicted molten volume fractions

volume increases almost linearly with time once the convective motion is established. For this flow regime, the results were correlated with a least-squares fit by the following equation

$$MVF = 1.915Ste^{0.629}\,t^{0.533}$$

CONCLUDING REMARS

A numerical study of natural convection dominated melting from below of a pure metal has been conducted. The effect of natural convection on solid-liquid phase change heat transfer under gravitationally unstable situations have been studied. Four Bénard convection cells appeared initially and grew in size as the melting progressed. These cells eventually merged to form two larger cells and, for a sufficiently large cavity, they yielded one large counterclockwise recirculating cell. These complex natural convection flows were found to significantly distort the solid-liquid interface and increase the melting rate. The predicted molten volume fractions were correlated in terms of the Stefan number and time.

Acknowledgements — The author gratefully acknowledges the financial support provided by the Natural Sciences and Engineering Research Council of Canada. The author is also very grateful to Miss Tremblay for typing this manuscript.

REFERENCES

[1] Crank, J. "Free and Moving Boundary Problems", Clarendon Press, 1984.

[2] Webb, B.W., Viskanta, R., "Analysis of Heat Transfer During Melting of a Pure Metal From an Isothermal Vertical Wall", Numer. Heat Transfer, vol. 9, pp 539–558, 1986.

[3] Brent, A.D., Voller, V.R., Reid, K.J., "Enthalpy-Porosity Technique for Modelling Convection-Diffusion Phase Change: Application to the Melting of a Pure Metal", Numer. Heat Transfer, vol. 13, pp 297–318, 1988.

[4] Lacroix, M., "Computation of Heat Tranfer During Melting ofa Pure Substance from an Isothermal Wall", Numer. Heat Transfer, Part B, vol. 15, pp 191–210, 1989.

[5] Gau, C., Viskanta, R., "Effect of Natural Convection on Solidification from Above and Melting from Below of a Pure Metal", Inr. J. Heat Mass Tranfer, vol. 28, No. 3, pp 573–587, 1985.

[6] Diaz, L.A., Viskanta, R., "Visualization of the Solid-Liquid Interface Morphology Formed by Natural Convection during Melting of a Solid from below", Int. Commun. Heat Mass Transfer, vo. 11, pp 35–43, 1984.

[7] Lacroix, M., "Analysis of Natural Convection Melting of a Pure Metal from Below", Numer. Heat Transfer, submitted, 1991.

[8] Wu, Y.K., Purd'homme, M., Nguyen, T.H., "Simulation Numérique de la Fusion à l'intérieur d'un Cylindre Adiabatique Chauffé par le Bas", Int. J. Heat Mass Transfer, submitted, 1990.

A Fast Implementation of the Isotherm Migration Method

A.S. Wood

Department of Mathematics, University of Bradford, Richmond Road, Bradford, West Yorkshire BD7 1DP, UK

ABSTRACT

In this paper we present a numerical scheme, based on a modified isotherm migration formulation, for Stefan problems with *arbitrary* conditions specified on the fixed boundary. An efficient finite difference technique is developed and the results presented show that the method can provide accurate simulations to general moving boundary problems. To facilitate initialising the numerical scheme and assessing its accuracy, a systematic algorithm for developing a double power series solution to the problem is also described. Finally, we consider its application to heat flow problems arising in the simulation of practical thermoelectric devices.

INTRODUCTION

The isotherm migration method (*IMM*) proposed by Chernousko [1] and Dix & Cizek [2] is a natural and effective approach to dealing with diffusion-like problems involving a moving concentration or phase-change boundary - the so-called Stefan problems. Since the original treatises by Stefan [3,4] on melting ice in the Arctic region, several authors have sucessfully based efficient numerical procedures on the *IMM* formulation - Crank & Phahle [5], Crank & Gupta [6], Durack & Wendroff [7], Crank & Crowley [8,9], Turland [10,11], and Wood [12,13]. Crank [14] also gives an excellent survey of the subject. The flexibility of the approach is evident in that *mushy* regions (with a melting/freezing range) are readily modelled without the usual loss of accuracy (associated with enthalpy-type methods) at points close to the moving interface.

The *IMM* approach interchanges the rôles of the usual dependent (temperature or concentration) and independent (spatial) variables so that we follow $x(U,t)$, the location of the isotherm U at time t. In many papers, test problems are proposed on which to demonstrate the prowess of a particular numerical process. For Stefan problems this has unerringly been a problem with (constant) Dirichlet conditions on the fixed boundary. In practice, however, conditions of

Neumann or Robin's type are more likely to represent the true behaviour of the temperature (concentration) at these boundaries. The consequence of such conditions is that the usually constant temperature range in the solution region is now time dependent and isotherms that we are tracking may well become undefined or, indeed, may be created. A *standard* computer package is likely to fail due to *overflow* at some point.

For pure conduction problems Wood [13] has shown that, with some modifications, the *IMM* method can be used effectively to simulate pure heat conduction problems with these boundary conditions. It is the purpose of this paper to extend the realm of applicability to Stefan problems. However, one difficulty concerning the flux-type boundary conditions treated in this paper is the determination of an analytic solution for the purposes of assessing the numerical scheme. Evans et al [15] and Kreith & Romie [16] made some headway and in this paper we describe a systematic scheme to obtain a double power series solution.

A MODEL PROBLEM

Consider a material in the semi-infinite region $x > 0$ to be initially ($t = 0$) liquid at its fusion temperature U_f. For time $t > 0$ one of the following boundary conditions is applied at $x = 0$.

(i) Dirichlet:

$$U = U_0 \,, \tag{1}$$

in which U_0 is a constant temperature such that $U_0 < U_f$.

(ii) Neumann:

$$K\frac{\partial U}{\partial x} = F \,, \tag{2}$$

in which K is thermal conductivity and $F > 0$ is a constant heat flux out of the region $x > 0$.

(iii) Robin's:

$$K\frac{\partial U}{\partial x} = HU \,, \tag{3}$$

in which $H > 0$ is a surface conductance.

In each case heat flows out of the material which subsequently begins to solidify. The process is characterised by a solid/liquid boundary $s(t)$ which moves away from $x = 0$. There is no heat flow in the liquid region and the temperature $U(x,t)$ in the solid region is governed by

$$\frac{\partial U}{\partial t} = \frac{\partial^2 U}{\partial x^2} \,, \quad 0 < x < s(t) \,, \; t > 0 \,. \tag{4}$$

At the interface $s(t)$ the temperature equals the fusion temperature,

$$U = U_f \,, \quad x = s(t) \,, \; t > 0 \,, \tag{5}$$

and by equating the latent heat liberated at the solidification front with the heat conducted away from the front into the solid region we obtain the Stefan condition

$$\frac{ds}{dt} = \frac{c}{L}\frac{\partial U}{\partial x}\,, \quad x = s(t)\,,\ t > 0\,, \tag{6}$$

in which L is latent heat of fusion and c is specific heat. The problems given by equations (1) – (6) can be related to the physical solution space (χ, τ) via the change of variables

$$x = \frac{\chi}{\ell}\,, \quad s = \frac{S}{\ell}\,, \quad t = \frac{\kappa\tau}{\ell^2}$$

in which κ is thermal diffusivity and ℓ is a notional length.

If the condition at $x = 0$ were a constant Dirichlet condition then the problem would admit the usual error function solution (see, for example, Carslaw & Jaeger [17], §11.2, p.285). For the Neumann or Robin's conditions we look for a double power series solution of the form

$$U(x,t) = \sum_{i=0}^{\infty}\sum_{j=0}^{\infty} \alpha_{i,j} x^i t^j\,, \quad s(t) = \sum_{j=0}^{\infty} b_j t^j\,.$$

Evans et al [15] (Neumann condition) determined the first few coefficients in these series and Kreith & Romie [16] (Robin's condition) presented some graphical results using the *thermal analyser*. It is difficult to go much beyond the first 2 or 3 terms by hand and some thought is given to a systematic solution to this difficulty in the next section.

SERIES SOLUTIONS OF THE HEAT EQUATION

For the temperature we assume a double power series solution of the form

$$U(x,t) = \sum_{m=0}^{\infty}\sum_{n=0}^{m}\Big\{\alpha_{2m,2n} + \alpha_{2m+1,2n+1}x\Big\}x^{2n}t^{m-n}\,.$$

Substituting this into the partial differential equation (4) and comparing coefficients of terms of the form $x^i t^j$ leads to the following recurrence relations for the $\alpha_{i,j}$.

$$\begin{aligned} \alpha_{2m,2n} &= \frac{(2n+2)(2n+1)}{m-n}\alpha_{2m,2n+2}\,, \\ \alpha_{2m+1,2n+1} &= \frac{(2n+3)(2n+2)}{m-n}\alpha_{2m+1,2n+3}\,. \end{aligned} \qquad m = 1,2,\ldots\,,\ n = m-1,\ldots,0\,. \tag{7}$$

Setting $\alpha_{2m,2m} = \alpha_{2m+1,2m+1} = 1$ then equation $U(x,t)$ takes the modified form

$$U(x,t) = \sum_{m=0}^{\infty}\sum_{n=0}^{m}\Big\{a_{2m}\alpha_{2m,2n} + a_{2m+1}\alpha_{2m+1,2n+1}x\Big\}x^{2n}t^{m-n}\,, \tag{8}$$

in which the $\alpha_{i,j}$ are *all* known from equation (7). It remains to determine $a_0, a_1, \ldots$ and the coefficients $b_0, b_1, \ldots$ in the power series for the moving boundary, namely

$$s(t) = \sum_{m=0}^{\infty} b_m t^m , \tag{9}$$

using the boundary conditions associated with the heat flow problem. Since $s(t) = 0$ then $b_0 = 0$.

The boundary conditions (5) and (6) are common to both the Neumann and Robin's problems. Differentiating equations (8) and (9) with respect to x and t, respectively, and substituting the resulting series into the Stefan condition (6), at which $x = s(t)$, we obtain the polynomial equation

$$\begin{aligned} pb_p t^{p-1} &= \frac{c}{L} a_1 \\ &+ \frac{c}{L} \sum_{m=1}^{\infty} \Big\{ a_{2m} \sum_{n=0}^{m-1} (2n+2)\alpha_{2m,2n+2} \Big(\sum_{j=1}^{\infty} b_j t^{j-1} \Big)^{2n+1} \\ &\qquad + a_{2m+1} \sum_{n=0}^{m} (2n+1)\alpha_{2m+1,2n+1} \Big(\sum_{j=1}^{\infty} b_j t^{j-1} \Big)^{2n} \Big\} t^{m+n} , \end{aligned} \tag{10}$$

for $p = 1, 2, \ldots$. Comparing coefficients of $O(t^{p-1})$ gives b_p.

We now substitute the series (8) into the temperature condition (5), at $x = s(t)$. Comparing $O(1)$ terms gives the value

$$a_0 = U_f \tag{11}$$

and the remaining polynomial equation is

$$\begin{aligned} 0 = a_1 b_p t^p &+ \sum_{m=1}^{\infty} \Big\{ a_{2m} \sum_{n=0}^{m} \alpha_{2m,2n} \Big(\sum_{j=1}^{\infty} b_j t^{j-1} \Big)^{2n} t^{m+n} \\ &+ a_{2m+1} \sum_{n=0}^{m} \alpha_{2m+1,2n+1} \Big(\sum_{j=1}^{\infty} b_j t^{j-1} \Big)^{2n+1} t^{m+n+1} \Big\} , \end{aligned} \tag{12}$$

for $p = 1, 2, \ldots$. Comparing coefficients of $O(t^p)$ gives the value a_{2p}. The Neumann and Robin's conditions at $x = 0$ determine the specific form of the ordered pairs (b_p, a_{2p}).

Neumann Condition (2)

Differentiating equation (8) with respect to x and substituting into condition (2) yields the following values, on comparing terms of $O(t^m)$.

$$\begin{aligned} O(1): &\qquad a_1 = \frac{F}{K} , \\ O(t^m): &\qquad a_{2m+1} = 0 , \quad m = 1, 2, \ldots . \end{aligned} \tag{13}$$

Hence,

$$U(x,t) = a_0 + a_1 x + \sum_{m=1}^{\infty} a_{2m} \sum_{n=0}^{m} \alpha_{2m,2n} x^{2n} t^{m-n} \tag{14}$$

and, in the appropriately modified forms of equations (10) and (12), comparing terms of $O(1)$ and $O(t)$, respectively, gives the values

$$b_1 = \frac{ca_1}{L} \quad \text{and} \quad a_2 = -\frac{a_1 b_1}{\alpha_{2,0}}. \tag{15}$$

For each $p = 2, 3, \ldots$, equations (10) and (12) now have the forms, respectively,

$$pb_p t^{p-1} = \frac{c}{L} \sum_{m=1}^{\infty} a_{2m} \sum_{n=0}^{m-1} (2n+2)\alpha_{2m,2n+2} \Big(\sum_{j=1}^{\infty} b_j t^{j-1}\Big)^{2n+1} t^{m+n}, \tag{16}$$

from which b_p is obtained ($O(t^{p-1})$ terms), and

$$0 = a_1 b_p t^p + \sum_{m=1}^{\infty} a_{2m} \sum_{n=0}^{m} \alpha_{2m,2n} \Big(\sum_{j=1}^{\infty} b_j t^{j-1}\Big)^{2n} t^{m+n}, \tag{17}$$

from which a_{2p} is obtained ($O(t^p)$ terms). In practical terms, to use equations (16) and (17) we need finite upper limits on the summations $\sum_m$ and $\sum_j$. A little work to determine how many terms are required so that (16) and (17) will contain *all* terms of $O(t^{p-1})$ and $O(t^p)$, respectively, yields the following polynomial equations ($p = 2, 3, \ldots$). For b_p

$$pb_p t^{p-1} = \frac{c}{L} \sum_{m=1}^{p-1} a_{2m} \sum_{n=0}^{n^*} (2n+2)\alpha_{2m,2n+2} \Big(\sum_{j=1}^{j^*} b_j t^{j-1}\Big)^{2n+1} t^{m+n}, \tag{18}$$

in which $n^* = \min\{p-1-m, m-1\}$ and $j^* = p - m - n$, and for a_{2p}

$$0 = a_1 b_p t^p + \sum_{m=1}^{p} a_{2m} \sum_{n=0}^{n^*} \alpha_{2m,2n} \Big(\sum_{j=1}^{j^*} b_j t^{j-1}\Big)^{2n} t^{m+n},$$

in which $n^* = \min\{p-m, m\}$ and $j^* = p - m - n + 1$. Additionally, for a given p, it is the term $m = p$ ($n = 0$) that contains the unknown a_{2p}. Terms $m = p$ ($n > 0$) are of order *greater than* $O(t^p)$ and terms $m = 1, \ldots, p-1$ ($n = 0$) are of order *less than* $O(t^p)$. Thus, for a_{2p}, we can write

$$0 = a_1 b_p t^p + \sum_{m=1}^{p-1} a_{2m} \sum_{n=0}^{n^*} \alpha_{2m,2n} \Big(\sum_{j=1}^{j^*} b_j t^{j-1}\Big)^{2n} t^{m+n} + a_{2p}\alpha_{2p,0} t^p. \tag{19}$$

Finally, we need to *multiply out* the $\sum_j$ terms. Consider, first, equation (18). The term $(\sum_j)^{2n+1}$ can be expanded using the multinomial formula

$$\left(b_1 + \ldots + b_{j^*} t^{j^*-1}\right)^{2n+1} = \sum_{\underline{q}} (2n+1; q_1, \ldots, q_{j^*}) \prod_{j=1}^{j^*} (b_j t^{j-1})^{q_j} \tag{20}$$

in which $\underline{q} = [q_1 \dots q_{j^*}]^T$. The sum is taken over all values of $q_1, \dots, q_{j^*}$ such that

$$\sum_{j=1}^{j^*} q_j = 2n + 1$$

and the multinomial coefficient is given by

$$(2n+1; q_1 \dots q_{j^*}) = \frac{(2n+1)!}{\prod_{j=1}^{j^*} q_j!}. \tag{21}$$

Since we require to pick out terms of $O(t^{p-1})$ then there is the additional condition that

$$\sum_{j=2}^{j^*} (j-1) q_j + m + n = p - 1 .$$

A similar story holds for equation (19) in which $2n+1$ is replaced by $2n$, $p-1$ is replaced by p and the appropriate value of j^* is used. An $O(n)$ routine is used to evaluate the coefficient in equation (21) and a non-recursive $O(j^{*n})$ process is used to traverse all elements of the sum in equation (20) (Wood [18]). The first few coefficients are

$$\begin{aligned}
a_0 &= U_f , \\
a_1 &= \frac{F}{K} , \\
a_2 &= -\frac{a_1 b_1}{\alpha_{2,0}} , \\
a_4 &= -\frac{a_1 b_2 + a_2 b_1^2 \alpha_{2,2}}{\alpha_{4,0}} , \\
a_6 &= -\frac{a_1 b_3 + 2 a_2 b_1 b_2 \alpha_{2,2} + a_4 b_1^2 \alpha_{4,2}}{\alpha_{6,0}}
\end{aligned}$$

and

$$\begin{aligned}
b_1 &= \frac{c a_1}{L} , \\
b_2 &= \frac{c a_2 b_1 \alpha_{2,2}}{L} , \\
b_3 &= \frac{2c(a_2 b_2 \alpha_{2,2} + a_4 b_1 \alpha_{4,2})}{3L} .
\end{aligned}$$

These certainly agree with the few coefficients for $s(t)$ given in Evans et al [15] and those presented in Carslaw & Jaeger [17], §11.3, p.292.

Robin's Condition (3)

A similar analysis to the above leads to the temperature series

$$U(x,t) = a_0 + a_1 x + \sum_{m=1}^{\infty} a_{2m} \sum_{n=0}^{m} \left\{ \alpha_{2m,2n} + \frac{H \alpha_{2m,0}}{K \alpha_{2m+1,1}} \alpha_{2m+1,2n+1} x \right\} x^{2n} t^{m-n} \tag{22}$$

and the initial coefficients

$$a_1 = \frac{Ha_0}{K}, \quad b_1 = \frac{ca_1}{L} \quad \text{and} \quad a_2 = -\frac{a_1 b_1}{\alpha_{2,0}}. \tag{23}$$

For $p = 2, 3, \ldots$, b_p is given by comparing terms of $O(t^{p-1})$ in

$$\begin{aligned} pb_p t^{p-1} &= \frac{c}{L} \sum_{m=1}^{p-1} a_{2m} \sum_{n=0}^{n_1^*} (2n+2)\alpha_{2m,2n+2} \Big(\sum_{j=1}^{j^*} b_j t^{j-1}\Big)^{2n+1} t^{m+n} \\ &+ \frac{cH}{LK} \sum_{m=1}^{p-2} a_{2m} \frac{\alpha_{2m,0}}{\alpha_{2m+1,1}} \sum_{n=1}^{n_2^*} (2n+1)\alpha_{2m+1,2n+1} \Big(\sum_{j=1}^{j^*} b_j t^{j-1}\Big)^{2n} t^{m+n} \\ &+ \frac{cH}{LK} a_{2p-2}\alpha_{2p-2,0} t^{p-1}, \end{aligned} \tag{24}$$

in which $n_1^* = \min\{p-1-m, m-1\}$, $n_2^* = \min\{p-1-m, m\}$ and $j^* = p-m-n$, and a_{2p} is given by comparing $O(t^p)$ terms in

$$\begin{aligned} 0 &= (a_1 b_p + a_{2p}\alpha_{2p,0})t^p \\ &+ \sum_{m=1}^{p-1} a_{2m} \sum_{n=1}^{n_1^*} \alpha_{2m,2n} \Big(\sum_{j=1}^{j_1^*} b_j t^{j-1}\Big)^{2n} t^{m+n} + a_{2p}\alpha_{2p,0} t^p \\ &+ \frac{H}{K} \sum_{m=1}^{p-1} a_{2m} \frac{\alpha_{2m,0}}{\alpha_{2m+1,1}} \sum_{n=0}^{n_2^*} \alpha_{2m+1,2n+1} \Big(\sum_{j=1}^{j_2^*} b_j t^{j-1}\Big)^{2n+1} t^{m+n+1}, \end{aligned} \tag{25}$$

in which $n_1^* = \min\{p-m, m\}$, $j_1^* = p-m-n+1$, $n_2^* = \min\{p-1-m, m\}$ and $j_2^* = p-m-n$. $s(t)$ is still defined by equation (9), albeit with different coefficient values. The first few values are

$$\begin{aligned} a_0 &= U_f, \\ a_1 &= \frac{Ha_0}{K}, \\ a_2 &= -\frac{a_1 b_1}{\alpha_{2,0}}, \\ a_4 &= -\frac{a_1 b_2 + a_2(b_1^2\alpha_{2,2} + Hb_1\alpha_{2,0}/K)}{\alpha_{4,0}}, \\ a_6 &= -\frac{1}{\alpha_{6,0}}\Big(a_1 b_3 + a_2\big(2b_1 b_2\alpha_{2,2} + \frac{H}{K}\alpha_{2,0}(b_2 b_1^3 \frac{\alpha_{3,3}}{\alpha_{3,1}})\big) + a_4(b_1^2\alpha_{4,2} + \frac{H}{K}b_1\alpha_{4,0})\Big) \end{aligned}$$

and

$$\begin{aligned} b_1 &= \frac{ca_1}{L}, \\ b_2 &= \frac{ca_2}{L}\Big(b_1\alpha_{2,2} + \frac{H\alpha_{2,0}}{2K}\Big), \\ b_3 &= \frac{c}{3L}\Big(a_2(2b_2\alpha_{2,2} + \frac{3Hb_1^2\alpha_{2,0}\alpha_{3,3}}{K\alpha_{3,1}}) + a_4(2b_1\alpha_{4,2} + \frac{H\alpha_{4,0}}{K})\Big). \end{aligned}$$

AN ISOTHERM MIGRATION FORMULATION

The *IMM* tracks selected isotherms and, as indicated in the introduction, requires a constant temperature range. Conditions (2) and (3) both lead to variable temperature ranges since $U(0,t)$ is a function of time. Further, if the heat flux F in condition (2) is zero then the resulting insulation condition transforms to an undefined gradient $\partial x/\partial U$ in the *IMM* formulation. Finally, for reasons of numerical expediency, it is usual to treat a non-dimensional problem. To take into account all these ideas the model problem is transformed using the relation

$$W(x,t) = \frac{U(x,t) - U_0(t)}{U_f - U_0(t)} + \frac{\varpi x}{s}, \tag{26}$$

in which $U_0(t) = U(0,t)$. The first term in equation (26) fixes the temperature range to $0 \le W \le 1+\varpi$ and the second term ensures a non-zero gradient at $x = 0$ if condition (1) is used (see equation (29) below). The dimensionless problem is

$$\frac{\partial W}{\partial t} = \frac{\partial^2 W}{\partial x^2} - \frac{\dot{U}_0[1 - W + \varpi x/s]}{U_f - U_0} - \frac{\varpi x \dot{s}}{s^2}, \quad 0 < x < s(t), \; t > 0, \tag{27}$$

where $\dot{} \equiv d/dt$, subject to one of

$$W = 0, \quad t > 0, \tag{28}$$

$$\frac{\partial W}{\partial x} = \frac{\mathcal{F}}{U_f - U_0} + \frac{\varpi}{s}, \quad t > 0, \tag{29}$$

or

$$\frac{\partial W}{\partial x} = \frac{\mathcal{H} U_0}{U_f - U_0} + \frac{\varpi}{s}, \quad t > 0, \tag{30}$$

on the fixed boundary, $x = 0$, in which $\mathcal{F} = F/K$ and $\mathcal{H} = H/K$. On the moving boundary $x = s(t)$ the conditions become

$$W = 1 + \varpi, \quad t > 0, \tag{31}$$

and

$$\frac{ds}{dt} = \beta\left(\frac{\partial W}{\partial x} - \frac{\varpi}{s}\right), \quad t > 0, \tag{32}$$

where $\beta = c(U_f - U_0)/L$ is the Stefan number - the ratio of sensible to latent heat. With a little effort, using the differential relations (Crank & Phahle [5])

$$\frac{\partial W}{\partial t} = -\left(\frac{\partial x}{\partial W}\right)^{-1} \frac{\partial x}{\partial t}, \quad \frac{\partial W}{\partial x} = \left(\frac{\partial x}{\partial W}\right)^{-1} \quad \text{and} \quad \frac{\partial^2 W}{\partial x^2} = -\left(\frac{\partial x}{\partial W}\right)^{-3} \frac{\partial^2 x}{\partial W^2},$$

equations (27) and (32) take the *IMM* form

$$\frac{\partial x}{\partial t} = \left(\frac{\partial x}{\partial W}\right)^{-2} \frac{\partial^2 x}{\partial W^2} r + \left[\frac{\dot{U}_0(1 - W + \varpi x/s)}{U_f - U_0} + \frac{\varpi x \dot{s}}{s^2}\right] \frac{\partial x}{\partial W},$$

$$0 < W < 1 + \varpi, \; t > 0, \tag{33}$$

and

$$\frac{ds}{dt} = \beta\left[\left(\frac{\partial x}{\partial W}\right)^{-1} - \frac{\varpi}{s}\right], \quad W = 1 + \varpi\,, \; t > 0\,, \tag{34}$$

in which $x(W,t)$ is the location of the isotherm W at time t and

$$\begin{aligned} x(0,t) &= 0\,, \\ x(1+\varpi,t) &= s(t)\,. \end{aligned} \quad t > 0\,, \tag{35}$$

The values of U_0 and $\dot{U}_0$ appearing in equation (33) can be estimated from difference replacements of the fixed-boundary conditions (see next section).

A NUMERICAL SOLUTION PROCESS

The temperature range $0 \leq W \leq 1 + \varpi$ is discretised into N elements of size $h = (1+\varpi)/N$ and time is subdivided into intervals of size k. An explicit finite difference replacement for equation (33) yields

$$\begin{aligned} x_i^{m+1} &= x_i^m + \frac{4k(x_{i+1}^m - 2x_i^m + x_{i-1}^m)}{(x_{i+1}^m - x_{i-1}^m)^2} \\ &+ k\left[\frac{\dot{U}_0^m(1 - W_i + \varpi x_i^m/s^m)}{U_f - U_0^m} + \frac{\varpi x_i^m \dot{s}^m}{(s^m)^2}\right]\frac{(x_{i+1}^m - x_{i-1}^m)}{2h}\,, \\ &1 \leq i \leq N-1\,,\; m \geq 0\,, \end{aligned} \tag{36}$$

in which $x_i^m \sim x(W_i, t_m)$, $W_i = ih$, $t_m = t_0 + mk$, t_0 is the time at which the numerical process is initialised, $U_0^m \sim U_0(t_m)$, $\dot{U}_0^m \sim \dot{U}_0(t_m)$, $s^m \sim s(t_m)$ and $\dot{s}^m \sim \dot{s}(t_m)$. The conditions (35) become

$$x_0^m = 0\,, \qquad x_N^m = s^m\,, \quad m \geq 0\,. \tag{37}$$

The boundary s and, in the Neumann and Robin's problems, the temperature U_0 can be advanced using the $O(k)$ replacements

$$s^{m+1} = s^m + k\dot{s}^m\,, \quad m \geq 0\,, \tag{38}$$

and

$$U_0^{m+1} = U_0^m + k\dot{U}_0^m\,, \quad m \geq 0\,. \tag{39}$$

We also need to advance $\dot{s}$ and, in the Neumann and Robin's problems, $\dot{U}_0$. An $O(h^2)$ replacement for the Stefan condition (34) gives

$$\dot{s}^{m+1} = \beta^{m+1}\left[\frac{2h}{(3s^{m+1} - 4x_{N-1}^{m+1} + x_{N-2}^{m+1})} - \frac{\varpi}{s^{m+1}}\right], \quad m \geq 0\,, \tag{40}$$

in which $\beta^{m+1} = c(U_f - U_0^{m+1})/L$. As $x \to 0$ so the partial differential equation (27) for W becomes, at $t = t_{m+1}$,

$$\dot{U}_0^{m+1} = (U_f - U_0^{m+1}) \lim_{x\to 0} \left.\frac{\partial^2 W}{\partial x^2}\right|_{t=t_{m+1}}, \quad m \geq 0\,. \tag{41}$$

At $x = 0$ we use a central difference replacement for the derivative $\partial^2 W/\partial x^2$ which introduces the fictitious value W_{-1}. From equations (29) and (30) a central difference replacement for $\partial W/\partial x$, at $x = 0$, can be used to eliminate W_{-1}. Thus, for equation (41), we obtain the following estimates for the second-order derivative.

Neumann:

$$\frac{2}{(x_1^{m+1})^2}\left[h - x_1^{m+1}\left(\frac{\mathcal{F}}{U_f - U_0^{m+1}} + \frac{\overline{\omega}}{s^{m+1}}\right)\right], \quad m \geq 0 .$$

Robin's:

$$\frac{2}{(x_1^{m+1})^2}\left[h - x_1^{m+1}\left(\frac{\mathcal{H}U_0^{m+1}}{U_f - U_0^{m+1}} + \frac{\overline{\omega}}{s^{m+1}}\right)\right], \quad m \geq 0 .$$

The explicit replacement (36) will have a restriction on the maximum size of the time step k. This arises from stability considerations. Typically, to ensure the correct physical movement of the isotherms, a virtual increase in x_i^m should produce a virtual increase in x_i^{m+1}. Thus, the coefficient of x_i^m in equation (36) must be positive leading to the requirement that

$$k = \min_{1 \leq i \leq N-1} \frac{(x_{i+1}^m - x_{i-1}^m)^2}{8 - \left(\frac{\dot{U}_0^m}{U_f - U_0^m} + \frac{\dot{s}^m}{s^m}\right)\frac{(x_{i+1}^m - x_{i-1}^m)^3 \overline{\omega}}{2hs^m}} . \tag{42}$$

To overcome this we could use an implicit replacement like the Crank-Nicolson algorithm. However, this will require the solution of a system of non-linear equations for $[x_1^{m+1} \ldots x_N^{m+1}]^T$ at *each* time step. Instead we shall make use of the hopscotch replacement proposed by Gourlay [19]. At each step, (36) is applied at nodes (W_i, t_m) for which $i + m$ is **even**, that is, at alternate spatial nodes. Then at the *same* time step, we apply an implicit replacement of equation (33) at nodes for which $i + m$ is **odd**. The values $x_{\pm 1}^{m+1}$ appearing in this replacement will have already been estimated from the explicit sweep and so the implicit replacement can be written in the *explicit* form

$$x_i^{m+1} = \frac{\delta_a^{m+1}}{\delta_b^{m+1}} \tag{43}$$

in which

$$\delta_a^{m+1} = x_i^m + 4k\frac{(x_{i+1}^{m+1} + x_{i-1}^{m+1})}{(x_{i+1}^{m+1} - x_{i-1}^{m+1})^2} + k\frac{\dot{U}_0^{m+1}(1 - W_i)(x_{i+1}^{m+1} - x_{i-1}^{m+1})}{2h(U_f - U_0^{m+1})}$$

and

$$\delta_b^{m+1} = 1 + \frac{8k}{(x_{i+1}^{m+1} - x_{i-1}^{m+1})^2} - k\left[\frac{\dot{U}_0^{m+1}}{(U_f - U_0^{m+1})} + \frac{\dot{s}^{m+1}}{s^{m+1}}\right]\frac{(x_{i+1}^{m+1} - x_{i-1}^{m+1})\overline{\omega}}{2hs^{m+1}} .$$

The hopscotch approach has the benefit of being an explicit-type algorithm, in the sense that no system of equations needbe solved at each time step, combined with a (theoretical) unconditional stability. This makes it an efficient numerical process.

RESULTS AND DISCUSSION

The numerical process is initialised at $t = t_0$ using the appropriate series solution, obtained earlier. For each W_i, x_i^0 is the solution of the equation

$$U(x_i^0, t_0) = U_0(t_0) + \big(U_f - U_0(t_0)\big)\left(W_i - \frac{\overline{\omega} x_i^0}{s_0}\right)$$

in which the initial value $t_0 = .5$ is used. The simulations were run to a time $t = 1$. Tables 1 and 2 show the results of applying the explicit and hopscotch schemes, respectively, to the Dirichlet problem with values $c = L = U_f = 1$ and $U_0 = 0$ (Wood [12]). A fictitious flux of $\overline{\omega} = 0$ was used.

Table 1. *Dirichlet* isotherm locations at $t = 1$ given by the explcit scheme.

Isotherm	Numerical Location			Exact
W	$h = .1$	$h = .05$	$h = .025$	Location
0.0	0.0	0.0	0.0	0.0
0.2	0.220822	0.220574	0.220507	0.220483
0.4	0.447204	0.446686	0.446544	0.446493
0.6	0.685757	0.684917	0.684685	0.684602
0.8	0.945778	0.944512	0.944160	0.944033
1.0	1.242759	1.240855	1.240319	1.240125

Table 2. *Dirichlet* isotherm locations at $t = 1$ given by the hopscotch scheme.

Isotherm	Numerical Location			Exact
W	$h = .1$	$h = .05$	$h = .025$	Location
0.0	0.0	0.0	0.0	0.0
0.2	0.220775	0.220563	0.220504	0.220483
0.4	0.447102	0.446661	0.446538	0.446493
0.6	0.685587	0.684874	0.684674	0.684602
0.8	0.945514	0.944446	0.944143	0.944033
1.0	1.242358	1.240754	1.240294	1.240125

Convergence to the analytic value is evident. Applying Richardson's extrapolation to the value of the weighted 1-norm error measure

$$\|\underline{e}\|_1 = \frac{1}{N}\sum_{i=1}^{N}\left|\frac{x(W_i, t_m) - x_i^m}{x(W_i, t_m)}\right|$$

yields convergence rates of, approximately, 1.85 (explicit) and 1.83 (hopscotch). These agree with the theoretical expectation of $O(h^2)$. Also, for these results, the heuristic time-step control (42) was applied leading to 115 ($N = 10$), 460 ($N = 20$) and 1840 ($N = 40$) time steps being required. If the control (42) is *turned off*, then we can use the hopscotch scheme with larger time-step sizes k. Table 3 shows the results of one simulation using a fixed value of $k = .005$ and $h = .05$ ($N = 20$).

Table 3. *Dirichlet* isotherm locations at $t = 1$ using hopscotch with fixed k.

Isotherm	Location	
W	Exact	Numerical
0.0	0.0	0.0
0.2	0.220483	0.220713
0.4	0.446493	0.446944
0.6	0.684602	0.685254
0.8	0.944033	0.944863
1.0	1.240125	1.241120

The above results were obtained with only 100 time steps, requiring considerably less CPU time, and yet still provide a 3-decimal-place solution. The efficiency of hopscotch is well demonstrated.

Table 4 displays results of applying the explicit scheme to the Neumann problem with values $c = F = K = 1$ and $L = 5$. A *flux* of $\bar{\omega} = 1$ was used.

Table 4. Error in the numerical predictions of the *Neumann* isotherm locations at $t = 1$ given by the explicit scheme.

Isotherm	Location		Rel. Error $\times 10^{-4}$		
W	Exact	Numerical	$h = .4$	$h = .2$	$h = .1$
0.0	0.0	0.0	0.0	0.0	0.0
0.4	0.03896	0.03896	0.1153	0.0289	0.0072
0.8	0.07806	0.07806	0.1159	0.0290	0.0073
1.2	0.11731	0.11731	0.1165	0.0292	0.0073
1.6	0.15670	0.15670	0.1171	0.0293	0.0073
2.0	0.19624	0.19624	0.1178	0.0295	0.0074

The above results display a convergence rate of about 2.0002 which agrees very closely with theoretical predictions. The heuristic time-step control (42) was again applied and led to 1295 ($N = 5$), 5191 ($N = 10$) and 20786 ($N = 20$) time steps being required. The resulting CPU times on a Cyber 180/830 machine, running under NOS, were 2.5s ($N = 5$), 21s ($N = 10$) and 167s ($N = 20$). The moderately large CPU times originate from the fact that the boundary $s(t)$ is moving very slowly. However, as seen from the data presented in table 5, the moving boundary is tracked with exceptional accuaracy even with a coarse mesh size - just 5 cells are used for these values. Thus, if a knowledge of the location of the boundary is of primary intereset, as opposed to the associated temperature distribution, then the 2.5s of CPU time mentioned above is very acceptable for the accuracy obtained.

Table 5. Location of the moving boundary $s(t)$.

t_m	$s(t_m)$	s^m	$\dot{s}(t_m$	$\dot{s}^m$
0.5	0.09903174	-	0.19618741	-
0.6	0.11864297	0.11864322	0.19546551	0.19546751
0.7	0.13819586	0.13819646	0.19475590	0.19475855
0.8	0.15758066	0.15758171	0.19406210	0.19406548
0.9	0.17705637	0.17705710	0.19337450	0.19337868
1.0	0.19624253	0.19624485	0.19270614	0.19271118

Further, the uniformity of the error distribution in table 4 is worth mentioning. Over the entire temperature range the relative error changes by about 2% and the order of accuracy at nodes adjacent to the moving boundary is comparable with that at *internal* points. This is in contrast with other flexible methods, such as the enthalpy formulation, in which the accuracy at nodes close to $s(t)$ can be 1 or 2 orders of magnitude worse than for internal nodes (for example, Wood et al [20]). This is in part due to the fact that difference replacements for the enthalpy formulation tend to artificially hold the temperature at the node immediately adjacent to $s(t)$ at the fusion temperature.

The data presented in table 6 show that problems covering a moderate spread of boundary flux values can be simulated with good results.

Table 6. Error $\|\underline{e}\|_1$ as a function of flux F.

F	$\|\underline{e}\|_1$
0.2	0.222×10^{-7}
0.4	0.347×10^{-6}
0.6	0.169×10^{-5}
0.8	0.508×10^{-5}
1.0	0.117×10^{-4}

In particular, as the boundary at $x = 0$ approaches an insulated state ($F = 0$) the method copes extremely well. This justifies the introduction of the fictitious flux $\overline{\omega}$. $\overline{\omega}$ may also be used to *tweek* the accuracy of the numerical predictions and for the Neumann problem with $N = 5$, $c = F = K = 1$ and $L = 5$, table 7 shows the weighted error measure $\|\underline{e}\|_1$ as a function of $\overline{\omega}$.

Table 7. Error $||\underline{e}||_1$ as a function of fictitiousflux $\overline{\omega}$.

$\overline{\omega}$	$\lVert\underline{e}\rVert_1 \times 10^{-4}$
0.0	0.1946
0.2	0.1688
0.4	0.1502
1.0	0.1165
2.0	0.0901
5.0	0.0635

Similar results on accuracy and behaviour can be obtained for the Robin's moving boundary problem.

To finish we mention the case in which there is no moving boundary and a fixed spatial region of solution, $0 < x < 1$. If the boundary conditions are of the form

$$\frac{\partial U}{\partial x} = 0\,, \quad x = 0\,,\ t > 0\,,$$
$$\frac{\partial U}{\partial x} = H(V - U)\,, \quad x = 1\,,\ t > 0\,,$$

then the resulting problem heat flow problem is a crude model of the thermistor (Westbrook [21]). For numerical results, the reader is refered to Wood [13,22].

In conclusion we have shown that the *IMM* method can be used to accurately simulate heat flow problems, with or without phase-change boundaries, and with a variety of *practical* boundary conditions. In addition, if the hopscotch difference replacement is used to numerically represent the resulting models then an efficient computational process can be designed.

REFERENCES

1. Chernousko, F.L. Solution of non-linear heat conduction problems in media with phase changes, Int. Chem. Engng., Vol.10, pp. 42-48, 1970.
2. Dix, R.C. and Cizek, J. The isotherm migration method for transient heat conduction analysis, Heat Transfer, Vol.1, pp. 1-10, Proc. 4th Int. Heat Transfer Conf., Paris, Versailles, 1970. Elsevier, Amsterdam, 1970.
3. Stefan, J. Über einige Probleme der Theorie der Wärmleitung, Sber. Akad. Wiss. Wien, Vol.98, pp. 473-484, 1889.
4. Stefan, J. Über die Theorie der Eisbildung, inbesondere über die Eisbildung im Polarmeere, Sber. Akad. Wiss. Wien, Vol.98, pp. 965-983, 1889. (Also in Ann. Phys. U. Chem., Vol.42, pp. 269-286, 1891)
5. Crank, J. and Phahle, R.D. Melting ice by the isotherm migration method, Bull. J. Inst. Maths. Applics., Vol.9, pp. 12-14, 1973.
6. Crank, J. and Gupta, R.S. Isotherm migration method in two dimensions, Int. J. Heat Mass Transfer, Vol.18, pp. 1101-1107, 1975.

7. Durack, D. and Wendroff, B. Computing a two-dimensional quench front, Nucl. Sci. Engng., Vol.64, pp. 187-191, 1977.

8. Crank, J. and Crowley, A.B. Isotherm migration along orthogonal flow lines, Int. J. Heat Mass Transfer, Vol.21, pp. 393-398, 1978.

9. Crank, J. and Crowley, A.B. On an implicit scheme for the isotherm migration method along orthogonal flow lines in two dimensions, Int. J. Heat Mass Transfer, Vol.22, pp. 1331-1337, 1979.

10. Turland, B.D. A general formulation of the isotherm migration method for reactor accident analysis, UKAEA (Culham) Preprint CLM-P582, 1979. (Also in Proc. 5th Int. Conf. Structural Mechanics in Reactor Technology, Berlin, 1979)

11. Turland, B.D. Finite difference front-tracking algorithms for Stefan problems arising in reactor safety studies, UKAEA (Culham) Preprint CLM-P662, 1982. (Also in Proc. Simp. Free Boundary Problems, Montecatini, Italy, 1981)

12. Wood, A.S. An unconditionally stable implementation of the isotherm migration method, NA Report 90-15, Univ. Bradford, 1990. (To appear in J. Numer. Heat Transfer (B), Vol.20, 1991)

13. Wood, A.S. A note on the use of the isotherm migration method, NA Report 90-16, Univ. Bradford, 1990. (To appear in J. Comp. Appl. Math., 1991)

14. Crank, J. Free and Moving Boundary Problems, Clarendon Press, Oxford, 1984.

15. Evans II, G.W., Isaacson, E. and MacDonald, J.K.L. Stefan-like problems, Quart. Appl. Math., Vol.8, pp. 312-319, 1950.

16. Kreith, F. and Romie, F.E. A study of the thermal diffusion equation with boundary conditions corresponding to solidification or melting of materials initially at the fusion temperature, Proc. Phys. Soc. (B), Vol.68, pp. 277-291, 1955.

17. Carslaw, H.S. and Jaeger, J.C. Conduction of Heat in Solids, 2nd Edn., Clarendon Press, Oxford, 1959.

18. Wood, A.S. Some comments on the multinomial distribution function, SOR Report 90-39, Univ. Bradford, 1990.

19. Gourlay, A.R. Hopscotch: A fast second-order partial differential equation solver, J. Inst. Maths. Applics., Vol.6, pp. 375-390, 1970.

20. Wood, A.S., Ritchie, S.I.M. and Bell, G.E. An efficient implementation of the enthalpy method, Int. J. Numer. Meths. Engng., Vol.17, pp. 301-305, 1981.

21. Westbrook, D.R. The thermistor: A problem in heat and current flow, Numer. Meths. PDEs, Vol.5, pp. 259-273, 1989.

22. Wood, A.S. Modelling the thermistor (Ed. M. Heliö), Proc. ECMI-90, Lahti, Finland, 1990. Kluwer, Stuttgart, 1991.

An Inverse Finite Element Method for Directly Formulated Free and Moving Boundary Problems

A.N. Alexandrou

Department of Mechanical Engineering, Worcester Polytechnic Institute, Worcester, MA 01609, USA

ABSTRACT

In free boundary problems *a priori* unknown geometric parameters need to be evaluated simultaneously with the primary field unknowns. These problems are highly nonlinear and difficult to handle computationally. Here an accurate and efficient method is presented which can appropriately handle the complexities associated with these problems. The method is based on a tesselation that is constructed by isolines as characteristic coordinates lines. Thus opposite sides of finite elements lie on isolines. The method allows the simultaneous determination of the location of the isolines with the primary unknowns. Its accuracy and efficiency is demonstrated in a number of model problems

INTRODUCTION

In free boundary problems, in addition to the field variables unknown geometric parameters need to be computed. Free boundary problems are highly nonlinear due to nonlinearities associated with *a priori* unknown geometric parameters, such as the melt/solid interphase in a solidification problem, and those due to the convective terms in the equations of motion (inertia) and energy (convection). Typical free boundary problems are those that include phase change such as solidification (or melting) and evaporation (or condensation). Therefore, without any loss of generality, most of the discussion presented in this work will be concentrated on problems associated with phase change and in particular with solidification.

In most materials, when the temperature reaches a transition temperature a phase change takes place, either by releasing or by absorbing heat. A characteristic feature of the problem is the presence of a demarcation line which separates the different phases. The boundary can be stationary or moving but in either case the location of the solidification front is *a priori* unknown, and it must be evaluated simultaneously with the governing equations. The presence of the solidification front is not merely an additional complexity to a non-linear problem. The solution has to account for the special boundary conditions that are associated with the latent heat of solidification and the change in material properties between the melt and solid phases. Obviously, this is difficult to achieve because the phases are defined by the solid/melt interphase which is *a priori* unknown.

Solidification has been studied extensively producing a plethora of information dealing with theoretical and numerical solutions to the problem. More recently, significant work has been done in relation to crystal growth. These processes are associated with the formation of single crystals which are the basis of the manufacturing of semiconductor devices [1] [2] [3]. Most numerical methods used to solve the problem are for Stefan-like problems where the governing equations (but not the boundary conditions) are linear. The solution to the linear problem is obtained using a variety of methods such as finite difference or finite elements. The location of the free boundary is obtained by relaxing one of the boundary conditions (such as the discontinuity in the thermal flux across the solidification front) and satisfying it iteratively after repeated solutions to the linear problem. The above methods are mostly limited to linear problems and because of the Picard iteration schemes they have slow convergence (or no convergence at all) and therefore, are costly to use or not suitable, for highly nonlinear problems. The simplest method dealing with a moving solidification front is the enthalpy method, [5]. The enthalpy method is used successfully by Voller and other investigators in one dimensional binary alloy solidification [5] [6]. However, the method cannot be easily implemented in problems involving two or three dimensions because of the more complicated search for the solidification temperature. The temperature search does not accommodate efficiently the non-linear terms and the material property changes in the melt and solid regions. The applicability of this method is limited, therefore, to one-dimensional and linear problems, [4].

The most promising methods to successfully handle phase change problems are those based on inverse formulations [4]. Using a set of orthogonal coordinates, inverse formulations appropriately transform the governing equation and make the space coordinates the dependent variables and the temperature the independent variable. Solutions to inverse problems have been successfully obtained for some solidification problems. One of the earliest inverse methods which it mathematically inverts the energy equation, the Isotherm Migration Method, [7], has been successfully implemented using finite difference methods in linear one-dimensional steady and unsteady problems [8] [9]. The applicability of purely inverse methods has been restricted to steady linear problems because in transient and nonlinear problems it is difficult to define an appropriate orthogonal coordinate system to invert the governing equations. Additional problems in the implementation of the methods are the variable nonlinearities in the problem and those introduced by the transformation.

In the method presented here the temperature at nodal points is *a priori* fixed and the problem is solved for the location and motion of the nodes. As the title reveals, the method is based on an inversely defined (but not mathematically inverted) problem. It is fundamentally different from the methods discussed above; it does not involve any variable transformation and thus it avoids all the complexities associated with such transformation and it can be applied to both steady and unsteady problems irrespective of the dimensionality of the problem.

In this paper the method and its accuracy will be discussed in detail in a variety of model problems. Two of the examples presented here are taken from [10]. These are simple problems and serve as excellent sample problems to illustrate the method. The method will be demonstrated in more diverse problems such as three-dimensional solidification and inverse airfoil design.

INVERSE FINITE ELEMENT METHOD

Consistent with the *inverse* approach, the Inverse Finite Element Method presented in this paper addresses the question: "At what location (at the nodes of the finite elements) does the variable, such as temperature, have a specific value"? Therefore, with the solution of the problem, the location and advancement of the solidification front is automatic.

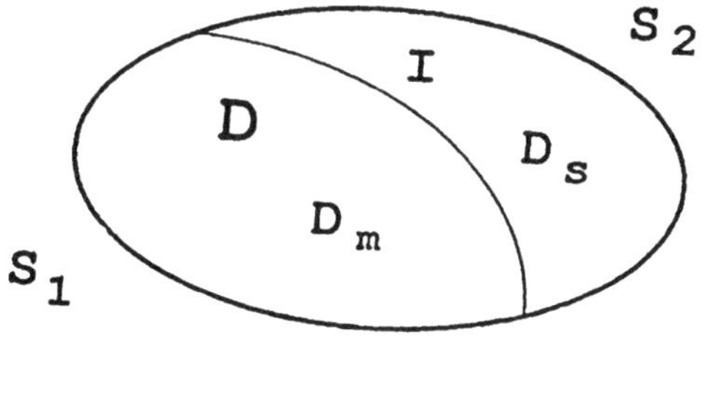

Figure 1: Typical Solidification Domain. Subdomains D_m and D_s correspond to the melt and solid regions and are separated by the solidification front, I.

Consider a typical solidification problem where the domain of interest, D, is bounded by S_1 and S_2 and separated by the solidification front, I, into the melt and solid regions, D_m, D_s, as shown in Figure 1. In the absence of melt flow, the problem is described by the energy equation. In dimensional form:

$$\rho_m c_m \frac{\partial T}{\partial t} = \kappa_m \nabla^2 T \tag{2}$$

$$\rho_s c_s \frac{\partial T}{\partial t} = \kappa_s \nabla^2 T \tag{3}$$

where ρ, c and κ are the density, specific heat and conductivity respectively which can be temperature dependent or constant. Subscripts m and s refer to the melt and solid phases respectively.

Boundary conditions:

$$T_m = T(S_1, t) \qquad \text{along} \qquad S_1$$

$$T_s = T(S_2, t) \qquad \text{along} \qquad S_2$$

The temperature along the solidification front is the solidification temperature, T_s. During phase change heat is released due to the latent heat of solidification, H. Therefore, in order to maintain local thermal equilibrium

$$\kappa_m \nabla T \cdot n - \kappa_s \nabla T \cdot n = \rho_m H \mathbf{U} \cdot n \tag{4}$$

where n is the outward unit normal in the melt region along the solidification front and $\mathbf{U}$ the velocity of the front.

Following a typical Finite Element Galerkin formulation and by invoking the divergence theorem, the problem can be discretized and solved on a finite computational mesh as:

$$R_E^i = \int\int \left[\phi_i \rho_m c_m \frac{\partial T}{\partial t} + \kappa_m \nabla T \nabla \phi_i\right] |J_c| d\xi d\eta + \underbrace{\int -\kappa_m \nabla T \cdot n \phi_i ds}_{I_1} \tag{5}$$

$$R_E^i = \int\int \left[\phi_i \rho_s c_s \frac{\partial T}{\partial t} + \kappa_s \nabla T \nabla \phi_i\right] |J_c| d\xi d\eta + \underbrace{\int -\kappa_s \nabla T \cdot n \phi_i ds}_{I_2} \tag{6}$$

where ϕ_i $(i = 1, N)$ are the basis functions, N the number of nodes per element and ξ and η the local coordinates after an isoparametric transformation.

If Derichlet boundary conditions, (specified temperature) are described along S_1 and S_2 the line integrals in the residual equations are set equal to zero. If the boundary conditions are of Neumann type (cooling by convection) the line integrals are replaced by:

$$I_1, I_2 = \int \phi_i h \left(T - T_\infty\right) ds$$

Due to the boundary condition along the solidification front the line intergrals are replaced by:

$$I_1 + I_2 = \int \phi_i \rho H \mathbf{U} \cdot n dS$$

In classical Finite Element solutions the evaluation of the integrals is performed numerically by Gauss integration and the resulting system of equations (5) - (6) is solved for the unknown nodal temperature using an iterative procedure such as full Newton iteration

$$\mathbf{q}^{(n+1)} = \mathbf{q}^{(n)} - \mathbf{Q}^{-1}\mathbf{R}\left(\mathbf{q}^{(n)}\right) \tag{7}$$

where $\mathbf{q}$ is the vector of the unknowns and $\mathbf{Q} = \frac{\partial \mathbf{R}}{\partial \mathbf{q}}$, the Jacobian of the residuals with respect to the unknown temperature.

The Finite Element solution procedure outlined above, however, cannot be applied to the described problem because the location of the solidification front and therefore, the melt and solid regions which have different material properties are not known. The Inverse Finite Element Method overcomes this difficulty by invoking an inverse thinking to the above classical formulation; it fixes the temperature at the nodal points and solves equations (5)-(6) for the location of the nodes. With this approach the "melt" and "solid" elements and the elements around the solidification front are automatically known. Therefore, the proper equations and boundary conditions are applied to each region. The solution is obtained in few iterations using a Newton iteration procedure by forming the Jacobian with respect to the coordinates of the nodal points rather than the temperature.

ONE-DIMENSIONAL FREE BOUNDARY TEST PROBLEM

Consider the simple one dimensional free boundary problem:

$$\begin{aligned} \frac{\partial^2 u}{\partial x^2} - 1 = 0 \quad & in \quad 0 \le x < \alpha, \\ u = 0 \quad & in \quad \alpha \le x < 1, \\ u \ge 0 \quad & in \quad 0 \le x < 1, \end{aligned} \tag{8}$$

$$u(0) = \tfrac{1}{4}, \qquad u(1) = 0.$$

This problem has the exact solution

$$u = \begin{cases} \frac{1}{2}(x-\alpha)\left(x - \dfrac{1}{2\alpha}\right) & in \quad 0 \le x < \alpha \\ 0 & in \quad \alpha \le x \le 1 \end{cases}, \tag{9}$$

provided u is continuous and non-negative in the interval $(0,1)$. The free boundary parameter, α, is found by imposing the condition that the first derivative of u is continuous. This yields $\alpha = 2^{-\frac{1}{2}}$.

Following a traditional Galerkin finite element formulation the variable u is expressed in terms of a quadratic finite element basis function

$$u = \sum_{i=1}^{9} \phi_i(\xi)\, u_i. \tag{10}$$

Each of the elements in the domain is mapped isoparametrically into a basic line element with limits $-1 \le \xi \le 1$, using the relation

$$x = \sum_{i=1}^{9} \phi_i(\xi)\, x_i. \tag{11}$$

The governing equation (Equation 8) is then discretized by forming the residual equation

$$R_i = \int \left(\frac{\partial^2 u}{\partial x^2} - 1\right) \phi_i dx = 0. \tag{12}$$

Finally, by invoking the divergence theorem and by changing the limits of integration to account for the isoparametric transformation the residual equation becomes

$$R_i = \int \left(\frac{\partial \phi_i}{\partial x} \frac{\partial u}{\partial x} - \phi_i \right) \frac{\partial x}{\partial \xi} d\xi = 0. \tag{13}$$

In a traditional "direct" FEM approach and for problems without free boundaries Equation 13 is solved for u on a fixed computational domain subject to the appropriate boundary conditions. Here, the "inverse" approach is invoked: the domain is discretized into N line elements and the variable u is specified at the nodal points so that the boundary conditions are satisfied. The actual variation of u along the internal points is arbitrary. It must be, however, a monotonically decreasing function from the left to the right boundary. The particular choice of the variation of u will be reflected in the refinement of the mesh which results with the final solution and it will affect the local accuracy of the solution; adjacent nodes with a large variation in u will be automatically spaced further apart with a less accurate local solution. Curve (a) in Figure 2 shows the linear variation of u used in this example. Curves (b) and (c) represent an intermediate and the converged solution solution respectively. In the analysis, the location of the nodes is treated as an initial guess (curve (a)) and a solution is sought for the coordinate x which satisfies Equation 13. Throughout the iteration procedure the nodal values of u are kept constant.

The solution of Equation 13 can be obtained by any appropriate iterative scheme. Here, the full Newton-Raphson method is used which ensures quadratic convergence.

$$\mathbf{x}^{(\mathbf{n}+1)} = \mathbf{x}^{(\mathbf{n})} - \mathbf{J}_\mathbf{c}^{-1} \mathbf{R}\left(\mathbf{x}^{(\mathbf{n})}\right),$$

where $\mathbf{x}$ is the vector of the unknown nodal locations and $\mathbf{J_c}$ the Jacobian of the residuals:

$$\frac{\partial R_i}{\partial x_j} = \int \left\{ \frac{\partial \phi_i}{\partial x} \frac{\partial u}{\partial x} - \phi_i \right\} \frac{\partial}{\partial x_j} \left(\frac{\partial x}{\partial \xi} \right) + \left(\frac{\partial x}{\partial \xi} \right) \left\{ \frac{\partial u}{\partial x} \frac{\partial}{\partial x_j} \left(\frac{\partial \phi_i}{\partial x} \right) - \frac{\partial \phi_i}{\partial x} \frac{\partial}{\partial x_j} \left(\frac{\partial u}{\partial x} \right) \right\} d\xi. \tag{14}$$

The derivatives with respect to the coordinates are easily evaluated using the isoparametric mapping (equations 10 and 11) :

$$\frac{\partial}{\partial x_j} \left(\frac{\partial (\cdot)}{\partial x} \right) = \frac{\partial (\cdot)}{\partial \xi} \left\{ - \frac{\dfrac{\partial \phi_j}{\partial \xi}}{\left(\dfrac{\partial x}{\partial \xi} \right)^2} \right\}, \tag{15}$$

$$\frac{\partial}{\partial x_j} \left(\frac{\partial x}{\partial \xi} \right) = \frac{\partial \phi_j}{\partial \xi}. \tag{16}$$

It is important to note here that the original governing equations have not been inverted. Rather an inverse thinking is used to solve the "direct" problem.

Obviously this feature resolves the mathematical difficulties introduced by the variable transformation of the purely inverse formulations.

In three to four iterations the solution converged within a tolerance of 10^{-6} to the exact solution, making the numerical solution indistinguishable from the exact solution (Figure 2). Such accuracy is expected away from the free boundary since the exact solution has the same quadratic variation as the finite element basis functions. However, the method predicted the location of the free boundary with the same accuracy.

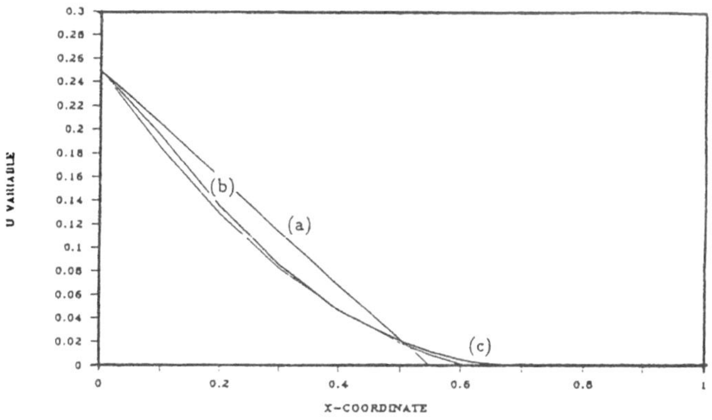

Figure 2: Initial guess (a), an intermediate, and the converged solution (b),(c). In three to four iterations the solution converges to the exact analytic solution.

ONE-DIMENSIONAL MOVING BOUNDARY TEST PROBLEM

The same approach is used for problems with moving boundaries such as the one shown in Figure 3. The liquid is at the melting temperature with the wall at a subcooled temperature. This problem is one of penetration depth, where the solidification front advances into the melted liquid. The objective is to determine the rate of penetration of the solid front as well as the temperature distribution in the solid.

Using the non-dimensional quantities

$$\theta = \frac{T - T_s}{T_s - T_w}, \quad \tau = \frac{t\alpha}{L^2}, \quad J = \frac{h_{sl}}{C_p(T_s - T_w)}, \quad x^* = \frac{x}{L},$$

and by dropping the superscript $*$, the energy equation reduces to

$$\frac{\partial \theta}{\partial \tau} = \frac{\partial^2 \theta}{\partial x^2}, \tag{17}$$

with the boundary conditions

$$\theta(0,\tau) = -1.0, \quad \theta(1,\tau) = 0.0, \quad \text{and} \quad \frac{\partial \theta}{\partial x} = J \frac{\partial x}{\partial \tau}\bigg|_{\mathbf{x = interphase}},$$

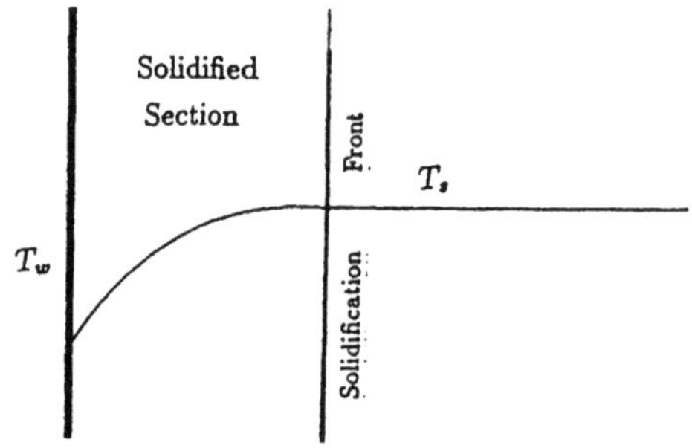

Figure 3: Schematic of one-dimensional solidification with the melt at the melting temperature and a subcooled wall.

where T_s and T_w are the solidification and wall temperature respectively. J is the Stefan number defined in terms of the specific heat C_p, the latent heat of solidification h_{sl}, and α and L the thermal diffusivity and a characteristic dimension respectively (all properties pertain to the solid phase). Equation 17 is linear but the problem is made nonlinear by the unknown location of the advancing solidification front.

The exact solution to the above problem is given as

$$\theta = \frac{1}{erf\,(\lambda)} erf\left[\frac{x}{2\,(\tau)^{\frac{1}{2}}},\right] \tag{18}$$

where λ is the solution of

$$\sqrt{\pi}\lambda \exp\left(\lambda^2\right) erf\,(\lambda) = \frac{1}{J}.$$

The location of the solid/melt interphase is given as

$$\delta\,(t) = 2\lambda\sqrt{\tau}.$$

Since the unsteady solution is obtained at selected nodal points which are defined with respect to a moving coordinate system (the mesh), the time derivative in the energy equation (Equation 17) must be rearranged according to

$$\frac{\partial\theta}{\partial\tau} = \frac{d\theta}{d\tau} - V_{mesh}\frac{\partial\theta}{\partial x} \tag{19}$$

However, since the temperature is kept constant at the nodal points (because of the inverse approach),

$$\frac{d\theta}{d\tau} = 0.$$

The velocity of the mesh, V_{mesh}, is defined in terms of the velocity of the nodal points as $\frac{dx}{d\tau}$. Here, a simple first order finite difference formula is used to numerically approximate V_{mesh} instead of a higher order method.

$$V_{mesh} = \frac{x_{t+dt} - x_t}{dt}$$

This is deemed necessary in order to separate the issues related to the accuracy and applicability of the method from those related to the accuracy of time integration schemes. Therefore, the results shown below can be improved further by changing the order of the integration scheme.

The finite element residual, after using the divergence theorem, reduces to

$$R_i = \int \left\{ \phi_i \left(V_{mesh} \frac{\partial \theta}{\partial x} \right) - \frac{\partial \phi_i}{\partial x} \frac{\partial \theta}{\partial x} \right\} \frac{\partial x}{\partial \xi} d\xi + \phi_i J \frac{dx}{d\tau} \bigg|_{x=\delta} . \tag{20}$$

The computational domain is discretized into N line elements and a temperature profile that satisfies the boundary conditions is assumed. The nonlinear Equation 20, is then solved for x_i, the location of the nodal points, using Newton iteration and the Jacobian

$$\frac{\partial R_i}{\partial x_j} = \int \phi_i \frac{\partial}{\partial x_j} \left(V_{mesh} \frac{\partial \theta}{\partial x} \frac{\partial x}{\partial \xi} \right) - \frac{\partial}{\partial x_j} \left(\frac{\partial \phi_i}{\partial x} \frac{\partial \theta}{\partial x} \frac{\partial x}{\partial \xi} \right) d\xi + \phi_i J \frac{\partial}{\partial x_j} \left(\frac{\partial x}{\partial \tau} \right) \bigg|_{x=\delta} \tag{21}$$

The analysis was started from an arbitrary initial time, τ_o with the initial condition taken from the analytic solution (Equation 18). The results for subsequent time steps are in very good agreement with the analytic solution, (Figure 4).

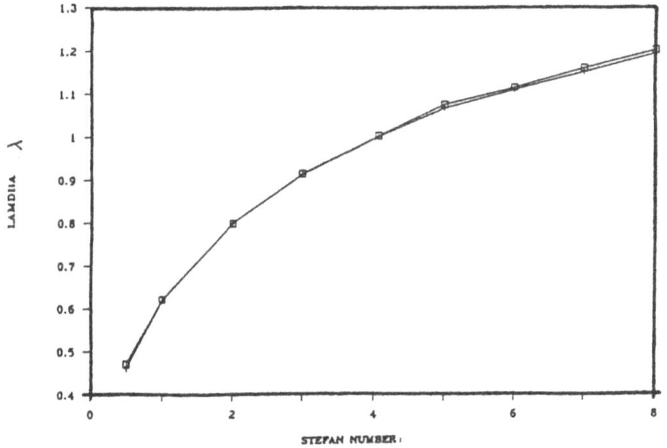

Figure 4: The numerical and exact analytic solution for the rate of penetration of the solidification front in a melt at melting temperature.

THREE DIMENSIONAL SOLIDIFICATION

One advantage of this inverse approach is the method's transparency to the dimensionality of the problem. Consider for example the three dimensional solidification problem shown in Figure 5.

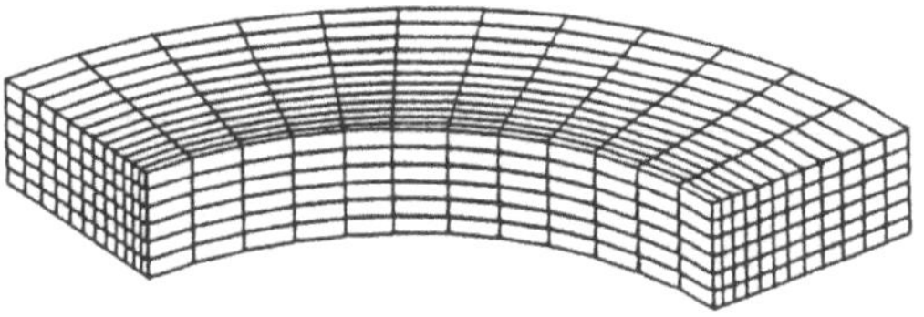

Figure 5: Physical domain of a three dimensional heat transfer problem

For a steady solution the inner and outer-radius are kept at constant temperatures T_1 and T_2 and the other surfaces are assumed to be adiabatic. If heat flux is specified along the inner or outer radial surfaces then the problem becomes unsteady. Using the IFEM temperature is fixed along planes normal to the radial direction and the solution is obtained for the location of the plane. If one of the planes is chosen to correspond to the solidification front then the solution will automatically yield the position of the solidification front. However, the solution procedure may result in a highly distorted mesh. Here a more general approach is presented which maintains the orthogonality of the mesh throughout the solution procedure.

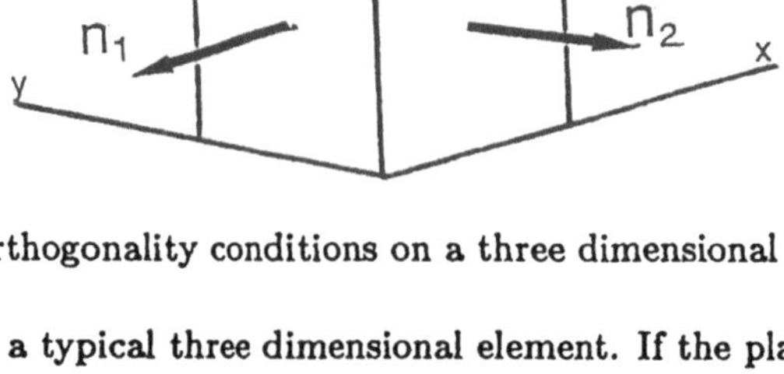

Figure 6: Orthogonality conditions on a three dimensional element

Figure 6 shows a typical three dimensional element. If the planes normal to the Z-direction are *a priori* fixed to be isothermal then ∇T is a vector normal to those planes. If n_1 and n_2 are the unit normal vectors on the other surfaces then, in addition to the energy equation in three dimensions

$$\nabla^2 T = 0 \tag{22}$$

we can require that:

$$n_1 \cdot \nabla T = 0 \tag{23}$$

and

$$(n_1 \times \nabla T) \cdot n_2 = |n_2|^2 \tag{24}$$

along the planes to the isothermal plane.

Equation(22) - Equation (24) and Equation (4) are solved for the coordinates x, y, z of each node. Figure (7) shows the IFEM solution procedure. Starting with an initial guess for the mesh, the governing equations are solved for the location of the nodes. In the final solution opposite planes of the elements will either correspond to isothermal planes or will be orthogonal to adjacent planes. This approach is general and can be used to many arbitrary geometries. However, it introduces two additional unknowns per node to solve for. This is not necessary in geometries where the isothermal planes are aligned with one of the major axes. In such case the orthogonality conditions are not needed and the solution includes only one unknown per node.

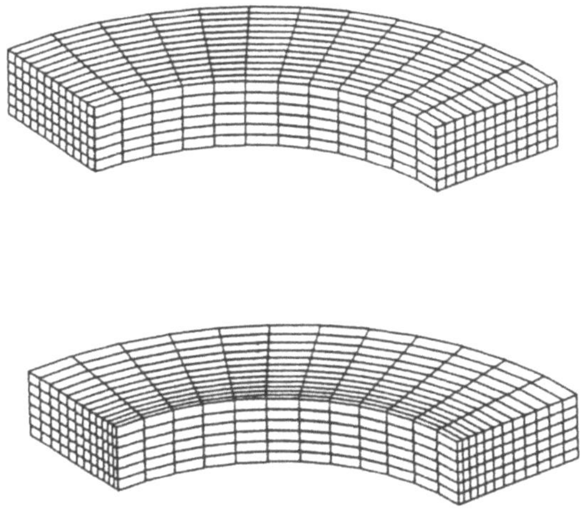

Figure 7: Initial guess (upper plot) and final mesh (lower plot) using the IFEM. During the solution procedure the temperature is kept constant and the solution is obtained for the location of the nodes.

INVERSE AIRFOIL DESIGN

The philosophy of the IFEM can also be used to Inverse Airfoil Design, a seemingly unrelated problem to solidification. In Inverse Airfoil Design the objective is to obtain an "optimum" airfoil section with desired aerodynamic characteristics such as pressure distribution (or lift). This problem highlights a critical issue in Engineering Analysis and Design; analysis is a small part of the design procedure since the analysis is performed on every design alternative until an "optimum" configuration is obtained. Here, the IFEM is applied in a design sense: the solution i sought for the airfoil shape which yields a pre-determined optimum pressure distribution.

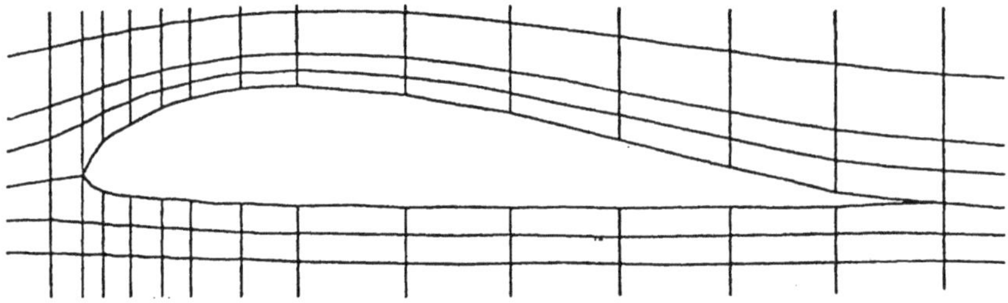

Figure 8: Inviscid flow past a two-dimensional airfoil.

Consider inviscid flow past an airfoil as shown in Figure. (8). Using the streamfunction Ψ and the irrotationality condition

$$\nabla^2 \Psi = 0 \tag{25}$$

Equation (25) can be solved for Ψ and then using

$$u = \frac{\Psi}{y} \qquad v = -\frac{\Psi}{x}$$

the velocity field can be determined.

Using the IFEM and a Galerkin formulation, and by applying the divergence theorem the residual equation is given as:

$$R^i = \int\int \nabla\phi_i \cdot \nabla\Psi - \underbrace{\int \phi_i \nabla\Psi \cdot n ds}_{I} \tag{26}$$

where n is the outward unit normal vector.

The line integral I is evaluated only along the boundaries that do not correspond to essential boundary conditions. The same integral along the boundary of the airfoil can be expressed in terms of the desired velocity distribution V_s as:

$$I = \int \phi_i V_s ds \tag{27}$$

Note that the velocity V_s along the boundary of the airfoil can be readily expressed in terms of the pressure through the Bernoulli equation.

Consistent with the inverse approach the streamfunction is specified everywhere in the domain and Equations (26) and (27) are solved for the Y-coordinate of the nodes. The shape of the airfoil which satisfies Equation (27) is then automatically determined.

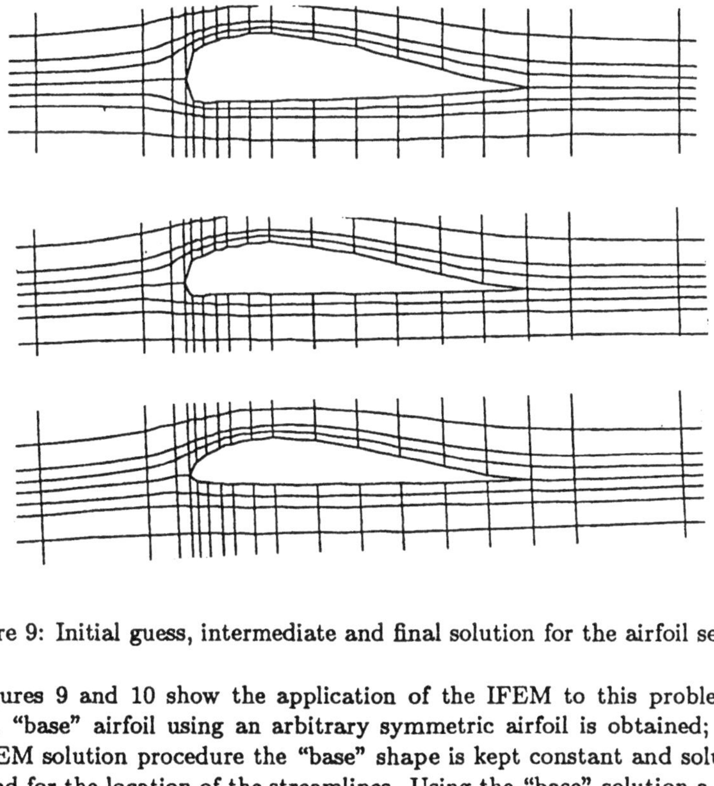

Figure 9: Initial guess, intermediate and final solution for the airfoil section

Figures 9 and 10 show the application of the IFEM to this problem. At first, a "base" airfoil using an arbitrary symmetric airfoil is obtained; during the IFEM solution procedure the "base" shape is kept constant and solution is obtained for the location of the streamlines. Using the "base" solution a desired velocity distribution V_s is specified along the airfoil boundary (shown in Fig. 10) and the problem is solved for the location of the streamlines including the streamline along the airfoil. Figure 10 shows the close agreement between the targeted velocity distribution shown by * and the final velocity along the airfoil (solid lines).

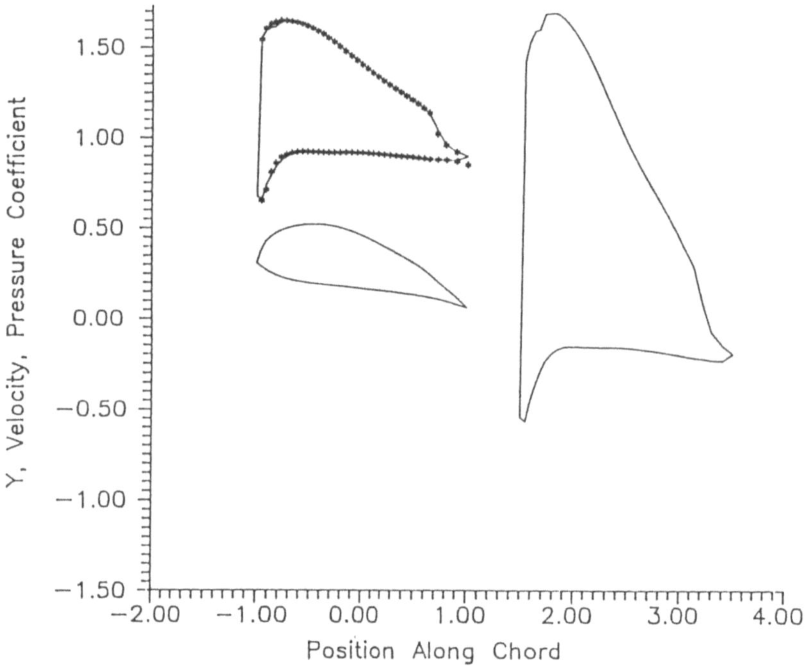

Figure 10: The final shape of the airfoil, the targeted (*) and final velocity (solid lines) and pressure coefficient.

CONCLUSIONS

The Inverse Finite Element Method is appropriate for problems with monotonically varying behavior. At its present form it cannot handle double-valued problems such problems with closed isotherms or isolines. However, it performs very well in most problems of practical and theoretical interest. The most important advantages of the method are the method's transparency to the complexity of the governing equations , the boundary conditions, the dimensionality of the problem or the complexity of the geometry. This method has been successfully used to solve systems integro-differential equations which are typical in analyses of viscoelastic polymers [12]. Very recently the method is used in two and three dimensional crystal growth problems

References

[1] Derby, J.J., and R.A. Brown 1985, "Finite Element Analysis of a Thermal-Capillary Model for Liquid Encapsulated Czochralski Growth", *J. Elec-*

troch. Soc. Solid-State Science and Technology, **132**, No. 2, 470.

[2] Cuvelier, C. and J.M. Driessen 1986, "Thermocapillary free boundaries in crystal growth", *J. Fluid Mech.*, **169**, 1.

[3] Langlois, W.E. 1982, "A parameter sensitivity study for Czochralski bulk flow of silicon", *J. Crystal Growth*,**56**, 15.

[4] Bell, G.E., 1985, "Inverse Formulations as a Method for Solving Certain Melting and Freezing Problems",*Numerical Methods in Heat Transfer*, John Wiley & Sons Ltd, Vol. III, Ch. 3.

[5] Voller, V.R. , 1986, "An implicit enthalpy solution for phase change problems: with application to a binary alloy solidification", *Appl. Math. Modelling*, **11**, 110.

[6] Voller, V.R. and M. Cross, 1981, "Accurate solutions of moving boundary problems using the enthalpy method", *Int. J. Heat Mass Transfer*, **24**, 545.

[7] Crank, J. and R.D. Phahle, 1973, "Melting ice by isotherm migration method", *Bull. J. Inst. Math. Applcs.*, **9**, 12.

[8] Talmon, Y., Davis, H.T. and L.E. Scriven, 1981, " Progressive freezing of composites analyzed by isotherm migration methods", *AIChE J.*, **27**, No. 6, 928.

[9] Talmon, Y., Davis, H.T. and L.E. Scriven, 1983, " Moving boundary problems in simple shapes solved by isotherm migration", *AIChE J.*, **29**, No. 5, 795.

[10] Alexandrou, A.N., 1989,"An Inverse Finite Element Method for Directly Formulated Free Boundary Problems", *Int. J. Num. Meth. in Eng.*,?, p.?.

[11] Alexandrou, A.N. and T.C. Papanastasiou, 1986, "An Inverse Finite Element Method with an Application to Extrusion with solidification", Submitted for publication.

[12] Alexandrou, A.N. and T.C. Papanastasiou, 1986, " Non-Isothermal Extrusion of Composite Materials", Solicited for publication, *J. Polymer Process. Society*

[13] Carslaw, H. and J.C. Jaeger, "Conduction of Heat in Solids", 296.

Finite Difference Enthalpy Methods for Dendritic Growth

K.-H. Tacke, A. Harnisch
Max-Planck-Institut für Eisenforschung, D-4000 Düsseldorf, Germany

ABSTRACT

The application of three-dimensional finite difference enthalpy methods to dendritic crystal growth is studied. A discrete expression for the curvature of the solid/liquid free boundary is given using the array of solid phase fractions. Specific enthalpy/temperature relations and an attachment condition are formulated to account for grid cell transitions. Dendritic structures are reproduced by the algorithm. Limitations of the approach are discussed.

INTRODUCTION

Dendrite growth is an important moving boundary problem, since many materials solidify in dendritic mode. Microstructures and material properties are influenced by the dendritic solidification. Dendritic growth is also of interest because ordered geometric patterns arise from the liquid state, which has no structure at the scale of the dendrite.

For the numerical simulation of dendrite evolution, diffusion equations have to be solved. The solid/liquid interface is a moving boundary of complex shape. This paper concentrates on solidification from a pure undercooled melt, where the formation of a dendritic structure is initiated with a solid nucleus (free dendrite growth). The growth of the structure is controlled by thermal diffusion. Latent heat is released when new solid material is formed, and has to be removed into the surrounding liquid by conduction to permit further growth. Heat removal is facilitated by surface areas as large as possible, which promotes the development of a branched structure. Capillarity is a counteracting factor limiting unrestricted branching.

The mathematical problem is not a standard moving boundary problem for the following reasons. First, solidification starts from an undercooled system, and thus from a metastable state. In finite difference approximations of phase change from metastable states an unusual phenomenon is encountered - the necessity of an *attachment condition*. Details will be given below. Second, dendritic growth involves a specific condition at the free boundary - the *Gibbs-Thomson equation*. To incorporate the Gibbs-Thomson equation a discrete expression of curvature is needed. The mechanisms generating the symmetry of the dendrite can be formulated as orientation-dependent factors of an extended Gibbs-Thomson equation. This requires a discrete expression of local interface orientation.

In this paper, a finite difference formulation is presented, which is a modification of the enthalpy method on the basis of the control volume technique [1]. The model has been previously developed for two-dimensional geometry [2] and is extended here to three dimensions.

Dendrite growth has been treated theoretically in a number of articles, among which [3, 4, 5, 6, 7, 8] are general references. Finite difference techniques for dendritic growth or undercooled systems have been described by Smith [9], Chorin [10], Rubinstein [11] and Solomon et al. [12]. Alternative methods for dendrite simulation are boundary layer or string models [13, 14, 15], phase field methods [16, 17], Monte Carlo simulation [18], diffusional aggregates [19] and BEM techniques [20, 21].

BASIC EQUATIONS

The growth of a dendritic crystal is modelled in a rectangular domain in cartesian coordinates. A uniform grid of cubic cells is used for the discretization. The control volume method implies that temperatures are calculated at the nodes, whereas enthalpies are taken to be averages over the control volumes. Heat flux densities through element interfaces are evaluated from the temperature differences between the elements adjacent to the interface. This gives the explicit finite difference form of the heat equation as

$$h_P^{k+1} = h_P^k + a\Big(\sum_{neighbours} u_{NB}^k - 6u_P^k \Big). \qquad (1)$$

Here, $h = H/\Delta H$ and $u = c_P(T - T_{m0})/\Delta H$ are the non-dimensional enthalpy and temperature with ΔH = latent heat of melting, c_P = specific heat capacity and T_{m0} = melting temperature at zero curvature. P is the node under consideration, NB refers to the neighbour cells E, W, N, S, H, L; k is the number of the time step. The factor $a = \lambda \Delta t/(\rho c_P \Delta x^2)$ is taken equal to 1/6, corresponding to the maximum time step for stability. λ is the thermal conductivity, ρ the density of the material. The numbers of grid cells in x_1, x_2 and x_3-direction are n_1, n_2, n_3. For the computations, non-dimensional spatial coordinates are used such that the cell side-length is unity.

The Gibbs-Thomson relation asserts that the temperature at a phase boundary deviates from equilibrium in proportion to local curvature, thereby introducing the effect of capillarity to the problem.

It is written in non-dimensional form as

$$u_m = -GK, \tag{2}$$

where $G = \gamma c_P/(\Delta H \Delta x \Delta s)$ is a dimensionless factor with γ = interfacial free energy, Δs = entropy of fusion. $K = \kappa \Delta x$ is the non-dimensional curvature.

H-T RELATIONS

In a time step, new enthalpies are computed from old temperatures using equation(1). The task is then to determine the temperature and, in the two-phase region, the solid phase fractions of all cells. In the conventional enthalpy method these functions can be simply taken from an H-T diagram, since T is a unique function of H.

The situation is more complex in the presence of metastable phases because the H-T relation becomes multivalued. We have to consider three types of grid cells, purely *solid* cells, purely *liquid* cells, and *two-phase* cells containing both solid and liquid material. Solid and liquid cells may be either in stable or metastable status. The pure solid substance at its melting point and zero curvature is assumed as reference state for enthalpy.

The relation between enthalpy and temperature is

$$h = u \tag{3}$$

for solid material (only sensible heat), and

$$h = u + 1 \tag{4}$$

for the liquid. The second term accounts for latent heat of fusion.

Two-phase cells are at the solid/liquid interface. Their temperature is therefore equal to the local interface temperature given by the Gibbs-Thomson equation (2). Thus, such cells have a position on a vertical line $u = u_m$ in the H-T diagram. The solid phase fraction of two-phase cells is determined from the *lever rule*

$$f = u + 1 - h, \tag{5}$$

which is derived from a heat balance stating that the element enthalpy must be equal to the sum of sensible and latent heat.

CELL STATUS AND ATTACHMENT CONDITION

How can an undercooled liquid cell become two-phase or solid ? Since we assume that no solid nuclei other than the initial seed are present, this transition can only happen when the evolving solid front reaches the cell or grows into the cell.

The mathematical criterion for transition is called the *attachment condition*. It can be formulated in various ways. For the present study, we use a simple heuristic attachment condition postulating that an undercooled liquid cell becomes a two-phase cell if at least one of its six directly adjacent neighbours is already fully solid.

Solidus and liquidus enthalpies with respect to the local melting temperature are calulated from

$$h_s = u_m \tag{6}$$

$$h_l = u_m + 1. \tag{7}$$

The new cell status is computed according to Table 1 which covers all possible transitions.

<table>
<tr><th></th><th>A.C.</th><th>$h \leq h_s$</th><th>$h_s < h < h_l$</th><th>$h \geq h_l$</th></tr>
<tr><td rowspan="2">$f = 0$</td><td>yes</td><td rowspan="2">(solid)</td><td>two-phase</td><td rowspan="2">liquid</td></tr>
<tr><td>no</td><td>liquid</td></tr>
<tr><td>$0 < f < 1$</td><td></td><td>solid</td><td>two-phase</td><td>liquid</td></tr>
<tr><td rowspan="2">$f = 1$</td><td>yes</td><td rowspan="2">solid</td><td>two-phase</td><td rowspan="2">(liquid)</td></tr>
<tr><td>no</td><td>solid</td></tr>
</table>

Table 1: Rules for cell status transitions, h = new enthalpy, f = old solid fraction, A.C. = attachment condition

DISCRETE EXPRESSION OF INTERFACE ATTRIBUTES

The array of solid phase fractions $\{f\}$ is used to estimate the curvature and other properties of the interface by introducing a hypothetical interpolation function $F(x_1, x_2, x_3)$. F is defined at all points in space and coincides with f at the nodes. The function F vanishes at points far from the structure and is equal to 1 in the interior. Near the boundary it assumes values between 0 and 1.

The equation

$$F(x_1, x_2, x_3) = \frac{1}{2} \tag{8}$$

defines a set of points which is taken as an approximation to the original boundary. An implicit representation of the boundary is thus given together with a discrete set of function values. The curvature is

$$K = \frac{2 \sum_{\substack{i,j=1 \\ i<j}}^{3} F_{x_i x_j} F_{x_i} F_{x_j} - \sum_{\substack{i,j,k=1 \\ i \neq j \neq k, j<k}}^{3} F_{x_i x_i} (F_{x_j}^2 + F_{x_k}^2)}{(\sum_{i=1}^{3} F_{x_i}^2)^{3/2}}. \tag{9}$$

Difference quotients are inserted into equation (9), e.g.

$$F_{x_1} = (f_E - f_W)/2 \tag{10}$$

$$F_{x_1 x_1} = (f_E - 2f_P + f_W) \tag{11}$$

$$F_{x_1 x_2} = (f_{NE} - f_{NW} + f_{SW} - f_{SE})/4. \tag{12}$$

The resulting expressions are independent of the exact nature of the hypothetical interpolation function.

Similar arguments may be used to derive discrete expressions for the outward normal vector, $(-F_{x_1}, -F_{x_2}, -F_{x_3})$, and the propagation speed of the interface in normal direction, $v_n = F_t/(\sum_i F_{x_i}^2)^{1/2}$. The latter expressions are not used in this study, but will become relevant when anisotropy is included.

COMPUTATION PROCEDURE

Initially, the material is assumed to be uniformly supercooled at temperature $u_0 < 0$. All grid cells are at enthalpy $h_0 = u_0 + 1$, and thus in metastable liquid state. To start the algorithm, eight grid cells at the centre of the domain are taken as the initial seed and are converted to two-phase state such that the enthalpy is preserved.

The time step of the numerical algorithm is summarized as follows. Assume old temperatures, enthalpies and solid fractions to be given. New enthalpies are obtained using equation (1). Local curvatures are calculated from the old solid fractions. Solidus and liquidus enthalpies are calculated with equations (6) and (7). The new status is determined from the attachment condition and Table 1. For single-phase cells, nodal temperatures are calculated from equations (3) or (4). For two-phase cells, equation (2) is used, and the new solid phase fraction is given by equation (5). At this point, the time step is complete.

The walls of the computation domain are assumed to be thermally insulated. This implies that an expanding structure will eventually be affected by the domain walls. As our main interest is free growth, a scheme to continuously extend or shift the computation domain in x_3-direction was developed. In this procedure, a new x_1-x_2 layer of grid cells filled with undercooled liquid is added on top of the structure whenever the boundary layer in front of the dendrite tip can be perceived at the topmost cell layer (temperature rise of 0.5 % of absolute initial temperature). The bottom x_1-x_2 layer is then discarded. The zone near the dendrite tip can thus be taken to be in free growth state, unaffected by the domain walls in all stages of the computation.

LIMITATIONS OF PRESENT MODEL

The main limitation of the current formulation is that the rectangular symmetry is caused by numerical grid effects rather than by modelling the underlying physical mechanisms. A more accurate model would apply an extended Gibbs-Thomson equation, e.g.

$$u_m = -G(\Theta)K - E(\Theta)v_n, \tag{13}$$

where $G(\Theta)$ and $E(\Theta)$ are orientation dependent factors [14]. The second term represents kinetic undercooling.

In our present model, the simplified attachment condition produces a canalization effect along the grid lines. Transition from a metastable to two-phase state can only occur adjacent to fully solid cells which favours growth in directions parallel to the coordinate axes. Additional work is in progress to improve the attachment condition and to incorporate equation (13).

COMPUTATION RESULTS

Figures 1 to 3 show steps in the evolution of a dendritic pattern. The x_1-x_2 layers of the domain shown in the figures have $60 \times 60 = 3600$ grid cells, although in the computations - because of symmetry - only a representative triangle on every x_1-x_2 layers was treated. We used $n_3 = 100$ layers in x_3-direction. The constant in equation (2) was taken as $G = 0.5 \cdot 10^{-4}$, the initial undercooling was $u_0 = -0.32$.

The structure in Figure 1 was obtained at time step $k = 100$. A rudimentary dendrite with primary arms is seen. This structure is further evolved in Figure 2, where secondary branches emerge. A fully developed structure, after 1000 time steps, is shown in Figure 3. It has distinct secondary arms and indications of tertiary branches.

Note that the scales of the graphics are not equal. The maximum extension of the structures shown in Figures 1 and 2 is approximately 20 and 38 grid cells respectively, whereas the fully developed dendrite of Figure 3 extends over a large part of the bottom x_1-x_2 layer. The first two structures are not affected by the domain walls. The third structure is considered to be free of wall effects in its upper part. Its initial seed and the other five primary arms have been removed from the view area by the domain shifting procedure. Figure 3 represents a quasi-stationary state with constant growth velocity. Apart from continuous formation of new sidebranches and their disappearance at the bottom, the overall structure remains unchanged when computation is carried on.

The graphics were generated as isosurfaces of the field of phase fractions $\{f\}$ at a value of $f = 1/2$. Using a triangulation technique, a set of connected triangles was generated and transferred to a visualization toolkit, where the geometric model was rendered and Gouraud shaded.

CONCLUSIONS

Three-dimensional dendritic behaviour has been reproduced by numerically integrating the time dependent thermal diffusion/phase change problem. The symmetry is controlled by the numerics rather than by expressions of anisotropy. If this effect can be sufficiently reduced by further development, a significant step in numerical microstructure simulation will be reached.

References

[1] S.V. Patankar, Numerical Heat Transfer and Fluid Flow, Hemisphere, New York, 1980

[2] K.-H. Tacke, Application of Finite Difference Enthalpy Methods to Dendritic Growth and Ripening, in Free Boundary Problems: Theory and Applications, (Ed. K.-H. Hoffmann, J. Sprekels), Vol. II, pp. 636-643, Pitman Research Notes in Mathematics, Longman, 1990

[3] J.S. Langer, Instabilities and pattern formation in crystal growth, Rev. Mod. Phys. 52 pp. 1-28, 1980

[4] W. Kurz, D.J. Fisher, Fundamentals of Solidification, Trans Tech Publications, Aedermannsdorf, 1984

[5] M.E. Glicksman, R.J. Schaefer, J.D. Ayers, Dendritic Growth - A Test of Theory, Metall. Trans. 7A pp. 1747-1759, 1976

[6] S.-C. Huang, M.E. Glicksman, Fundamentals of Dendritic Solidification, I. Steady-State Tip Growth, II. Development of Sidebranch Structure, Acta Metallurgica 29 pp. 701-734, 1981

[7] M.E. Glicksman, Free Dendritic Growth, Mat. Sc. Engng. 65 pp. 45-55, 1984

[8] E. Ben-Jacob, P. Garik, Ordered Shapes in Nonequilibrium Growth, Physica D 38 pp. 16-28, 1989

[9] J.B. Smith, Shape Instabilities and Pattern Formation in Solidification: A New Method for Numerical Solution of the Moving Boundary Problem, J. Comput. Phys. 39 pp. 112-127, 1981

[10] A.J. Chorin, Curvature and Solidification, J. Comput. Phys. 58 pp. 472-490, 1985

[11] L. Rubinstein, Mathematical modelling of the growth of a monocomponent crystal from its melt in a non-homogeneous temperature field, in Free boundary problems: applications and theory, (Ed. A. Bossavit, A. Damlamian, M. Fremond), Vol III, pp. 166-178, Pitman, Boston, 1985

[12] A.D. Solomon, V. Alexiades, D.G. Wilson, G. Geist, J. Kerper, An Enthalpy Method for the Solidification of a Supercooled Liquid, Oak Ridge National Laboratory Report ORNL-6217, 1986

[13] E. Ben-Jacob, N. Goldenfeld, J.S. Langer, G. Schön, Boundary-layer model of pattern formation in solidification, Physical Review A 29 pp. 330-340, 1984

[14] E. Ben-Jacob, N. Goldenfeld, J.S. Langer, G. Schön, String Models of Interfacial Pattern Formation, Physica D 12 pp. 245-252, 1984

[15] J.S. Langer, Dynamics of Dendritic Pattern Formation, Mat. Sc. Engng. 65 pp. 37-44, 1984

[16] G. Caginalp, J.T. Lin, A Numerical Analysis of an Anisotropic Phase Field Model, Univ. of Pittsburgh, Report, 1986

[17] G. Caginalp, An Analysis of a Phase Field Model of a Free Boundary, Arch. Rational Mechanics and Analysis 92 pp. 205-245, 1986

[18] Y. Saito, T. Ueta, Lattice-Gas Model Simulation of Crystal Shapes, J. Crystal Growth 99 pp. 171-174, 1990

[19] R.C. Ball, M.J. Blunt, O. Rath Spivak, Diffusion-controlled growth, in Fractals in the Natural Sciences (Ed. M. Fleischmann, D.J. Tildesley, R.C. Ball), pp. 123-132, Princeton University Press, 1989

[20] Y. Saito, G. Goldbeck-Wood, H. Müller-Krumbhaar, Numerical simulation of dendritic growth, Physical Review A 38 pp. 2148-2157, 1988

[21] C. Misbah, H. Müller-Krumbhaar, Y. Saito, Dendritic Growth and Directional Solidification, J. Crystal Growth 99 pp. 156-160, 1990

Figure 1: Dendritic structure at time step $k = 100$. The maximum diameter of the structure is approximately 20 grid cells.

Figure 2: Dendritic structure at time step $k = 200$. The maximum diameter of the structure is approximately 38 grid cells.

Figure 3: Fully developed structure at time step $k = 1000$. Only the upper half of the computation domain ($i_3 = 50 - 100$) is used for graphic representation. The height of the structure is approximately 40 grid cells.

Numerical Model of Heat Transfer Processes in Solidifying and Cooling Steel Ingot (on the basis of BEM)

B. Mochnacki, E. Majchrzak, A. Kapusta
Silesian Technical University, 44-100 Gliwice, Zwyciestwa 42, Poland

ABSTRACT

One of the possibilities of the BEM utilization in numerical modeling of foundry processes is the coupling of a typical BEM algorithm for linear parabolic equations with a certain procedure called the Alternating Phase Truncation Method (APTM) presented by J.Rogers, A.Berger and M.Ciment in [1,2]. The solution obtained in this way by E.Majchrzak is described in [3,4]. In this paper the essential generalization of the APTM will be shown. The algorithm proposed by the authors of this paper allows to solve the problems of temperature field identification in case of solidification in an interval of temperature and also in case of "mixed" solidification when the process proceeds partly in a constant temperature and in part in an interval of temperature (e.g. cast iron). It must be pointed out that the proposed generalization can be used in numerical computations of a wide class of non-linear problems in which the thermophysical parameters of the material are functions of temperature.

INTRODUCTION

Solidification of metals and alloys can proceed in different ways. From the practical and also mathematical point of view two basic models are considered

(A) solidification in a constant temperature,

(B) solidification in an interval of temperature.

In case (A) the two changing with time sub-areas (e.g. liquid state $D_2(t)$ and solid one $D_1(t)$) are analysed. Then a non-steady temperature field $T(x,t)$ is described by a system of partial differential equations in the form

$$
\begin{aligned}
x\in D_1(t): &\quad c_1\rho_1\frac{\partial T_1}{\partial t} = \operatorname{div}\,(\lambda_1 \operatorname{grad}\, T_1) \\
x\in D_2(t): &\quad c_2\rho_2\frac{\partial T_2}{\partial t} = \operatorname{div}\,(\lambda_2 \operatorname{grad}\, T_2)
\end{aligned}
\tag{1}
$$

where c_e, ρ_e, λ_e, e=1,2 are the thermophysical parameters of sub-areas $D_e(t)$ (specific heat, mass density and thermal conductivity.
The mathematical description is supplemented by the following boundary and initial conditions (compare Figure 1)

$$x \in \Gamma_0: \qquad \Phi(T, n \circ \mathrm{grad}\ T) = 0\ , \tag{2}$$

$$x \in \xi_1: \qquad \begin{cases} -\lambda_2 n \circ \mathrm{grad}\ T_2 = -\lambda_1 n \circ \mathrm{grad}\ T_1 + \rho_1 L_1 v_1 \\ T_2 = T_1 = \hat{T}_1 \end{cases}, \tag{3}$$

$$t=0: \qquad T_e(x,0) = T_{oe}(x) \quad , \ e=1,2, \tag{4}$$

at the same time $n \circ \mathrm{grad}T$ is a normal derivative, L-latent heat, v-rate of solidification in normal direction, T_1-solidification point.

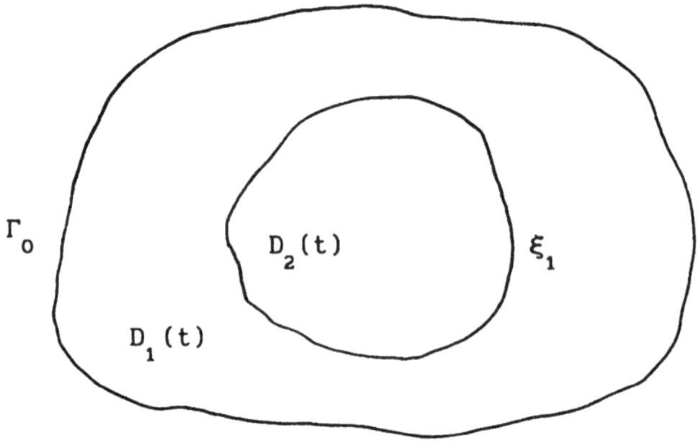

Fig. 1. The sub-areas D_2 and D_1

In case (B) we consider a multiphase object (e.g. steel casting area-Figure 2) in which liquid $D_3(t)$, transcient $D_2(t)$ and solid $D_1(t)$ sub-areas should be distinguished. The transcient zone is limited by the transistory positions of isotherms T_L and T_S (the beginning and the end of solidification). The latent heat is emitted in the interval $\langle T_S, T_L \rangle$ and the temperature field in this sub-area is a sourced one. It is possible to introduce a certain parameter called substitute thermal capacity C_2 [5] to the mathematical description and then the system of energy equations is of the form

$$x \in D_e(t): \quad C(T)\rho_e \frac{\partial T_e}{\partial t} = \mathrm{div}\,(\lambda_e \,\mathrm{grad}\, T_e)\ , \quad e=1,2,3 \tag{5}$$

where

$$C(T) = \begin{cases} c_3 \ , & T>T_L \\ C_2=c_2+c^{sp} \ , & T\in\langle T_S, T_L\rangle \\ c_1 \ , & T<T_S \end{cases} \tag{6}$$

The component c^{sp} (spectral latent heat) can be determined as a constant value $c^{sp}=L/(T_L-T_S)$ but one can find a more complex hypothesis concerning the considered parameter [6].

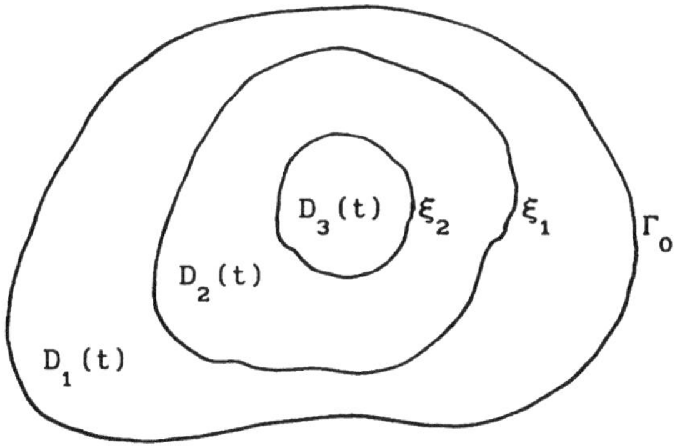

Fig. 2. The sub-areas D_3, D_2 and D_1

On the surfaces corresponding to the isotherms T_L and T_S the condition of temperature field and heat flux continuity should be assumed:

$$x\in\xi_1: \quad \begin{cases} -\lambda_2 n\circ \mathrm{grad}\, T_2 = -\lambda_1 n\circ\mathrm{grad}\, T_1 \\ T_2 = T_1 = T_S \end{cases} , \tag{7}$$

$$x\in\xi_2: \quad \begin{cases} -\lambda_3 n\circ \mathrm{grad}\, T_3 = -\lambda_2 n\circ\mathrm{grad}\, T_2 \\ T_3 = T_2 = T_L \end{cases} . \tag{8}$$

On the outer surface of the system the boundary condition (2) is given, and for t=0, e=1,2,3 the condition of the form (3).

On the basis of these two models it is possible to construct more complex problems, for example, the solidification and cooling processes in the volume of multiphase material $D_1(t) \cup D_2(t) \cup \ldots \cup D_{m+1}(t)$ with m moving boundaries $\xi_1, \xi_2, \ldots, \xi_m$ (compare Figure 3) on which latent heats $L_1, L_2, \ldots, L_m$ are emitted. Then

$$\begin{cases} x \in D_e(t): & c_e \rho_e \dfrac{\partial T_e}{\partial t} = \operatorname{div} (\lambda_e \operatorname{grad} T_e) , \quad e=1,2,\ldots,m+1, \qquad (9) \\ x \in \xi_e: & \begin{cases} -\lambda_{e+1} n \circ \operatorname{grad} T_{e+1} = -\lambda_e n \circ \operatorname{grad} T_e + \rho_e L_e v_e \\ T_{e+1} = T_e = \hat{T}_e \end{cases} , \qquad (10) \\ & \qquad e=1,2,\ldots,m \\ x \in \Gamma_0: & \Phi(T, n \circ \operatorname{grad} T) = 0 , \qquad (11) \\ t=0: & T_e(x,0) = T_{oe}(x) , \quad e=1,2,\ldots,m+1. \qquad (12) \end{cases}$$

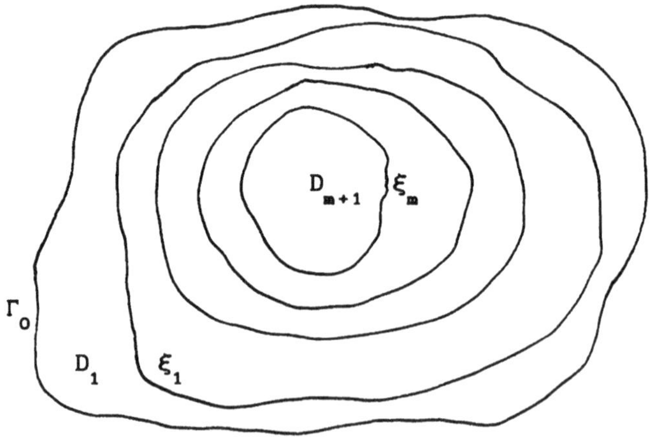

Fig. 3. The sub-areas D_e, e=1,2,...m+1

The next generalization consists in the division of transcient zone into parts $D_2(t), D_3(t) \ldots \cup D_m(t)$, at the same time index e=1 identifies the solid sub-area and e=m+1 the liquid one. In this case we formulate the following mathematical model

$$\begin{cases} x\in D_e(t): \quad C(T)\rho_e \dfrac{\partial T_e}{\partial t} = \mathrm{div}\,(\lambda_e \,\mathrm{grad}\, T_e)\,, \quad e=1,2,\dots,m+1, & (13) \\[2ex] C(T) = \begin{cases} c_{m+1}\,, & T\geq T_{m+1}=T_L \\ c_e + c_e^{sp}\,, & T\in\langle \hat{T}_{e-1}, \hat{T}_e)\,, \quad e=2,\dots,m \\ c_1\,, & T<T_1=T_S \end{cases} & (14) \\[4ex] x\in\xi_e: \quad \begin{cases} -\lambda_{e+1}\,n\circ\mathrm{grad}\, T_{e+1} = -\lambda_e\,n\circ\mathrm{grad}\, T_e \\[2ex] T_{e+1} = T_e = \hat{T}_e \end{cases} & (15) \\ \qquad\qquad\qquad\qquad\qquad\qquad e=1,2,\dots,m \\ x\in\Gamma_0: \quad \Phi(T, n\circ\mathrm{grad}\, T) = 0\,, & (16) \\[1ex] t=0: \quad T_e(x,0) = T_{oe}(x)\,, \quad e=1,2,\dots,m+1. & (17) \end{cases}$$

In the most general case it is possible to consider the model being the combination of the two previously presented, namely

$$\begin{cases} x\in D_e(t): \quad C(T)\rho_e \dfrac{\partial T_e}{\partial t} = \mathrm{div}\,(\lambda_e \,\mathrm{grad}\, T_e)\,, \quad e=1,2,\dots,m+1, & (18) \\[2ex] C(T) = \begin{cases} c_{m+1}\,, & T\geq T_{m+1}=T_L \\ c_e + c_e^{sp}\,, & T\in\langle \hat{T}_{e-1}, \hat{T}_e)\,, \quad e=2,\dots,m \\ c_1\,, & T<T_1=T_S \end{cases} & (19) \\[4ex] x\in\xi_e: \quad \begin{cases} -\lambda_{e+1}\,n\circ\mathrm{grad}\, T_{e+1} = -\lambda_e\,n\circ\mathrm{grad}\, T_e + L_e\rho_e v_e\,, \\[2ex] T_{e+1} = T_e = \hat{T}_e\,. \end{cases} & (20) \\ \qquad\qquad\qquad\qquad\qquad\qquad e=1,2,\dots,m \\ x\in\Gamma_0: \quad \Phi(T, n\circ\mathrm{grad}\, T) = 0\,, & (21) \\[1ex] t=0: \quad T_e(x,0) = T_{oe}(x)\,, \quad e=1,2,\dots,m+1. & (22) \end{cases}$$

It should be pointed out that considering the practical foundry problems one can distinguish such boundaries ξ_e on which $L_e=0$ (reduction of the Stefan condition to condition (15)) and also such sub-areas $D_e(t)$ with $c_e^{sp}=0$. This situation has place among others during the cast iron solidification when the process in part proceeds in the constant temperature and partly in the interval of temperature.
The generalization of the APTM discussed in this paper allows to solve (by numerical methods) the problems described above efficiently and in a rather simple way.

GENERALIZED ALTERNATING PHASE TRUNCATION METHOD (GAPTM)

A presentation of the APTM and its generalizations requires a modification of the mathematical models described previously by introducing to the considerations the physical enthalpy of alloy

$$H(T) = \int_0^T C(\mu)\rho(\mu)d\mu + \sum_{e=1}^{m} \rho_e L_e \eta_e(T) \ , \tag{23}$$

where $\rho(T)=\rho_e$, $T\in\langle\hat{T}_{e-1},\hat{T}_e)$ and

$$\eta(T)=\begin{cases} 0, & T\le\hat{T}_e \\ 1, & T>\hat{T}_e \end{cases} .$$

The enthalpy distributions are shown in Figure 4.

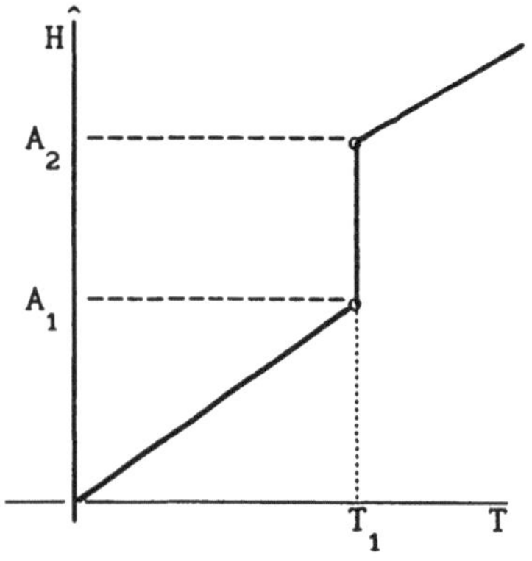

Fig. 4. 1. Model (A)

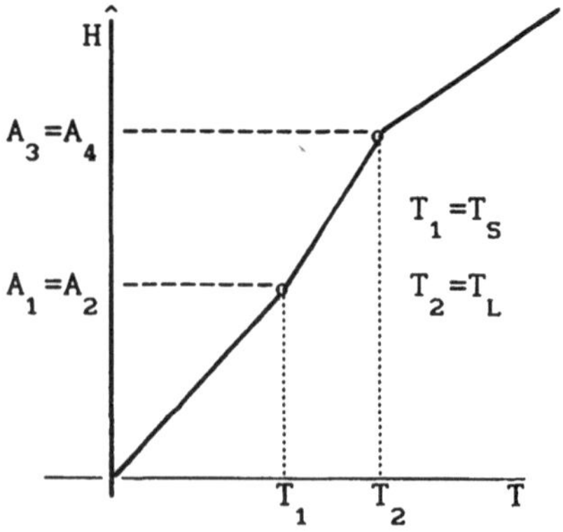

Fig. 4. 2. Model (B)

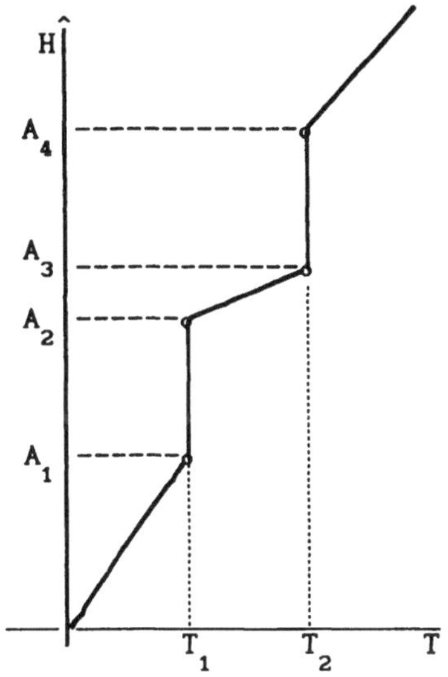

Fig. 4.3. Generalized model (A)

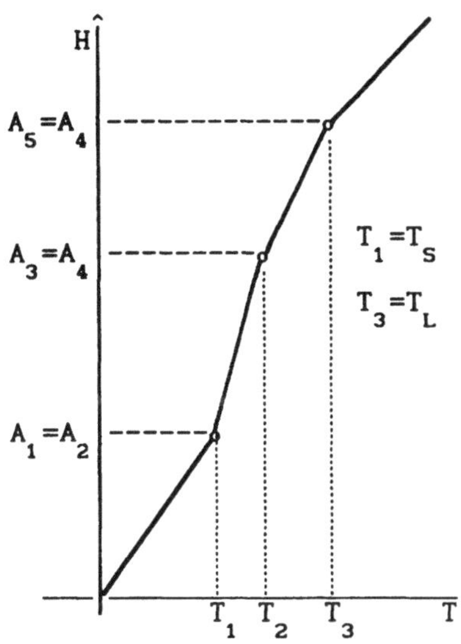

Fig. 4.3. Generalized model (B)

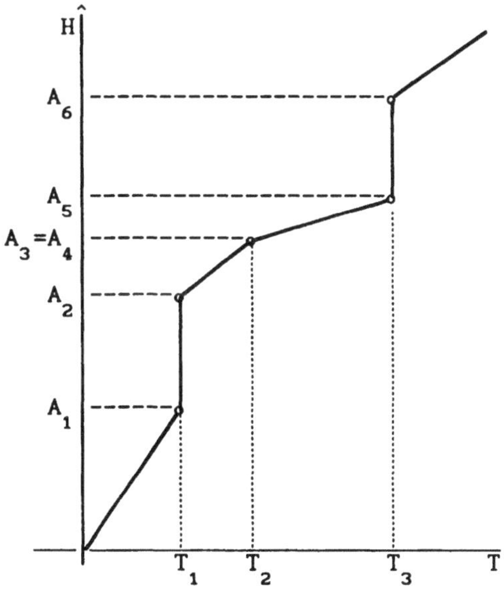

Fig. 4.3. Mixed model

So the mathematical model (18)-(22) transformed to the enthalpy convention is of the form

$$
\begin{cases}
x \in D_e(t): & \dfrac{\partial H_e}{\partial t} = \operatorname{div}(a_e \operatorname{grad} H_e), \quad e=1,2,\ldots,m+1, \quad (24) \\
& a_e = \lambda_e / C(T)\rho(T), \\
x \in \xi_e: & \begin{cases} -a_{e+1} n \circ \operatorname{grad} H_{e+1} = -a_e n \circ \operatorname{grad} H_e + L_e \rho_e v_e \\ H_{e+1} = A_{2e}, \; H_e = A_{2e-1}, \; A_{2e} = A_{2e-1} + L_e \end{cases}, \quad (25) \\
& e=1,2,\ldots,m \\
x \in \Gamma_0: & \hat{\Phi}(H, n \circ \operatorname{grad} H) = 0, \quad (26) \\
t=0: & H_e(x,0) = H_{oe}(x), \quad e=1,2,\ldots,m+1. \quad (27)
\end{cases}
$$

Let us assume that t_0 and $t_1 = t_0 + \Delta t$ denote two distinguished points of time lattice Δ_t and that the enthalpy distribution $H(x, t_0)$ in the considered area D is known.
The standard APTM algorithm [1,2] concerns the model (A) and consists of two stages. The final effect of its use is the determination of the enthalpy field for the time level t_1.
During the first stage such a quantity of enthalpy is added to points of the area D that whole area is formally reduced to the liquid state, in other words the following pseudo-initial condition should be assumed

$$V_2(x, t_0) = \max \{A_2, H(x, t_0)\} . \qquad (28)$$

Utilizing the numerical methods (e.g. BEM) we find the enthalpy distribution at the moment t_1 i.e. $\hat{V}_2(x, t_1)$. The first stage of calculations ends when the previously added enthalpy is substracted

$$V_2(x, t_1) = \hat{V}_2(x, t_1) + H(x, t_0) - V_2(x, t_0) . \qquad (29)$$

In the second stage the following pseudo-initial condition is constructed

$$V_1(x, t_0) = \min \{A_1, V_2(x, t_1)\} . \qquad (30)$$

In this way the area D becomes a homogenous one (solid state).
The enthalpy distribution $V_1(x, t_1)$ is again calculated and the final enthalpy values in the area D results from the formula

$$H(x,t_1) = \hat{V}_1(x,t_1) + V_2(x,t_1) - V_1(x,t_0) . \qquad (31)$$

The proof of the method's correctness is shown in [2].

In paper [7] B.Mochnacki and A.Kapusta discussed the generalization of this algorithm for the case shown in Figure 4.3. After further investigations it appeared that it is possible to adapt APTM for computations of solidification processes corresponding to Figures 4.2, 4.4 and 4.5.

Let us consider an arbitrary stage "e" ($e \neq m+1$, $e \neq 1$). The enthalpy distribution $V_{e+1}(x,t_1)$ for phase "e+1" is known. A reduction of the area D to the phase "e" corresponds to acceptation of the following initial conditions

$$V_e(x,t_0) = \min\{A_{2e-1}, \max[A_{2e-2}, V_{e+1}(x,t_1)]\} . \qquad (32)$$

If $\hat{V}_e(x,t_1)$ denotes the enthalpy field calculated in this stage then the final result of computations arises from the formula

$$V_e(x,t_1) = \hat{V}_e(x,t_1) + V_{e+1}(x,t_1) - V_e(x,t_0) . \qquad (33)$$

The solution $V_e(x,t_1)$ constitutes the basis for the next step of the GAPTM.

A certain modification of the algorithm has place in case of the first ($e=m+1$) and the last ($e=1$) stages of computations. Considering the first stage the following pseudo-initial condition should be assumed

$$V_{m+1}(x,t_0) = \max\{A_{2m}, H(x,t_0)\} . \qquad (34)$$

At the same time the correction (33) is of the form

$$V_{m+1}(x,t_1) = \hat{V}_{m+1}(x,t_1) + H(x,t_0) - V_{m+1}(x,t_0) . \qquad (35)$$

At the last stage

$$V_1(x,t_0) = \min\{A_1, V_2(x,t_1)\} . \qquad (36)$$

and

$$H(x,t_1) = \hat{V}_1(x,t_1) + V_2(x,t_1) - V_1(x,t_0) . \qquad (37)$$

The solution $H(x,t_1)$ obtained at the last stage constitutes the searched enthalpy field at the moment $t=t_1$.

The transition from t_0 to t_1 requires solving "m+1" linear problems in conventionally homogenous areas $D=D_e$, $e=1,2,...,m+1$.

On the other hand, however, the typical and well known BEM algorithm for the linear Fourier's equation can be used and it is an essential advantage of the GAPTM.

The details of numerical model are not presented here because the problems of the BEM utilization in an approximate solution of linear boundary-initial problems for homogenous areas were described many times.

EXAMPLES OF NUMERICAL SIMULATIONS

The first example concerns the identification of a non-steady temperature field in the volume of low carbon steel casting (the plate with thickness G=0.095m). The thermophysical parameters of casting sub-areas are as follows

- liquid state ($T \geq 1500^{\circ}C$): $c_3=820$[J/kgK], $\rho_3=700$[kg/m^3], $\lambda_3=20$[W/mK],
- transcient zone ($1470<T<1500$): $c_2=700$, $c_2^{sp}=6300$, $\rho_2=7200$, $\lambda_2=25$,
- solid state ($T \leq 1470$): $c_1=650$, $\rho_1=7500$, $\lambda_1=35$.

The pouring temperature (initial condition) was assumed as $T_3(x,0)=T_{03}(x)=1550$, for x=0: no grad $T(0,t)=0$, for x=G: $T(G,t)=800$.

In the area of the considered object 10 one-dimensional linear internal cells and two boundary points were distinguished.

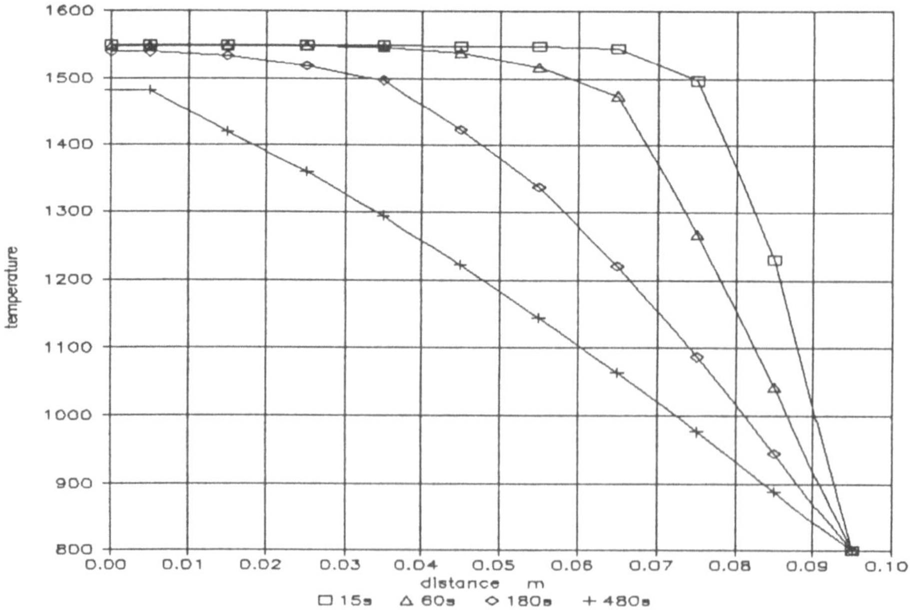

Fig.5. Solution on the basis of BEM-GAPTM (full lines) and FDM (points)

The GAPTM algorithm was coupled with the 1st scheme of the BEM for linear parabolic equations. The obtained results are shown in Figure 5. The same problem was solved by a repeatedly verified computer program for non-steady and non-linear solidification problems (on the basis of the FDM).
The comparison of numerical solutions is presented in Figure 5 and as it can be noticed the results are practically the same.

The second example is more "utilitarian". The aim of computations was the analysis of the solidification process in the central cross section of a rectangular steel ingot produced in the steel-works "Sendzimir" in Cracow.

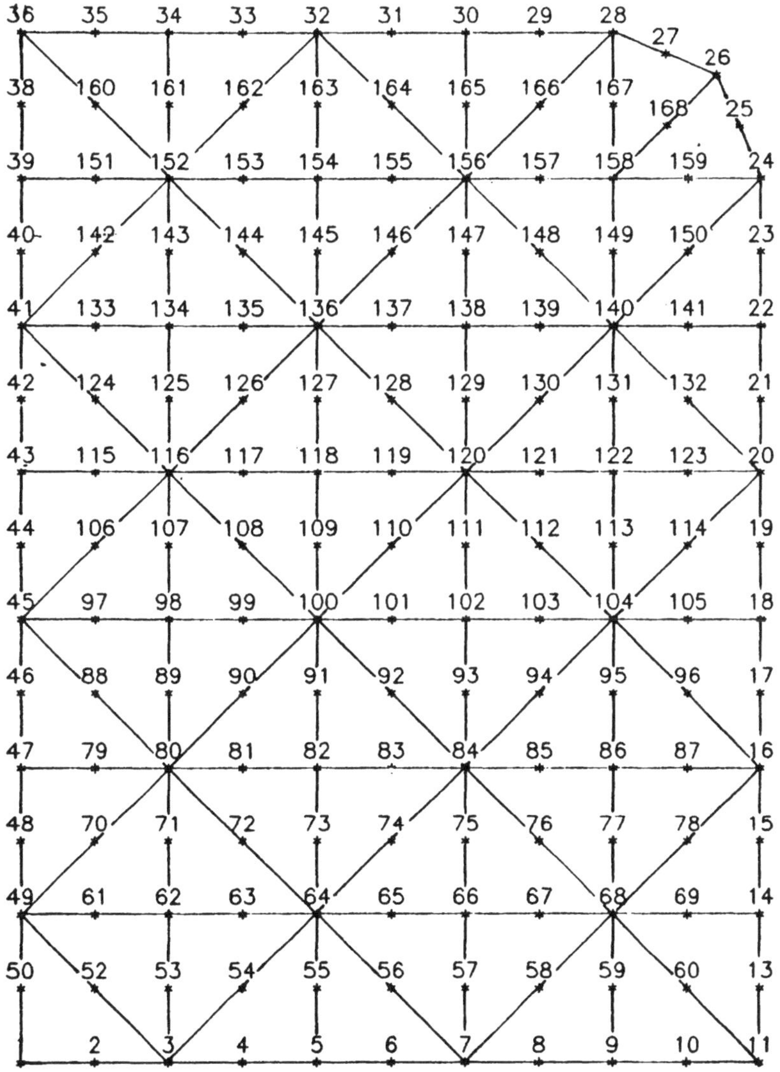

Fig. 6. Discretization of ingot area

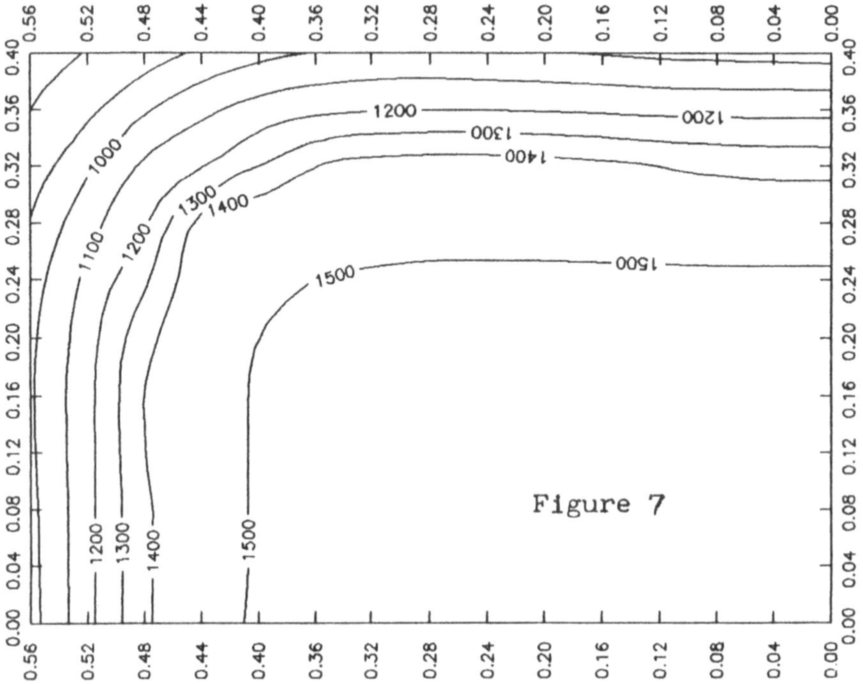

Figure 7

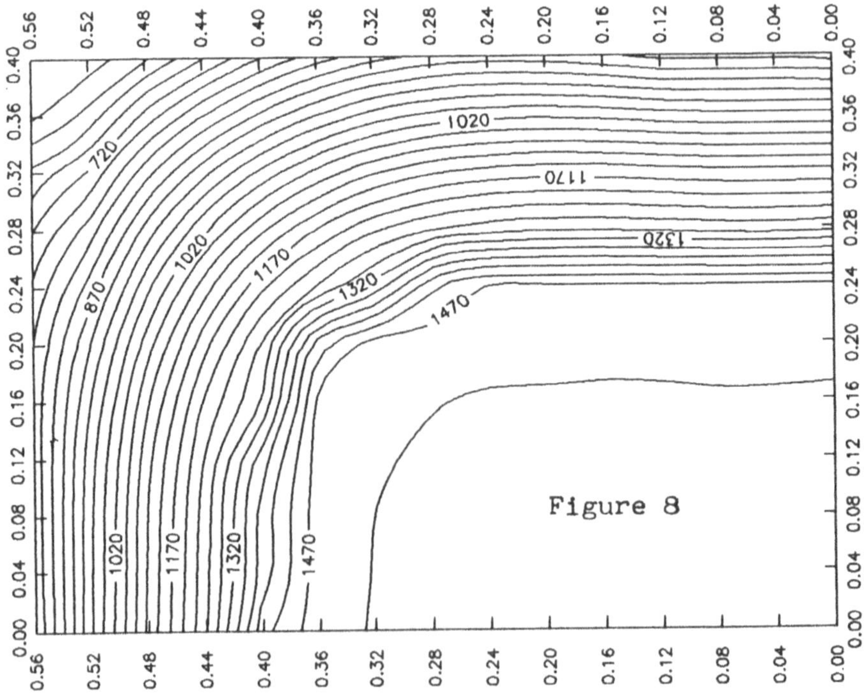

Figure 8

The quarter of the section of the dimensions 0.61*0.38[m] (the corner radius r=0.06[m]) was considered. The discretization of the object is shown in Figure 6. The following thermophysical parameters of steel were introduced: $\lambda_1=\lambda_2=\lambda_3=35$, $c_3=820$, $\rho_3=7200$, $c_2=735$, $C_2=7714$, $\rho_2=7320$, $c_1=650$, $\rho_1=7500$, $T_{30}=1550$, $T_L=1505$, $T_S=1470$. The boundary conditions along the axis of symmetry: n∘grad T=0, the heat flux q(x,t) on the outer surface of ingot was assumed [8]. The local and instantaneous values of q(x,t) were determined on the basis of measurements done in the steel-works "Sendzimir".

In Figures 7 and 8 the temperature fields for t=1800[s] and t=3600[s] are shown. The disturbances in the course of the isotherms particulary visible in Figure 8 are not a result of calculation errors but of the graphical procedure imperfection. For small temperature gradients and triangle meshes the procedure of isolines drawing is not excessively precise.

REFERENCES

1. Rogers, J., Berger, A. and Ciment, M., Numerical Solution of a Diffusion Consumption Problem with a Free Boundary, SIAM J. Num. Anal. Vol.12, pp. 645-659, 1975.

2. Rogers, J., Berger, A. and Ciment, M., The Alternating Phase Truncation Method for a Stefan Problem, SIAM J. Num. Anal. Vol.16, pp. 562-587, 1979.

3. Majchrzak, E., Numerical Simulation of Solidification by Boundary Elements Method, pp.208-212, Proceedings of the Int. Int. Conf. FOCOMP'88, Sofia, Bulgaria, 1988.

4. Majchrzak, E., Numerical Simulation of Continuous Casting Solidification by Boundary Elements Method, Eng. Anal. with Boundary Elements' (in print)

5. Mochnacki, B., Mathematical Methods in the Thermal Theory of Foundry, Chapter XVIII, Eng. Handbook-Foundry, Vol.2, WNT, Warsaw, 1986 (in Polish).

6. Mochnacki, B., Substitute Thermal Capacity of Metal Solidifying in an Interval of Temperature, Bull. of the Pol.Ac. of Sc., Techn. Sc., Vol.32, No 3-4, pp.127-143, 1984.

7. Mochnacki, B., Kapusta, A., The Analysis of Heat Transfer Processes in the Cylindrical Radial Continuous Casting Volume, Bull. of the Pol.Ac. of Sc., Techn. Sc., Vol.36, No 5-6, pp.309-320, 1988.

8. Kawaler, J., Mochnacki, B., Majchrzak, E. et al, The Optimization of the Ingot Mould W01 Shape, Report for st.pl. "Sendzimir", Cracow, 1989-1990 (not published).

SECTION 2: METAL CASTING AND WELDING

Spread in Hot Rolling: an Optimal Control Approach

P. Montmitonnet, A. Bern (*), J.L. Chenot

Ecole des Mines de Paris, Centre de Mise en Forme des Matériaux, Sophia Antipolis, 06565 Valbonne Cédex, France

(*) now at Aluminium Péchiney - Rhénalu, BP 16, 68600 Neuf-Brisach, France

ABSTRACT

Some forming processes mainly consist of steady-state phases, and should preferably be modelled by steady state approaches to save computing time. However, some difficulties occur as the geometry of free surfaces is not known. An optimal control method for determination of these free surfaces is presented and its inclusion in a finite element code is described. It consists of finding the free surface through which the material flux vanishes. The problem of contact of the bulging surface with rolls is discussed. Numerical results are substantiated by comparison with rolling experiments on model material slabs. Finally, a numerical study of the influence of material properties and rolling parameters is performed in the case of edge rolling.

INTRODUCTION

The surface of a workpiece being formed can be divided into:
- the contact surface through which the forming forces are transmitted by the tools.
- the free surface, which is generally free both from a mechanical (zero stress vector) and a geometrical point of view: its position is not fixed.

The latter is most important in forming processes where it determines the external shape of the formed product, one of its major requirements. In rolling, the main free surface effect is called spread, i.e. an increase of the width of the rolled stock (figure 1), in a process which should be primarily a reduction of height resulting in elongation. In flat rolling, this phenomenon leads to a problem in width control; in shape rolling, filling the groove or not decides whether or not the final shape will correspond to the designed shape.

Most rolling operations mainly consist of a steady state phase. We will restrict our analysis to this situation: we do not investigate end effects with such poetic names as "fishtail" or "cat's tongue".

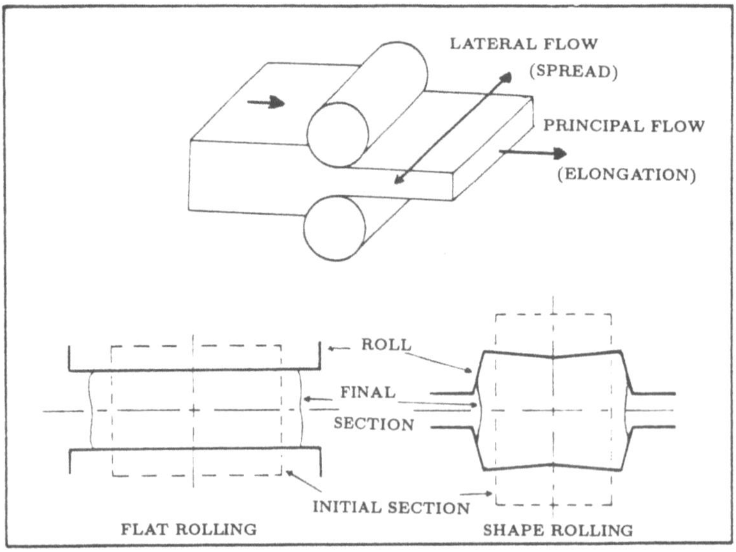

Figure 1 : description of spread in rolling.

Because of the huge variety of geometries, only very general methods can be thought of to model spread. Much work has been devoted to Finite Element modelling of rolling[1-8]. The most straightforward technique to predict spread and bulging uses the incremental approach, although spread is primarily a steady state effect. Several papers describe steady state formulations [4-8], which is a natural and cheaper way of solving the problem: only one computation of a velocity field, instead of up to several hundred increments.

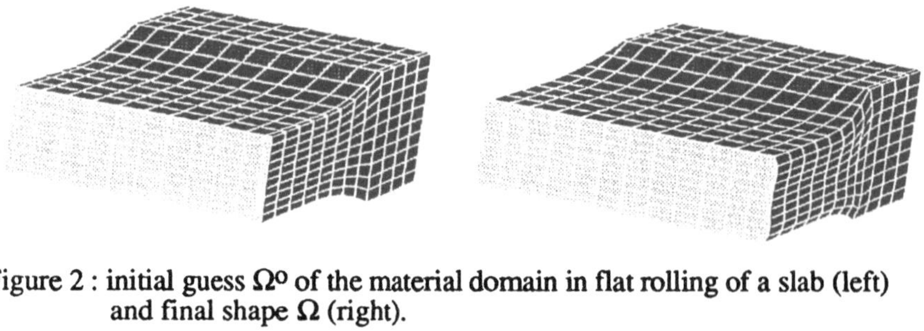

Figure 2 : initial guess Ω^0 of the material domain in flat rolling of a slab (left) and final shape Ω (right).

However, a difficulty appears for steady state computations. Contrary to the incremental approach, one has to suppose the workpiece already engaged in the roll gap (figure 2). Moreover, it is impossible to perform the computation of the correct velocity field on the real, bulged geometry, which is a result of the computation. One should hence initialize the domain Ω occupied by the rolled material. This initial guess Ω^0 is generally based on the intersection of the rolls and a cylinder, the cross section of which is that of the product before entry (Ω^0 hence results from a no-spread hypothesis).

As an approximate geometry has been assumed, the resulting velocity field violates the boundary condition on a stationary free surface $\partial\Omega_f$:

$$\mathbf{v}.\mathbf{n} = 0 \text{ at each point of } \partial\Omega_f \tag{1}$$

where $\mathbf{n}$ is the outer normal and $\mathbf{v}$ the velocity field. Hence, an iterative technique has to be used: an approximation $\mathbf{v}^{(i)}$ of the velocity field is computed on a certain approximation $\Omega^{(i)}$ of Ω . This velocity field is used to modify $\Omega^{(i)}$ into $\Omega^{(i+1)}$, and the cycle is repeated until no significant variation is found. Several techniques have been proposed to modify the domain:

The streamline method

Eq.(1) is commonly written:

$$\frac{dx}{v_x} = \frac{dy}{v_y} = \frac{dz}{v_z} \tag{2}$$

which represents the equations of streamlines. The velocity vector $\mathbf{v}(v_x,v_y,v_z)$ of points of the free surface must have been previously computed (see next paragraph); solving eq.(2) gives the streamline. In a discrete, finite element approach, eq.(2) can be written (for instance with an explicit integration scheme):

$$y_i = y_{i-1} + (x_i - x_{i-1}) \frac{v_y(x_{i-1})}{v_x(x_{i-1})}$$

$$z_i = z_{i-1} + (x_i - x_{i-1}) \frac{v_z(x_{i-1})}{v_x(x_{i-1})} \tag{3}$$

computed by recurrence, starting from a point (x_0,y_0,z_0) the position of which is unambiguously known. The x_i's are chosen positions in the flow direction. This technique has been used by Li and Kobayashi [4] for computing spread in slab rolling, by Mori and Osakada [5] in shape rolling and by Yamada et al.[6] for tube rolling.

Galerkin's formulation of the stationarity equation

Lee et al. [7] have applied Galerkin's method to eq.(1), writing:

$$W = \int_{\partial\Omega_F} N\, \mathbf{v}(\mathbf{x})\, \mathbf{n}(\mathbf{x})\, dS \tag{4}$$

where N's are the shape functions. Discretizing and writing that residual W should be zero results in a system:

$$[\mathbf{K}]\,(\Delta\mathbf{X}_F) = \mathbf{f} \tag{5}$$

which can be linearized assuming that the velocity field changes little when going to next approximation of Ω . $\Delta\mathbf{X}_F$ are the unknown displacements of free surface nodes (along $\mathbf{n}$), and $[\mathbf{K}]$ and $\mathbf{f}$ are given in full detail in [7]. This model was applied to computation of spread in flat rolling of billets.

FORMULATION WITH AN OPTIMAL CONTROL APPROACH

We formulate the following problem [8-12]:

"Find the free surface $\partial\Omega_F$ such that

$$J(\mathbf{v},\partial\Omega_F) = \int_{\partial\Omega_F} (\mathbf{v}.\mathbf{n})^2 \, dS \qquad (6)$$

is minimal, knowing that **v** is the velocity field solution of equilibrium equations and rheological equations on the domain Ω bounded by $\partial\Omega_F$, i.e. minimizing the functional (flow formulation):

$$\Phi(\mathbf{v}) = \int_{\Omega} \frac{K}{m+1} [\sqrt{3}\,\dot{\bar{\varepsilon}}]^{m+1} \, dV + \int_{\partial\Omega_C} \frac{\alpha K}{p+1} \|\mathbf{v}_g\|^{p+1} \, dS + \int_{\Omega} \frac{\rho_p K}{2} [\mathrm{div}(\mathbf{v})]^2 \, dV - \int_{\partial\Omega_T} \mathbf{T}.\mathbf{v} \, dS \text{ "} \qquad (7)$$

* $\partial\Omega_C$ is the contact surface, $\partial\Omega_T$ the part of the external surface on which a component of the stress vector **T** is prescribed. The fourth term of functional (7) thus allows to include back and front tensions if necessary.

* As for the rheological behaviour, an incompressible viscoplastic NORTON-HOFF law is used:

$$s = 2K \, [\sqrt{3}\,\dot{\bar{\varepsilon}}]^{m-1} \, \dot{\varepsilon} \qquad (8)$$

The corresponding viscoplastic potential gives the first term in eq.(7). s is the deviatoric stress, $\dot{\varepsilon}$ the strain rate tensor, $\dot{\bar{\varepsilon}}$ the equivalent strain rate. K is the material consistency and m the strain rate sensitivity parameter; as will be illustrated in paragraph 4, they may depend on temperature and strain

$$\bar{\varepsilon} = \int_{path} \dot{\bar{\varepsilon}} \, dt$$

* friction law:

$$\tau = -\alpha K \|\mathbf{v}_g\|^p \frac{\mathbf{v}_g}{\|\mathbf{v}_g\|} \qquad (9)$$

τ is the friction stress vector, $\mathbf{v}_g$ the sliding velocity; α and p are two friction coefficients. The friction potential is the second term of functional Φ.

* incompressibility is enforced by the penalty method (penalty coefficient ρ_p) giving the third term in eq.(8).

We therefore have to solve the following system after finite element discretization:

$$\frac{\partial \Phi}{\partial V}(V, X_F) = 0 \tag{10a}$$

$$\frac{dJ}{dX_F} = \frac{\partial J}{\partial V}\frac{\partial V}{\partial X_F} + \frac{\partial J}{\partial X_F} = 0 \tag{10b}$$

where the unknowns include the nodal velocity vector **V** and nodal positions on free surface X_F . In fact, each part of the system (eq(10a) and eq(10b)) is solved separately, and coupling is enforced by iterative solution of the two decoupled equations. Each equation is solved by Newton-Raphson's method, which involves the second derivatives of **Φ** and J. These imply tedious calculations, for instance:

$$\frac{d^2J}{dX_F^2} = \frac{\partial^2 J}{\partial X_F^2} + 2\left(\frac{\partial^2 J}{\partial V \partial X_F}\right)\frac{\partial V}{\partial X_F} + \frac{\partial J}{\partial V}\frac{\partial^2 V}{\partial X_F^2} + \left(\frac{\partial V}{\partial X_F}\right)^2 \frac{\partial^2 J}{\partial V^2} \tag{11}$$

However, a numerical study [10,12] has shown that it is sufficient to keep the term $\frac{\partial^2 J}{\partial X_F^2}$ in eq(10b), and also to solve a simplified equation:

$$\frac{\partial J}{\partial X_F} = 0 \quad \text{instead of} \quad \frac{dJ}{dX_F} = 0 \tag{12}$$

as both $\frac{\partial J}{\partial X_F}$ and $\frac{\partial J}{\partial V}$ vanish together at the solution **V.n** = 0.

Hence, the flow chart of the computation is established as follows:

1 - Initialize the shape of the product (initial meshing).
2 - initialize **V** by the newtonian solution V_0 (m=p=1 in eq.(7)).
3 - iterative computation of **V** with m and p ≠ 1; as long as $\frac{\|\Delta V_j\|}{\|V\|} > \varepsilon$,

$$V_{j+1} = V_j + \Delta V_j = V_j - \left(\frac{\partial^2 \Phi}{\partial V^2}\right)^{-1}\left(\frac{\partial \Phi}{\partial V}\right)$$

j is the iteration number.

4 - modify the positions of the nodes of the free surface:

$$X_F^{(i+1)} = X_F^{(i)} + \lambda\, \Delta X_F^{(i)} = X_F^{(i)} - \lambda \left(\frac{\partial^2 J}{\partial X_F^2}\right)^{-1} \cdot \frac{\partial J}{\partial X_F}$$

λ : parameter for under incrementation: find λ_{min} minimizing $J(X_F^{(i+1)}[\lambda_{min}])$

5 - test if $\| \Delta X_F^{(i)} \| < \varepsilon$; if not, i=i+1 , branch to 2

SPACE DISCRETISATION

Meshing .
Eqs.(10a) and (10b) are discretized by the finite element method. Throughout this paper, trilinear 8-node isoparametric hexahedral elements are used.

The extension of domain Ω is an important parameter. For truly steady state, Ω should extend to infinity in the main flow direction, which is of course impossible in practice. Intuitively, the region in which the spread takes place is limited to the roll gap. But it is much more difficult to precise quantitatively how far the domain should be extended to obtain a stable solution. A guideline for this determination is that the velocity field should stabilize at both ends of the mesh. Generally, if L is the roll gap length, a length L of material should be added before and after the roll gap, as pictured in figure 2.

The 2D meshing of the free surface is that induced by the restriction to $\partial\Omega_F$ of the 3D meshing of Ω.

Mode of displacement of the free surface nodes
Changes of the free surface are unambiguously defined by displacing the nodes in one direction only [7,12]. Only a displacement along normal $\mathbf{n}$ can provide a general and efficient method for finding the free surface. This technique will be used throughout in the following; the displacement direction of node i is the normal $\mathbf{n}_i$ projected in the Oyz plane (plane of cross section of the product, Ox is the rolling direction). The solution of the discrete system derived from eq.(10b) must hence be done in a local coordinate system $(\mathbf{t}_{1i},\mathbf{n}_i,\mathbf{t}_{2i})$, where the $\mathbf{t}$'s are vectors tangent to the surface at point i. The solution is afterwards transferred back into the (Oxyz) coordinate system.

Changes in contact conditions
The "free" surface is allowed to spread only to a certain extent, limited in shape rolling by the presence of the flank of the groove (figure 1). This conditional freedom requires special attention as the motion of free surface nodes must be restricted.

Definitions . The groove section is discretized into linear segments $[S_i]$. The line bearing segment $[S_i]$ is called (D_i). The normal to the groove will be called $\mathbf{N}$ ($\mathbf{N}_i$ for the normal to segment $[S_i]$). The nodal normal to the mesh is called $\mathbf{n}$. It should not be mistaken for the normal to the surface elements, defined at integration points and used in eq(6). The nodal normal is averaged over neighbouring elements.

Displacement of a node which is not in contact at the beginning of the iteration.
Solution of eq(10b) gives a displacement Δn along the normal, and the position of a node is then updated by:

$$\mathbf{X}' = \mathbf{X} + (\Delta n)\,\mathbf{n} \qquad (13)$$

If M' (fig. 3) is found inside the tool, it means that the point should come into contact; as it violates the non-penetration condition, its displacement is stopped when it intersects the groove (point M_1),

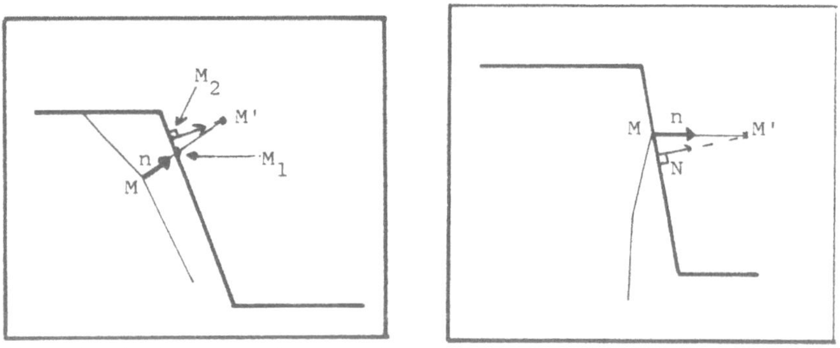

Figure 3 : deplacement of free surface nodes and contact with the rolls.
left: points initially not in contact
right: points initially in contact

<u>Displacement of a node which is in contact at the beginning of the iteration</u> . They may of course not penetrate the tool, but are allowed to slide along the groove surface in the Oyz plane. After some tests, the procedure pictured in fig3b has been derived. It consists of allowing a displacement along the normal **n** to the mesh (M moves to M'), then reprojecting M' on the groove surface along the groove normal **N**. As there might be some ambiguities in certain cases, an iterative procedure is used (figure 4):

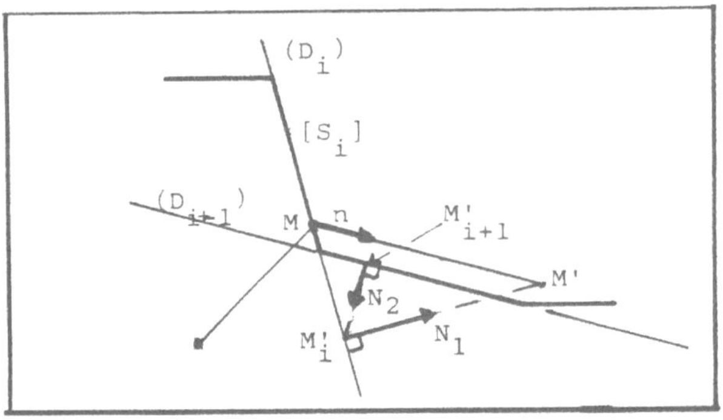

Figure 4 : iterative procedure for projecting nodes which have penetrated the roll.

(i) M (initially on segment $[S_i]$) moves to M' along **n**, normal to the mesh in M. M' (inside the roll) is projected along $\mathbf{N_i}$ on the line (D_i) bearing the segment $[S_i]$, which gives point M'_i ;
(ii) if M'_i is outside $[S_i]$, it is then projected similarly on line (D_{i+1}) or (D_{i-1}), along normal $N_{i\pm1}$, giving point $M'_{i\pm1}$.
(iii) the process is repeated until projection on (D_{i+p}) falls inside segment $[S_{i+p}]$.

APPLICATION TO SPREAD PREDICTION IN ROLLING

Experiments

Simulation experiments on plasticine have been carried out on an experimental rolling mill. Aluminium rolls, diameter 200 mm, were used. Rolling was performed at low speed (about 0.03 m/s), and with talc as a lubricant. Two geometries were tested:

- a 192 mm wide x 110 mm thick slab was rolled at 35% reduction (final thickness 72 mm); it should exhibit high spread.
- a 221 mm wide x 26 mm thick slab, 17% reduction (final thickness 21,5 mm); little spread is expected.

The plasticine used in these experiments is viscoplastic. From rheological measurements, we take m = 0,10. It is also slightly work-softening: in eq(8), K depends on strain:

$$K = K_0 \, \bar{\varepsilon}^{\,-0,03} \tag{14}$$

Computed results

Figure 5 and 6 show views of the two meshes before and after computation, and comparisons between computed and measured shapes; strain coupling according to eq(14) is assumed, m = 0,1. Friction coefficient α =1,3 K was used to fit the experiments. We have shown [12] that the computed free surface is remarkably unsensitive to mesh refinement, so that there is almost no effect of discretisation in the results presented.The thick slab, as expected, undergoes high spread; its lateral surface takes on a shape known as "diabolo" or "double bulge", i.e. a concave profile, with maximum spread just below the contact surface. The thinner slab, on the contrary, shows little spread (less than 1%) and a convex lateral section.

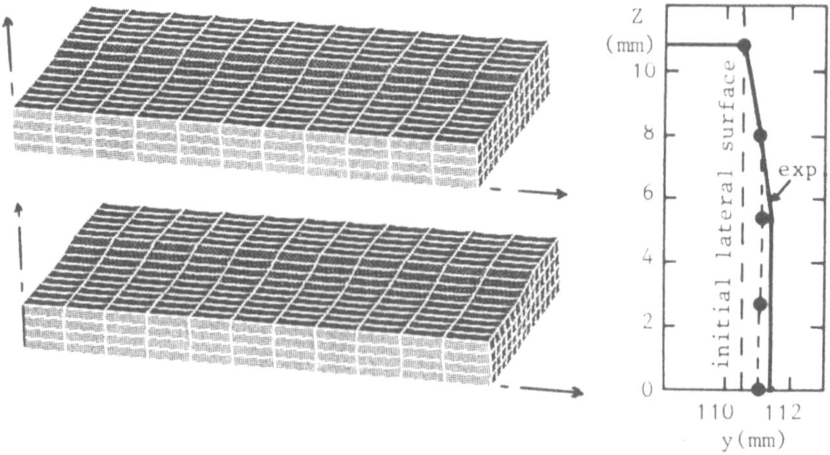

Figure 5 : initial mesh and bulged form for a thin slab.
Right: comparison of computed and experimental lateral profiles.

Quite a good agreement between experiments and computations is found; the discrepancy is nowhere greater than 0,5 mm, which remains within experimental errors.

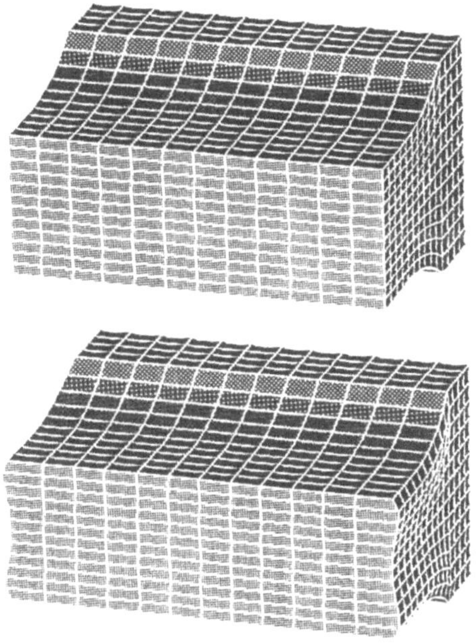

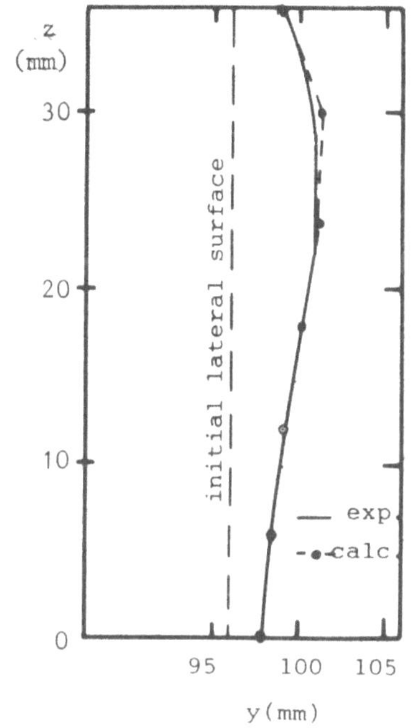

Figure 6 : comparison of experimental and computed lateral surface profiles, thick slab.

SENSITIVITY OF SPREAD TO MATERIAL PROPERTIES

Edge rolling and the dog-bone effect

Edge rolling consists of a reduction of the width of a slab by lateral rolls with vertical axes. Edge rolling is made necessary by excessive spread in previous flat rolling, thickness reducing passes. It has been selected for this study because of the spectacular phenomenon called "dogbone" (figure 7). Most of the time, the lateral rolls are grooved to limit this dogbone.

Dogbone is caused by a localization of the deformation in the vicinity of the lateral contact, due to the great width / thickness ratio. The central part undergoes little deformation, hence a small elongation. As a result, the elongation of the sides, which are much more deformed, is restricted, and spread flow (here in the thickness direction) is dominant.

A number of factors, both geometrical and physical, affect strain localization, and should therefore influence the amplitude of dogbone. We will investigate the effect of friction, strain rate sensitivity, work hardening or softening properties, and tensions at slab ends. Tensions are not likely to exist in this situation, but their effect is included for illustration, as it is important for other types of hot rolled products (bars or strip rolling).

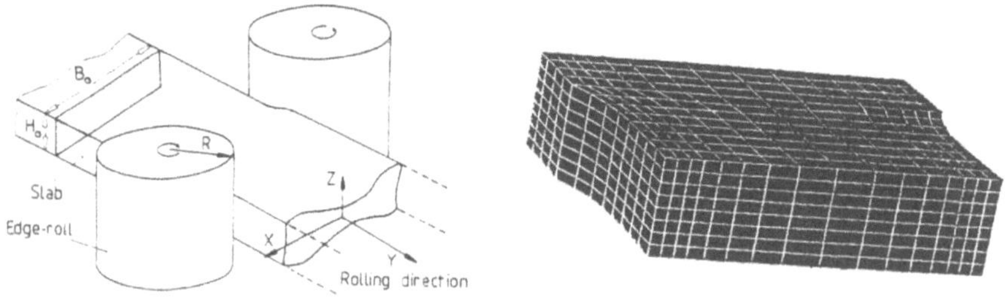

Figure 7 : schematic view of edge rolling (above, after [13]) and mesh used in the following computations.

Material properties

The rheology is described by eq(8), and friction by eq(9). The parameters for the reference case are:

- strain rate sensitivity m = 0.2 , a common value for hot steel at 1200°C
- consistency K is constant; as all terms in eq(7) are proportional to K, K=1 is assumed.
- for friction, $\alpha = 1.0$, p = 0.1
- no tensions

Materials properties are varied one by one from these reference conditions:

- m = 0.05
- K is made strain dependent:

$$K(\bar{\varepsilon}) = 2.7 \exp\left(-\frac{0.1}{\bar{\varepsilon}+0.1}\right) \quad \text{work hardening} \qquad (15a)$$

$$K(\bar{\varepsilon}) = \exp\left(\frac{0.1}{\bar{\varepsilon}+0.1}\right) \quad \text{work softening} \qquad (15b)$$

- friction is increased:
  * $\alpha = 1.3$, p = 0.1
  * sticking friction, i.e. $\mathbf{v} = \mathbf{v}_{roll}$ for all points in contact. This is a very high (in fact unrealistically high) friction.

- tension stresses (either front or back tension) 0.5 K in magnitude are included.

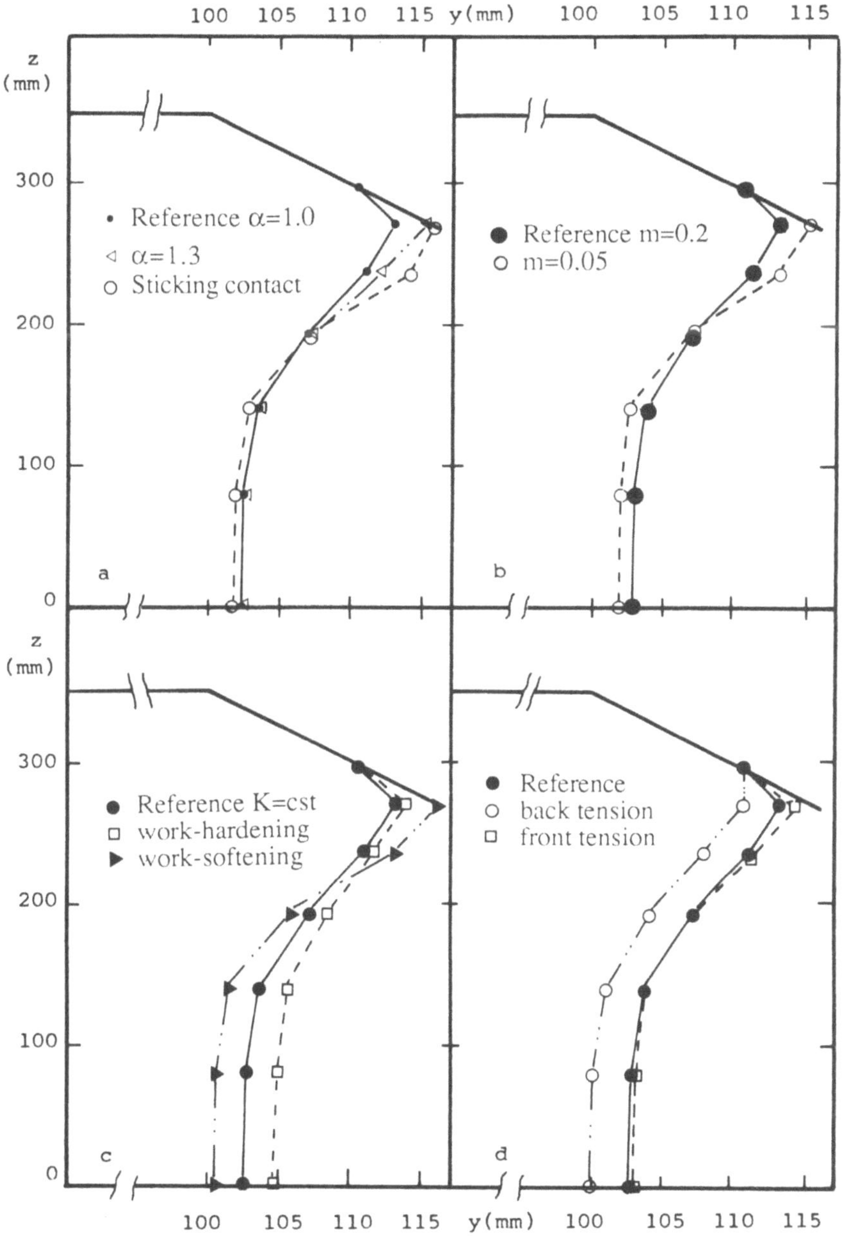

Figure 8 : influence of material properties and rolling parameters on dogbone.
a - influence of friction
b - influence of strain rate sensitivity
c - influence of strain hardening or softening
d - influence of front or back tension

Discussion

The mesh pictured in figure 7 has been used; it has been verified on a refined mesh that no significant discretization effect was present. The operation consists of rolling a 1000 mm wide x 200 mm thick slab down to 700 mm in width, with grooved rolls (12° flank angle, 100 mm deep groove); the roll diameter is 600 mm. Figure 8 presents the influence of the abovementioned parameters. Note that the y-direction is dilated 10 times with respect to the z-direction, in order to make the differences more conspicuous. Only 1/4th of the system is shown owing to symmetries. The dogbone forms at the groove / slab contact, but thickening extends to the central part of the slab.

The higher friction is, the greater the dogbone (figure 8a). On the contrary, thickening of the slab centre decreases slightly with increasing friction.

It is well known that decreasing the strain rate sensitivity (m=0.05 instead of 0.2) enhances strain localization. This effect is clearly demonstrated in figure 8b: with m=0.05, the dogbone is bigger and thickening of the central part is smaller. The same can be said of the strain hardening or softening tendancy (figure 8c); strain hardening disfavors localization and extends thickening towards the slab centre, whereas strain softening promotes it: spread in the central part almost vanishes.

Finally, figure 8d illustrates the effect of front and back tensions. Front tension is shown to have little effect on spread in the present case (this is not a general conclusion). Back tension severely decreases spread, both in the dogbone region and in the central part. In the case presented, a little thickness reduction is even predicted in the central part, but the tensions imposed are rather high.

CONCLUSION

The method presented in this paper for modelling rolling has also been applied to a diversity of situations, including extrudate swell in polymer flows, or bulging of continuously cast slabs. It has proved sufficiently robust, and has been applied to a great number of quite different rolling passes; some limitations exist, for instance bad convergence for very high reductions. It is slightly more expensive than streamline integration, but avoids some shortcomings of the latter connected with the determination of contact surfaces.

ACKNOWLEDGEMENT

The code used in this study was developed with financial support from Aluminium Pechiney and IRSID (Usinor - Sacilor company).

REFERENCES

[1] C. Liu, P. Hartley, C.E.N. Sturgess, G.W. Rowe, Elastic plastic finite element modelling of cold rolling of strip, Int. J. Mech. Sci. Vol. 27, 7-8, pp 531-541, 1985

[2] R. Shivpuri, P.C. Chou, a comparative study of slab, upper bound and finite element methods for predicting force and torque in cold rolling, Int. J. Mach. Tools Manuf. Vol. 29, 3, pp 305-322, 1989

[3] K. Mori, K. Osakada, H. Mikaido, T. Naoi, Y. Aburatoni, Finite element simulation of non steady state deformation in rolling of slabs and plates, Proc. 4th Int. Steel Rolling Conf., DEAUVILLE, June 1987. IRSID and ATS, Paris, 1987.

[4] G.J. Li, S. Kobayashi, Spread analysis in rolling by rigid plastic finite element method, in NUMIFORM 82 (ed. Wood et al.), Proc. 1st Conf. Numerical Methods for Industrial Forming Processes, Swansea, 1982. Pineridge Press, Swansea,1982

[5] K. Mori, K. Osakada, Finite element simulation of 3D deformation in shape rolling, in NUMIFORM 89 (ed. Thompson et al.), Proc. 3rd Conf. Numerical Methods for Industrial Forming Processes, Fort Collins, 1989. A.A. Balkema, Rotterdam, 1989.

[6] K. Yamada, S. Ogawa, S. Hamauzu, T. Kikuma, 3D analysis of mandrel rolling using rigid plastic finite element method, NUMIFORM 89 (ed. Thompson et al.), Proc. 3rd Conf. Numerical Methods for Industrial Forming Processes, Fort Collins, 1989. A.A. Balkema, Rotterdam, 1989.

[7] Y.S. Lee, P.R. Dawson, T.B. Dewhurst, Bulge prediction in steady state bar rolling process, in NUMIFORM 89 (ed. Thompson et al.), Proc. 3rd Conf. Numerical Methods for Industrial Forming Processes, Fort Collins, 1989. A.A. Balkema, Rotterdam, 1989.

[8] J.L. Chenot, P. Montmitonnet, P. Buessler, F. Fau , Finite element computation of spread in hot flat and shape rolling with a steady state approach. Accepted for publication in Eng. Comp.

[9] Zolesio, J.P. , Numerical algorithms and existence results for Bernouilli-like steady free boundary problems. Technical Report, CRMA 1183, Montreal University , 1983

[10] A. Bern, Contribution à la modélisation par éléments finis des surfaces libres en régime stationnaire, Doctors Thesis, Ecole des Mines de Paris, 1987 (in French).

[11] A. Bern, J.L. Chenot, Y. Demay, J.P. Zolesio, Numerical computation of free boundaries in non-newtonian stationary flows, in Proc. Finite Element Methods in Flow problems, Antibes, June 1986. SIAM, 1986.

[12] J.L. Chenot, P. Montmitonnet, A. Bern, C. Bertrand-Corsini, A method for determining free surfaces in steady state finite element computations. Accepted for publication in Comp. Meth. Appl. Mech. Eng.

[13] H.J. Huisman, J. Huetink, a combined eulerian-lagrangian 3D finite element analysis of edge rolling, J. Mech. Working Techn. Vol. 11, p 333, 1985.

Newtonian Fluid Computations in Lagrangian Variables with a Remeshing Technique. Application to the Filling Stage of the Diecasting Process

F. Muttin, T. Coupez, M. Bellet, J.L. Chenot

Ecole Nationale Supérieure des Mines de Paris, Centre de Mise en Forme des Matériaux, Sophia Antipolis, 06560 Valbonne, France

ABSTRACT

The Navier Stokes incompressible model is used to describe 2-dimensional metal casting flow. In a die-filling simulation the free moving boundary must be known precisely. For this reason the Lagrangian formulation is chosen instead of the Eulerian one. The computation domain and the velocity field are approached by means of the finite element method. A non-structured mesh is built by using quadratic six-noded triangular elements. In time the mesh degenerates. A remeshing technique, based on the Delaunay algorithm, provides a new computation mesh. As for the application, we present the computation of a die- filling case with low Reynolds number.

1. Introduction.

From the practician's point of view, the numerical simulation of the filling stage in the casting process is of major interest for many reasons :

- the prediction of defects, such as the incomplete filling of some regions or the merging of metal free surfaces at too low a temperature,
- the possible optimisation of filling conditions, especially the gate systems,
- the determination of the temperature field at the end of the filling, in order to define consistent initial conditions for a thermomechanical model of the solidification stage,
- and more generally, to be able to study the feasibility of new operating conditions or new processes.

Such industrial ambitious objectives have induced the development of various numerical models using different approaches. The finite difference and finite volume method have been first worked out and remain the more frequently used, with many variations : Marker And Cell (MAC) technique [1], Sola-Vof method... [2-4]. Their main drawbacks are :

- a spatial description which is mandatory structured and regular,
- the approximations by finite differences of the derivatives included in the Navier Stokes equations,
- a coarse description of the external surfaces of the computation domains : mould/fluid interface, free surface of the fluid.

The finite element method presents the major advantage of the more precise integral formulation, associated to a more general spatial discretisation. The method can be adapted to either Eulerian (the mesh is fixed in space) [5-6] or Lagrangian formulation (the mesh is then convected with the material flow). The latter is original and has been developed by very few authors for the simulation of fluid flow [7-9]. The better representation of free surfaces ("front tracking" procedures can be avoided), and also the easy introduction of surface tensions make it especially attractive.

However, in the case of very large deformations, such as those encountered in die filling, this method can give satisfying results only when it is associated to a remeshing technique of the fluid domain, in order to regenerate periodically elements having a correct shape ratio. Ramaswamy gives simple examples of a remeshing technique with application to wave propagation [9] with moderate distorsions. In this paper, we are mainly concerned with a general method for automatic remeshing to cope with the fluid flow simulation in the casting process. The flow is considered isothermal. In the first part, the Lagrangian formulation of Navier Stokes equations is briefly outlined. Then, the capability of the remeshing technique is demonstrated with an example.

2. Theoretical formulation.

2.1 Navier Stokes equation in Lagrangian variables.

The motion of a liquid melt is well governed by a Newtonian incompressible fluid flow model. To respect precisely the motion of the fluid free surfaces the Lagrangian variables are used. This consists of taking the time t and the particle coordinates x_0 at an initial time, as variables . Let $x(x_0,t)$ be the coordinates at time t of the particle initially at x_0 . The velocity and the acceleration of the particle may be expressed with the following partial derivatives

$$V = \frac{\partial x}{\partial t} \tag{1}$$

$$\dot{V} = \frac{\partial V}{\partial t} \tag{2}$$

The Navier Stokes equations postulate the conservation of linear momentum and mass in any point of the body at any time. In Lagrangian variables they take the following form

$$\rho \frac{\partial V}{\partial t} = \nabla.\sigma + \rho g \tag{3}$$

$$\nabla.V = 0 \tag{4}$$

where the stress tensor σ satisfy the linear constitutive law

$$\sigma = - p I + 2\mu D \tag{5}$$

in which I is the unit tensor and D is the tensor of rates of deformation

$$D(V) = \frac{1}{2}(\nabla V + \nabla V^t) \qquad (6)$$

the superscript t denoting the transpose of a tensor. In the above equations ρ is the density, μ the dynamic viscosity, g is the gravity and p is the hydrostatic pressure.

In the momentum equation (3) we can note the simplification of the inertial term characteristic of the Lagrangian formulation. While in the Eulerian one a non linear convective term must be added.

The mechanical problem needs boundary conditions. They take the following form [10]

$$\sigma . n = \tau \frac{d\dot{t}}{ds} \qquad \text{on } \Gamma_f \qquad (7)$$

$$V = 0 \qquad \text{on } \Gamma_b \qquad (8)$$

where Γ_f is the free surface , Γ_b the fluid-mould interface, n the unit normal at the free surface, $s(x_0,t)$ the curvilinear coordinates on the free boundary at time t , $\dot{t}$ is the unit tangent at the free surface, and τ is the surface tension stress.

The momentum equation is a first order partial derivative equation which needs an initial value for the velocity. Commonly we will take

$$V = 0 \qquad \text{in } \Omega_0 \qquad (9)$$

where Ω_0 is the open which represent the initial position of the fluid.

2.2 Weak formulation.

We adopt a perturbation (penalization) method, proposed by Temam[11], to treat the continuum equation (4)

$$\varepsilon p + \nabla.V = 0 \qquad \varepsilon \to 0^+ \qquad (10)$$

It is then possible to eliminate the pressure p from the momentum equation (3). The elimination of p is made to simplify the formulation and save in the sequel computational cost.

A weak formulation of the momentum equation may be obtain by multiplying it by any virtual velocity field V^* , which satisfy the boundary condition (8), and by summing it over Ω_0 . By using (10) we obtain

$$\forall\ V^*(+B.C.) \qquad \rho\int_{\Omega_0} \frac{\partial V}{\partial t}.V^*\ d\Omega_0 = \frac{1}{\varepsilon}\int_{\Omega_0} \nabla(\nabla.V).V^*\ d\Omega_0$$

$$+\ 2\,\mu\int_{\Omega_0} (\nabla.D(V)).V^*\ d\Omega_0 + \rho\int_{\Omega_0} g.V^*\ d\Omega_0 \qquad (11)$$

By means of integrations by part and by using equations (10) (5) and (7) we obtain while V^* is null on Γ_b

$$\forall\ V^*(+B.C.) \qquad \rho\int_{\Omega_0} \frac{\partial V}{\partial t}.V^*\ d\Omega_0 = -\frac{1}{\varepsilon}\int_{\Omega_0} \nabla.V\ \nabla.V^*\ d\Omega_0$$

$$-\ 2\,\mu\int_{\Omega_0} tr\,(\,D(V)D(V^*)\,)\ d\Omega_0 + \rho\int_{\Omega_0} g.V^*\ d\Omega_0\ -\tau\int_{\Gamma_f} \vec{t}.\frac{dV^*}{d\,s}\ d\Gamma_f \quad (12)$$

where tr is the trace operator.

Neglecting inertial and surface tension effects we will obtain an elasticity problem to be solved, where the constants $1/\varepsilon$ and 2μ are the Lamé coefficients.

2.3 Time integration with a Lagrangian updating.

Now we look at the integration scheme. The time length is divided in N time steps $\Delta t = T/N$. Let us denote $t_i = i\Delta t$ an intermediate time.

The first step consists to find at time t_1 the location $x(x_0,t_1)$, the velocity $V(x_0,t_1)$ and the acceleration $\frac{\partial V}{\partial t}(x_0,t_1)$ of any fluid particle initially located in x_0 with the velocity $V(x_0,0)$ and the acceleration $\frac{\partial V}{\partial t}(x_0,0)$. We have three unknowns. We search for them as the solution of equation (12) which satisfy also the two following integration rules:

$$V(x_0,t_1) = V(x_0,0) + \Delta t\left\{(1-\theta)\frac{\partial V}{\partial t}(x_0,0) + \theta\,\frac{\partial V}{\partial t}(x_0,t_1)\right\} \qquad (13)$$

$$x(x_0,t_1) = x(x_0,0) + \Delta t\ V(x_0,t_1) + \frac{\Delta t^2}{2}\left\{(1-\theta)\frac{\partial V}{\partial t}(x_0,0) + \theta\,\frac{\partial V}{\partial t}(x_0,t_1)\right\} \quad (14)$$

where θ is the explicit/implicit parameter, which belongs to $[0,1]$. This scheme is Newmark [12] type. Note that the momentum equation (12) is not set on Ω_1 that means an explicit treatment of the momentum conservation. Equation (13) is implicit because it links the unknown velocity with the unknown acceleration. The updating of the particle coordinates by means of (14) will be made after the simultaneous solution of equations (12)-(13).

When we have found $x(x_0,t_1)$, the velocity $V(x_0,t_1)$ and the acceleration $\frac{\partial V}{\partial t}(x_0,t_1)$, we start to compute the second time step by a Lagrangian updating.

At time t_1 a new Lagrangian coordinates $\widetilde{x_0}$ is defined. Location, velocity and acceleration of particles are reinitialized :

$$\widetilde{x_0} = x(x_0,t_1) \tag{15}$$

$$V(\widetilde{x_0},0) = V(x_0,t_1) \tag{16}$$

$$\frac{\partial V}{\partial t}(\widetilde{x_0},0) = \frac{\partial V}{\partial t}(x_0,t_1) \tag{17}$$

and the new unknowns $x(x_0,t_2)$, $V(x_0,t_2)$, $\frac{\partial V}{\partial t}(x_0,t_2)$ are noted:

$$x(x_0,t_2) = x(\widetilde{x_0},t_1) \tag{18}$$

$$V(x_0,t_2) = V(\widetilde{x_0},t_1) \tag{19}$$

$$\frac{\partial V}{\partial t}(x_0,t_2) = \frac{\partial V}{\partial t}(\widetilde{x_0},t_1) \tag{20}$$

They are the solution of relations (13)-(14) in which x_0 is replaced by $\widetilde{x_0}$ and of relation (12) set on $\Omega_{\Delta t}$.

We repeat the algorithm while $i.\Delta t<T$. This updated Lagrangian formulation is a sequence of computations on successive current configurations.

2.4 Space discretization by finite elements.

Now we look at space and velocity field approximations. The open Ω_0 is approached by a set of n_e elementary cells Ω_e . Each point x_0 of Ω_e is located by means of the nodes location $\{x_{0,e}\}$ and the matrix $[N]$ of the shape functions.

$$x_0 = [N] \; \{x_{0,e}\} \tag{21}$$

Isoparametric element will be used in the sequel. Let us denote by $\{v_{1,e}\}$ the unknown velocities of the cell nodes, by $\{v_{0,e}\}$ and $\{\dot{v}_{0,e}\}$ respectively their velocity and acceleration at the previous time step. In each cell starting from these vectors we can obtain the asssociated field by means of the matrix $[N]$.

Let $<\nabla.N>$ be the vector which links the divergence of a vector field in any point of Ω_e in term of nodal values

$$\nabla.V = <\nabla.N> \{v_{1,e}\} \tag{22}$$

Let $[DN]$ be the matrix which links in vectorial form the strain rate tensor in term of nodal velocities

$$\{D(V)\} = [DN] \{v_{1,e}\} \tag{23}$$

The free boundary Γ_f of Ω_0 is approached by n_b element sides Γ_b . A boundary particle is located by means of the matrix $[\tilde{N}]$

$$x_0(s) = [\tilde{N}(s)] \{x_{0,b}\} \tag{24}$$

where s is the curvilinear coordinate defined on Γ_b and $\{x_{0,b}\}$ the nodal points of Γ_b . The unit tangent at Γ_b in s is denoted by $\vec{t_b}(s)$.

2.5 Momentum equation in terms of a linear system.

We use for the virtual velocity V^* the same approximation as for the velocity field (Galerkin method). Using the above approximations (21)-(24), and the numerical integration scheme (13), the momentum equation (12) becomes the linear system

$$C\ V = F \tag{25}$$

where the vector V is the assemblage of the free nodal velocity components

$$V = A_{e=1}^{n_e} \{v_{1,e}\} \tag{26}$$

C is the matrix defined by

$$C = A_{e=1}^{n_e} (C_\rho^e + C_\varepsilon^e + C_\mu^e) \tag{27}$$

F is the vector of the applied forces

$$F = A_{e=1}^{n_e} (F_\rho^e + F_g^e) + A_{b=1}^{n_b} F_\tau^b \tag{28}$$

The elementary mass matrix C^e_ρ is defined by

$$C^e_\rho = \frac{\rho}{\theta\,\Delta t}\int_{\Omega_e} [N]^t[N]\; d\Omega_e \tag{29}$$

C^e_ε the elementary matrix which comes from the incompressibility is set by

$$C^e_\varepsilon = \frac{1}{\varepsilon}\int_{\Omega_e} \langle\nabla.N\rangle^t \langle\nabla.N\rangle\; d\Omega_e \tag{30}$$

The viscous elementary matrix takes the following form

$$C^e_\mu = 2\,\mu\int_{\Omega_e} [DN]^t\,[DN]\; d\Omega_e \tag{31}$$

The inertial forces are given by

$$\langle F^e_\rho\rangle = \frac{\rho}{\theta}\left\{\frac{1}{\Delta t}\langle v_{0,e}\rangle + (1-\theta)\,\langle \dot{v}_{0,e}\rangle\right\}\int_{\Omega_e} [N]^t\,[N]\; d\Omega_e \tag{32}$$

The gravity forces is done by

$$\langle F^e_g\rangle = \rho\;\langle g\rangle\int_{\Omega_e} [N]\; d\Omega_e \tag{33}$$

The forces given by surface tension are

$$\langle F^b_\tau\rangle = \tau\int_{\Gamma_b} \langle\vec{t}_b(s)\rangle\left[\frac{d\tilde{N}}{ds}(s)\right] ds \tag{34}$$

The momentum equation is linear while an implicit treatment of the inertial term in term of velocity. Hughes et al. [13] use a lumped mass matrix (diagonal) while here we use the full mass matrix (Galerkin one).

In the sequel we use the six node triangular quadratic element. We use the four Gauss points integration rule to evaluate the different elementary matrix and vectors defined by equations (29)-(34). The matrix of incompressibility is exactly integrated. If a reduced integration is used, with the one node formula, some numerical problems occur, like zig-zag pattern velocity field.

3. Automatic remeshing.

3.1 Metal forming calculation and remeshing.

A lot of work has been presented on the mesh generation problem and a number of algorithms of varying degrees of automation have been proposed in two dimensions for a variety of applications. For metal forming calculation, the remeshing step was concerned primarily with remedying to mesh distorsions and consequently was only understood as reshaping an existing finite element mesh. A radical alternative to the stuctured mesh updating is the use of triangles. This element gives unstructured meshes more flexible in treating complex geometries while retaining an appropriate mesh point distribution. Moreover several triangle mesh generation schemes are now available and can be used for general purpose. More particulary we are concerned with large deformation occuring in metal forming or unsteady flow and our problem is to create several meshes for only one calculation.

The mesh generator presented here, was developed to solve the remeshing stage, in Lagrangian finite element analysis (updated Lagrangian description or incremental scheme) and more particularily for the numerical simulation of the forging process [14]. The drawback of classical reshaping method is finally the lack of flexibility of the structured grid because of the constraint of the connectivity. Another difficulty well solved with the help of unstructured grid is the dynamic mesh adaptation to the contact problem.

Obviously, the calculation presented here needs a fully automatic remeshing procedure and shows the effectiveness of our mesh generation scheme which seems very general and well adapted for any formulation using a convected mesh (Lagrangian element) and a time integration scheme.

3.2 The remeshing scheme.

The remeshing main procedure verifies at each step of calculation the distorsion of elements and the accuracy of the mesh boundary (penetration of the lines into the die or other obstacles, free surfaces curvature ...) and decides itself to remesh or not according to prescribed tolerances.

In case of remeshing, a new boundary is deduced from the old mesh fitting perfectly the external geometric constraint (contact adaptivity). A generalized algorithm of the Delaunay triangulation ([15],[16],[17]) is used to create a mesh fitting exactly the prescribed boundary without internal node. At this stage, points can be created into elements according to the method proposed in [16] and [17]. The mesh is refined by adding internal nodes in order to obtain elements having the best possible shape. It is easy to show that this condition is sufficient to finish the nodes generation algorithm.

Two procedures are used to perfect the mesh. The first one is a topological operation : exchanging the diagonal between two adjacent triangles [18], and the second one is a mesh regularisation using a Laplace algorithm (internal nodes are moved to the barycenter of neightbouring nodes). A combination of these two procedures in a global and iterative algorithm gives a mesh very close to the optimal one. At last, the time depending parameters are interpolated from the old grid to the new one and the calculation may continue.

Each stage of that remeshing scheme is fully automatic and does not need user intervention (although it is possible to change the tolerances and the mesh size parameters during computation). Moreover the history of remeshing is automatically stored and usable by the post processing system as a time depending solution.

4. Die filling example.

We look at a filling case by injection. Figure 1 presents the dimension of the mould. The fluid is initially located in the channel of injection with zero initial velocity and acceleration.

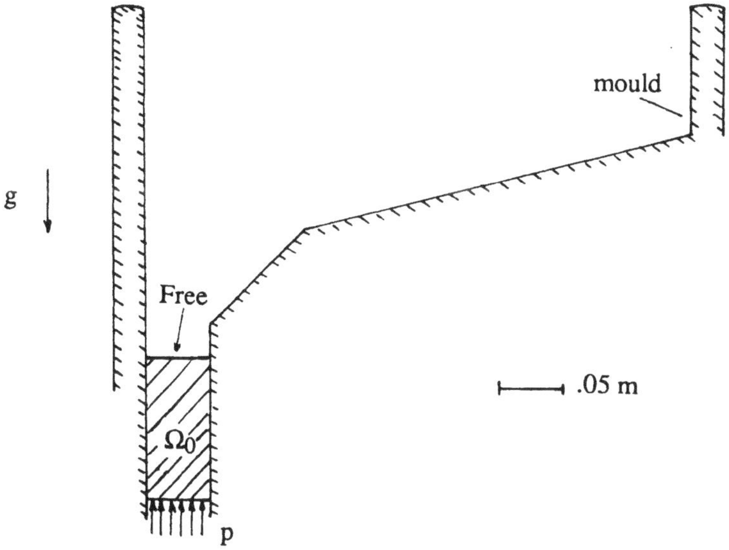

Figure 1. Initial configuration and mould dimensions.

The upper bound of the fluid is a free surface. We suppose a perfect adhesion between the fluid and the die. The lower bound is submitted to a constant pressure $p=85000\ N.m^{-2}$. To simulate material injection the location of the nodes on the pressurized bound are not updated. The fluid is a steel in semisolid state. Its viscosity is $100\ N.m^{-2}.s$, the density is $7000\ Kg.m^{-3}$, the gravity is $g=10\ m.s^{-2}$, the surface tension is equal to $0.9\ N.m^{-2}$.

The computation is stopped after 1860 time steps of $2\ 10^{-3}$ s. That requires 132 remeshings of the fluid. The CPU time is around 2 hours on DN10000 Apollo computer.

Figures 2 - 3 show the time sequence of liquid filling the mould and meshes. After 2.5 s. the fluid reaches the top of the mould. Figures 4 - 5 show velocity fields. The velocity is maximum in the injection channel. Its main value is $1.5\ m.s^{-1}$, giving a Reynolds number of 5.25 in the channel. We can note on Figures 2 - 3 that the meshes degenerate essentially where injection is made. An arbitrary Lagragian-Eulerian formulation (ALE) may be interesting in the channel to reduce the mesh distorsions. We can note the smoothness of the free flow surface which is a consequence of surface tension effects.

t = 0.079 s

t = 0.47 s

t = 1.76 s

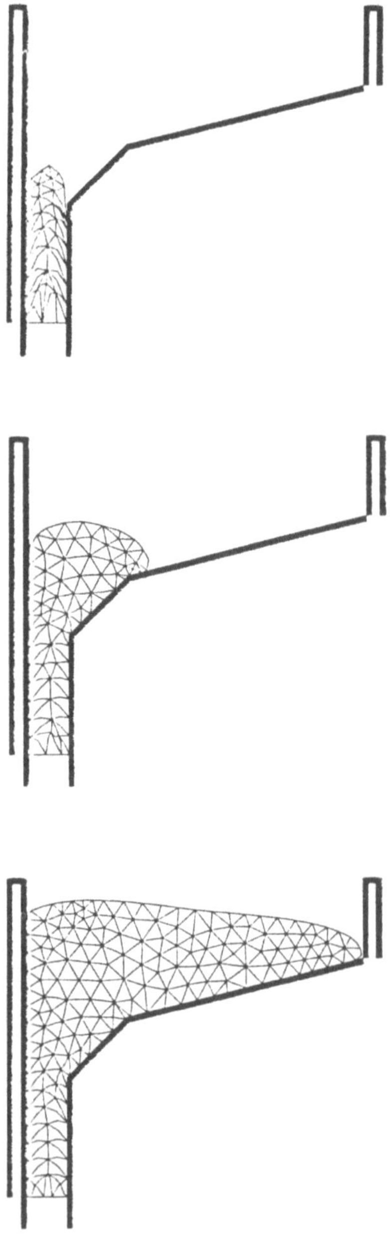

Figure 2. Computed mesh (the plot are time .079 .47 1.76 s.)

t = 2.52 s

t = 3.32 s

t = 3.87 s

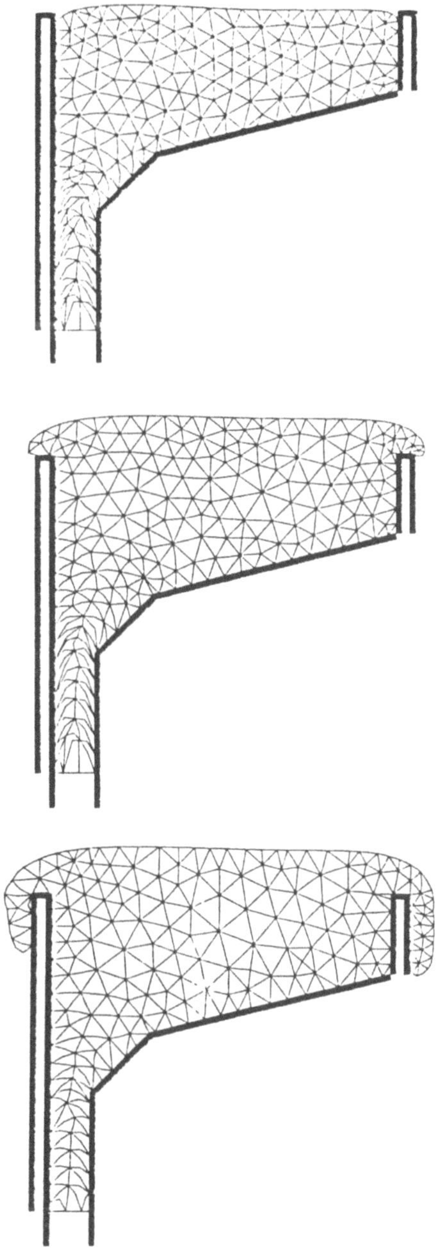

Figure 3. Computed mesh (the plot are time 2.52 3.32 3.87 s.)

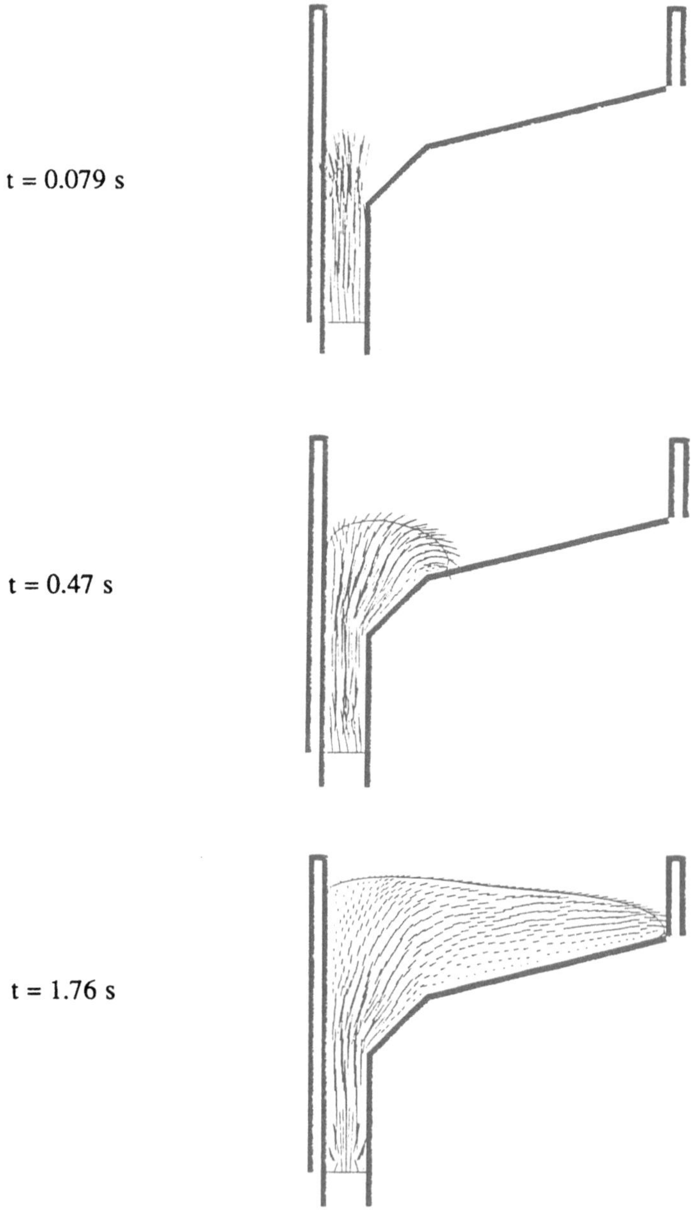

Figure 4. Computed velocity vectors (the plot are time .079 .47 1.76 s.)

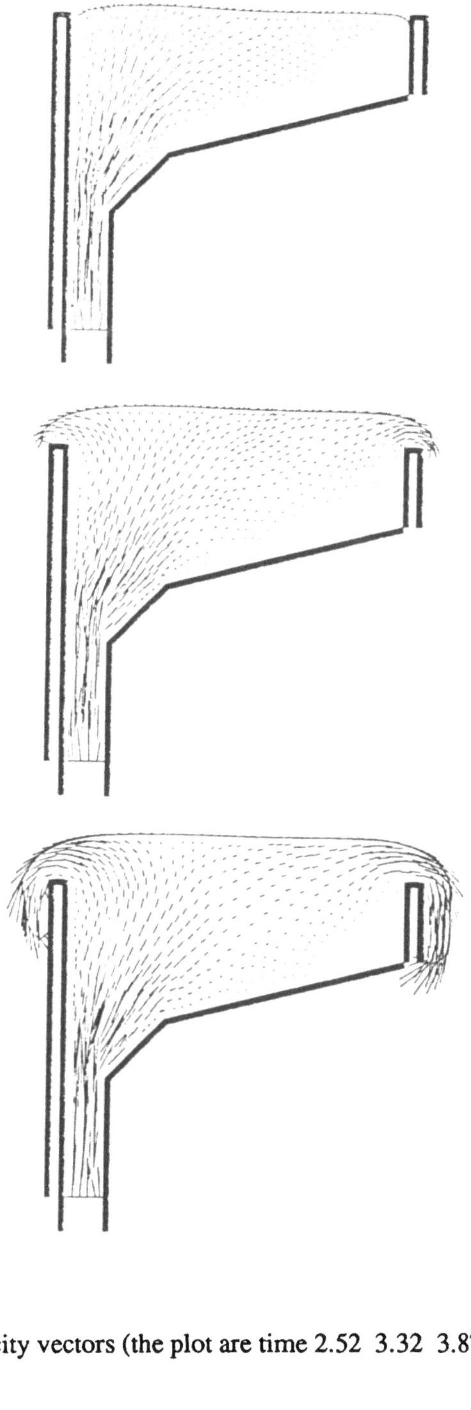

t = 2.52 s

t = 3.32 s

t = 3.87 s

Figure 5. Computed velocity vectors (the plot are time 2.52 3.32 3.87 s.)

5. Conclusion.

Our updated Lagrangian formulation of the Navier-Stokes equations is attractive. It allows to treat practical cases with low computation cost, the viscosity being aroud 100 Pa.s. Numerical comparisons of our solutions with analytical results and with other numerical schemes are in current development.

The automatic remeshing algorithm permit us to treat mould geometries and filling conditions of industrial complexity. With the automatic remeshing procedure there are no theoretical limitations on the flow geometry and its time-length.

Acknowledgments.

We wish to acknowledge financial supports provided by Creusot-Loire Industrie and by Agence Française pour la Maitrise de l'Energie.

References.

[1] F. H. Harlow, J. E. Welch, "Numerical Study of Large-Amplitude Free-Surface Motion" Phys. Fluids, **9** , 842, 1966.

[2] W.S. Hwang, R.A. Stoehr, "Modelling of fluid flow", ASM Metals Handbook, **14**, 9th edition, 867, 1988.

[3] C.W. Hirt, B.D. Nichols, "Volume of fluid (VOF) method for the dynamics of free boundaries", J. Comp. Phys., **39**, 210, 1981.

[4] W.T. Shah, "Generic casting modelling", AGARD Conference Proceedings n°426, Aerospace materials process modelling, 16-1, 1987.

[5] G. Dhatt, D.M. Gao, A.B. Cheikh, "A finite element simulation of metal flow in moulds", Int. J. Num. Meth. Eng., **30**, 821, 1990.

[6] B. Minaie, K.A. Stelson, V.R. Voller, "Fluid flow and solidification modeling", Modeling of materials processing, MD-vol **3**, ASME, New-York, 35, 1987.

[7] P. Bach, O. Hassager, "An algorithm for the use of the Lagrangian specification in Newtonian fluid mechanics and applications to free-surface flow", J. Fluid Mech., **152**, 179, 1985.

[8] F. Muttin, M. Bellet, J.L. Chenot, "Numerical simulation of inertia effects in Newtonian flows. Application to the filling stage of the die casting process", Proc. Int. Conf. on the Development of Industrial Forming Technology, Lisbonne (Port.), p.1-47, 1990.

[9] B. Ramaswamy, "Numerical simulation of unsteady viscous free surface flow", J. Comp. Physics, **90**, 396, 1990.

[10] K. J. Ruschak, " A Method for Incorporating Free Boundaries with Surface Tension in Finite Element Fluid-Flow Simulator", Int. J. Num. Meth. Engng, **15**, 639-648, 1980.

[11] R. Temam, Bull. Soc. Math. Fr. **96**, 115-152, 1968.

[12] N. M. Newmark, "A Method of Computation of Structural Dynamics" ASCE Journal of Engng. Mech. Division, **85**, 67-94, 1959.

[13] T. J. R. Hughes, W. K. Liu, A. Brooks, "Finite Element Analysis of Incompressible Viscous Flows by the Penalty Function Formulation" J. Comput. Phys. **30**, 1-60, 1979.

[14] C. Corsini, E. Wey, T. Coupez, J. L.Chenot, "Numerical simulation of the forging process with automatic remeshing" European conference on new advances in computational structural mechanics, Giens, France, 1991.

[15] D.F. Watson, "Computing the n-dimensional Delaunay tesselation with applications to Voronoi polytope", Computer Journal **24**, 2, 1981.

[16] F. Hermeline, "Triangulation automatique d'un polyèdre en dimension n" R.A.I.R.O. Analyse Numérique, **16** n°3, 211-242, 1982.

[17] J.L. Coulomb, " 2D and 3D mesh generation. Experiment with the Delaunay's tesselation", Conference on 'Automated mesh generation and adaptation', Grenoble, 1-2 october 1987.

[18] J.Y. Talon, "Algorithmes d'amélioration de maillages pour éléments finis", Conference on 'Automated mesh generation and adaptation', Grenoble, 1-2 october 1987.

Utilization of the Boundary Element Method for Numerical Analysis of Thermal Processes in the Casting-Mould System

E. Majchrzak

Foundry Institute, Silesian Technical University, 44-100 Gliwice, Towarowa 7, Poland

ABSTRACT

In this paper the boundary element method is applied to numerical simulation of solidification and cooling processes in the area of complex casting produced in a sand mould. Any metal solidifying in an interval of temperature (e.g. cast steel) is considered. The numerical solution has been obtained on the basis of the BEM algorithm for parabolic equations supplemented by certain procedures correcting the instantaneous temperature field in the casting volume, at the same time two different BEM schemes for the casting and mould sub-areas have been used, respectively. The examples presented in the final part of the paper concern computations of the solidification process in the cross-section of rectangular "T" thermal centre (2D problem).

INTRODUCTION

Let us consider a metal whose liquid state correponds to temperatures $T>T_L$ and solid one to the temperatures $T<T_S$ where T_L, T_S are the liquid and solid points. The sub-area in which $T_S<T<T_L$ is called the transcient zone. Thermal processes in the volume of solidifying and cooling metal are described by a system of differential equations

$$x \in D_m(t): \qquad c_m \rho_m \frac{\partial T_m}{\partial t} = \lambda_m \nabla^2 T_m + q_{vm} \; , \quad m=1,2,3, \qquad (1)$$

where m=1, 2, 3 corresponds to the liquid, transcient and solid sub-areas $D_m(t)$, and c_m, ρ_m, λ_m are the thermophysical parameters of sub-areas (specific heat, mass density, and thermal conductivity), q_{vm}-capacity of internal heat sources in $D_m(t)$ sub-areas, at the same time $q_{vm}=0$ for $m \neq 2$.

It can be shown that the system of equations (1) can be transformed to the form [1]

$$x \in D_m(t): \qquad C_m \rho_m \frac{\partial T_m}{\partial t} = \lambda_m \nabla^2 T_m \ , \ m=1,2,3, \tag{2}$$

where C_m is the substitute thermal capacity and $C_m = c_m$ for $m \neq 2$, $C_2 = c_2 - L\frac{dS}{dT}$, L is the latent heat, S-a function describing the volumetric solid state fraction at the point $x \in D_2(t)$. Assuming, that S is a linear function of the form

$$S = \frac{T_L - T}{T_L - T_S} \tag{3}$$

we obtain

$$C_m(t) = \begin{cases} c_1 \ , & T > T_L \ , \quad (m=1) \\ c_2 + \dfrac{L}{T_L - T_S}, & T \in \langle T_S, T_L \rangle \ , \quad (m=2) \ . \\ c_3 \ , & T < T_S \ , \quad (m=3) \end{cases} \tag{4}$$

The temperature field in the mould sub-area is determined by the equation

$$x \in D_4(t): \qquad c_4 \rho_4 \frac{\partial T_4}{\partial t} = \lambda_4 \nabla^2 T_4 \ . \tag{5}$$

The system of equations(2),(5) is supplemented by the following boundary and initial conditions (compare Figure 1)

$$x \in \Gamma_{12}(t): \begin{cases} T_1 = T_2 = T_L \\ \lambda_1 \dfrac{\partial T_1}{\partial n} = \lambda_2 \dfrac{\partial T_2}{\partial n} \end{cases}, \qquad x \in \Gamma_{23}(t): \begin{cases} T_2 = T_3 = T_S \\ \lambda_2 \dfrac{\partial T_2}{\partial n} = \lambda_3 \dfrac{\partial T_3}{\partial n} \end{cases}, \tag{6}$$

$$x \in \Gamma_c(t): \begin{cases} T_m = T_4 \\ \lambda_m \dfrac{\partial T_m}{\partial n} = \lambda_4 \dfrac{\partial T_4}{\partial n} \end{cases}, \quad m=1 \cup 2 \cup 3, \tag{7}$$

$$x \in \Gamma_{40}: \quad -\lambda_4 \frac{\partial T_4}{\partial n} = \alpha(T_4 - T_\infty), \tag{8}$$

$$t=0: \quad T_1 = T_p, \qquad T_4 = T_0 , \tag{9}$$

where $\frac{\partial T}{\partial n}$ is a normal derivative, α-heat transfer coefficient, T_∞-ambient temperature, T_p-pouring temperature, T_0-initial temperature in the mould sub-area.

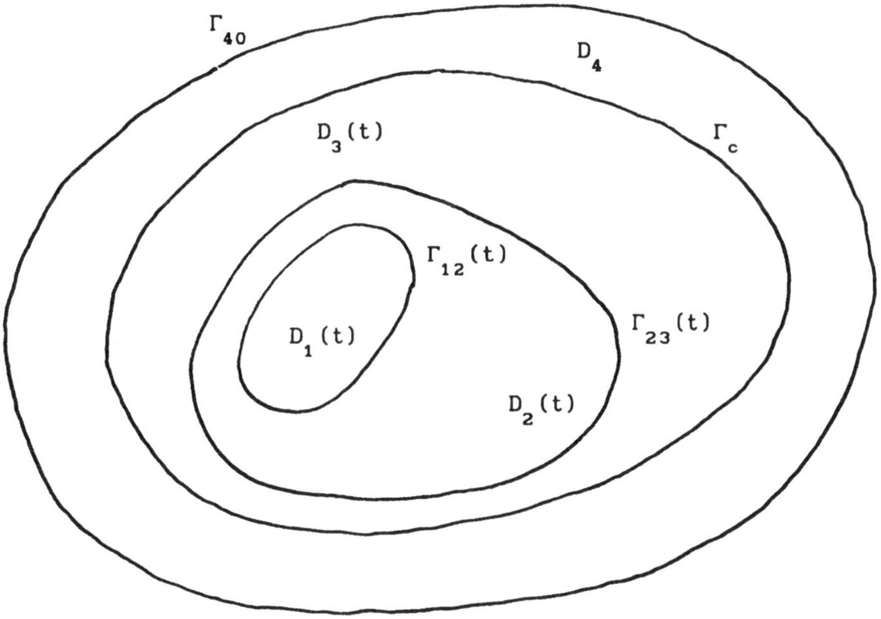

Fig. 1. The considered areas

In the numerical realization the system casting-mould has been reduced to the system molten metal-mould and the adequate computations have been corrected in a way that takes into account real conditions of solidification and cooling in the heterogenous casting area.
Details of this procedure will be presented in the next chapter.

CORRECTION OF TEMPERATURE FIELD IN THE CASTING AREA

Consider, now, two identical areas D and $\bar{D}$. The thermophysical parameters of D are equal to c, ρ, λ whereas for $\bar{D}$ area $\bar{c}$, $\bar{\rho}$, $\bar{\lambda}$. At the same time it is assumed that $\lambda=\bar{\lambda}$. The boundary conditions given on contours Γ_I, Γ_{II} and $\bar{\Gamma}_I$, $\bar{\Gamma}_{II}$ are the same - compare Figure 2.

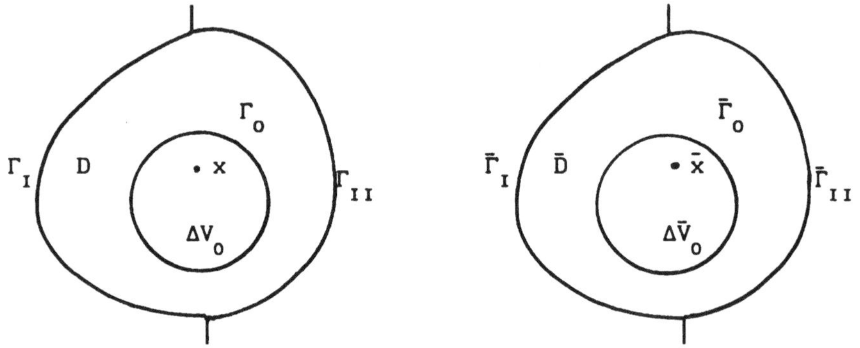

Fig.2. The areas D and $\bar{D}$

The temperature field in D and $\bar{D}$ determine the equations

$$c\rho\frac{\partial T}{\partial t} = \lambda\nabla^2 T \,, \qquad \bar{c}\bar{\rho}\frac{\partial \bar{T}}{\partial t} = \lambda\nabla^2\bar{T}. \tag{10}$$

A new variable is introduced

$$\tau = c\rho t/\bar{c}\bar{\rho}. \tag{11}$$

Because $\frac{\partial \bar{T}}{\partial t} = \frac{\partial \bar{T}}{\partial \tau}\frac{d\tau}{dt} = c\rho\frac{\partial \bar{T}}{\partial \tau}/\bar{c}\bar{\rho}$ therefore equation (11b) can be transformed to the form

$$c\rho\frac{\partial \bar{T}}{\partial \tau} = \lambda\nabla^2\bar{T}. \tag{12}$$

It should be pointed out that in formal substance equations describing heat transfer processes in D and $\bar{D}$ areas are identical although the time in D and $\bar{D}$ "runs" in a different way.

In Figure 3 the cooling curves in selected points $x \in D$ and $\bar{x} \in \bar{D}$ are shown. Slope of the chords marked in Figure 3 are the same, which means that

$$\frac{\Delta T}{\Delta t} = \frac{\Delta \bar{T}}{\Delta \tau} = \frac{\bar{c}\bar{\rho}\Delta\bar{T}}{c\rho\Delta t}. \tag{13}$$

and $c\rho\Delta T = \bar{c}\bar{\rho}\Delta\bar{T}$.

So, if T_1^{f-1} denotes the temperature at point x_1 at moment t^{f-1} whereas T_1^f temperature at the same point for $t=t^f$ then assuming the identical initial (pseudo-initial) conditions for D and $\bar{D}$: $T_i^{f-1} = T_i^f$ we obtain

$$c\rho(T_i^f - T_i^{f-1}) = \bar{c}\bar{\rho}(\bar{T}_i^f - T_i^{f-1}) \tag{14}$$

and it is possible to transform the solution for area D to the discrete solution for $\bar{D}$ calculating $\bar{T}_i^f$ from the formula

$$\bar{T}_i^f = T_i^{f-1} - c\rho(T_i^f - T_i^{f-1})/\bar{c}\bar{\rho}. \tag{15}$$

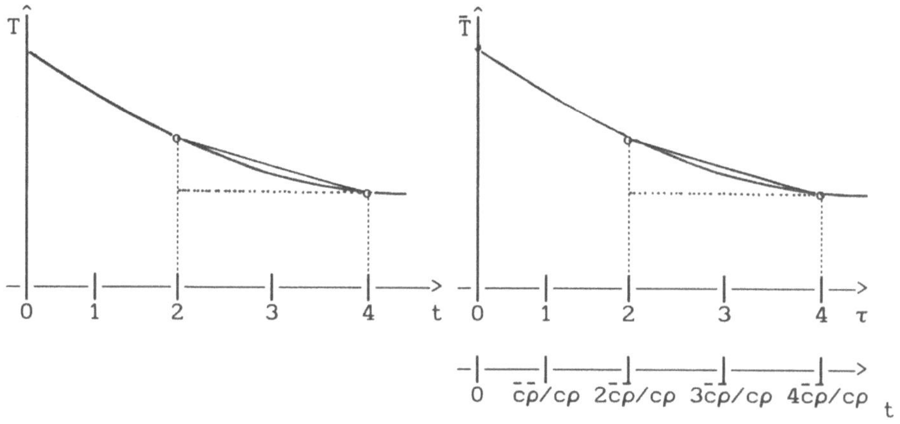

Fig. 3. Cooling curves at points x and $\bar{x}$

The same procedure can be reduced on the basis of energy balances for neighbourhood of points x and $\bar{x}$ (compare [2]).
In case of metal solidifying in an interval of temperature one ought to consider the following transitions (Figure 4)

$$\{1\}\ T_i^{f-1} > T_L,\ T_i^f > T_L, \qquad \{2\}\ T_i^{f-1} > T_L,\ T_S < T_i^f < T_L,$$

$$\{3\}\ T_i^{f-1} > T_L,\ T_i^f < T_S, \qquad \{4\}\ T_S < T_i^{f-1} < T_L,\ T_S < T_i^f < T_L, \tag{16}$$

$$\{5\}\ T_S < T_i^{f-1} < T_L,\ T_i^f < T_S, \qquad \{6\}\ T_i^{f-1} < T_S,\ T_i^f < T_S.$$

It should be pointed out that temperature T_i^f has been found for a homogenous area (molten metal) with parameters c_1 and ρ_1.
For the points fulfilling condition {1} temperature values are not corrected.

In cases {2}-{6}:

$$\{2\}\quad \bar{T}_i^f = T_L - c_1\rho_1(T_L - T_i^f)/c_2\rho_2,$$

$$\{3\}\quad \bar{T}_i^f = T_S + L\rho_2/c_3\rho_3 - c_1\rho_1(T_L - T_i^f)/c_3\rho_3,$$

$$\{4\}\quad \bar{T}_i^f = T_i^{f-1} - c_1\rho_1(T_i^{f-1} - T_i^f)/c_2\rho_2, \tag{17}$$

$$\{5\}\quad \bar{T}_i^f = T_S + L\rho_2/c_3\rho_3 - c_2\rho_2(T_L - T_i^{f-1})/c_3\rho_3 - c_1\rho_1(T_i^{f-1} - T_i^f)/c_3\rho_3,$$

$$\{6\}\quad \bar{T}_i^f = T_i^{f-1} - c_1\rho_1(T_i^{f-1} - T_i^f)/c_3\rho_3,$$

where $\bar{T}_i^f$ is the corrected value of temperature.

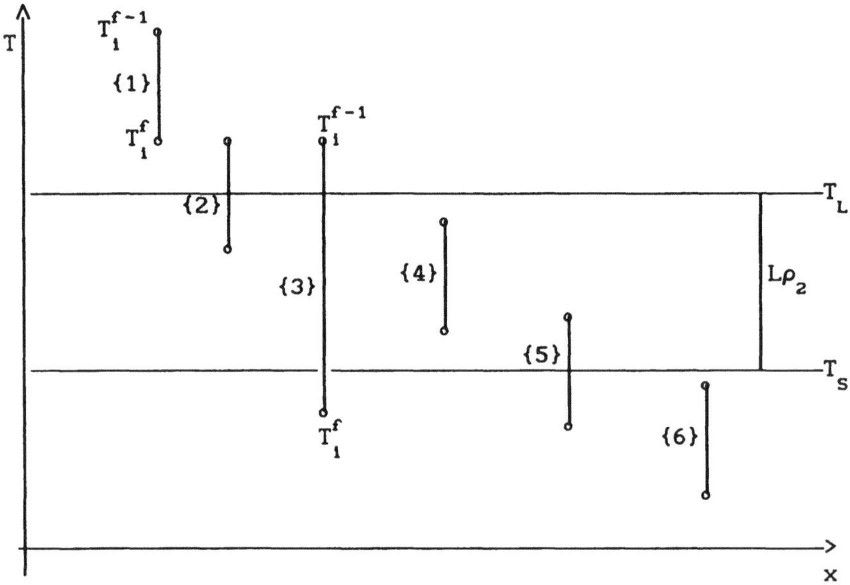

Fig.4. Correction of temperature field

Temperature field $\bar{T}_i^f$ obtained in this way creates pseudo-tial condition for the next step of computations.

COUPLING OF 1ST AND 2ND SCHEME OF BEM

As it is presented above, the non-steady temperature field in the casting volume can be found on the basis of numerical solution concerning homogenous area D (molten metal). The adequate algorithm must be coupled with the procedure identifying the temperature field in the volume of mould (interaction between D and D_4 is determined by the 4th kind boundary condition).
Adequate integral equations for the considered system are as

follows [3]

$$\begin{cases} c(\xi)T_1(\xi,t^F) + \dfrac{1}{\rho_1 c_1}\displaystyle\int_{t^0}^{t^F}\int_{\Gamma_1} T_1^*(\xi,x,t^F,t)q_1(x,t)d\Gamma_1(x)dt = \\ =\dfrac{1}{\rho_1 c_1}\displaystyle\int_{t^0}^{t^F}\int_{\Gamma_1} q_1^*(\xi,x,t^F,t)T_1(x,t)d\Gamma_1(x)dt+\int_D T_1^*(\xi,x,t^F,t^0)T_1(x,t^0)dD(x) \\ c(\xi)T_4(\xi,t^F) + \dfrac{1}{\rho_4 c_4}\displaystyle\int_{t^0}^{t^F}\int_{\Gamma_4} T_4^*(\xi,x,t^F,t)q_4(x,t)d\Gamma_4(x)dt = \\ =\dfrac{1}{\rho_4 c_4}\displaystyle\int_{t^0}^{t^F}\int_{\Gamma_4} q_4^*(\xi,x,t^F,t)T_4(x,t)d\Gamma_4(x)dt+\int_{D_4} T_4^*(\xi,x,t^F,t^0)T_4(x,t^0)dD_4, \end{cases} \tag{18}$$

where $[t^0, t^F]$ is the analysed interval of time, T_1^*, T_4^* - fundamental solutions. For the objects oriented in rectangular coordinate system they are of the form

$$\begin{cases} T_1^*(\xi,x,t^F,t) = \begin{cases} \dfrac{1}{4\pi a_1(t^F-t)}\exp\left(-\dfrac{r^2}{4a_1(t^F-t)}\right), & t\le t^F \\ 0, & t>t^F \end{cases}, \\ T_4^*(\xi,x,t^F,t) = \begin{cases} \dfrac{1}{4\pi a_4(t^F-t)}\exp\left(-\dfrac{r^2}{4a_4(t^F-t)}\right), & t\le t^F \\ 0, & t>t^F \end{cases}, \end{cases} \tag{19}$$

at the same time $r=\sqrt{(x_1-\xi_1)^2+(x_2-\xi_2)^2}$ is the distance from the considered point $x=(x_1,x_2)$ to point $\xi=(\xi_1,\xi_2)$ where the concetrated heat source is applied, $a_1=\lambda_1/c_1\rho_1$, $a_4=\lambda_4/c_4\rho_4$ and $q_1^*=-\lambda_1\dfrac{\partial T_1^*}{\partial n}$, $q_4^*=-\lambda_4\dfrac{\partial T_4^*}{\partial n}$.

Sub-areas D and D_4 are characterized by essentially different thermophysical parameters and it is difficult to obtain by the BEM sufficiently precise numerical solution (the choice of a suitable interval of time). In the paper [4] C.P.Hong, T.Umeda and Y.Kimura proposed to introduce to the algorithm the substitute thermal diffusivity of D_4 area instead of the actual value of a_4 and to correct the temperatures in the mould volume in every step of time. The author of this article realized numerical experiments in this range and the results of computations were not fully satisfactory. However, it turned out that the idea of coupling two schemes of the BEM, namely the 1st scheme for D and the 2nd for D_4 (with formally assumed initial condition in the form $x \in D: T(x,0)=T_p-T_0, x \in D_4: T(x,0)=0$) assures a very good accuracy of numerical solution (compare further chapters).
It is assumed that t^{f-1} and t^f denote two distinguished points of the time lattice Δ_t: $0=<t^0<t^1<\ldots<t^{f-1}<t^f<\ldots t^F<\infty$. Boundary Γ_1 of area D is divided into N_1 boundary constant elements Γ_{1j}, $j=1,\ldots N_1$, boundary Γ_4 of area D is divided into N_4 boundary constant elements Γ_{4j}, $j=1,\ldots N_4$, at the same time interior D is divided into L internal cells D_l and then a numerical approximation of Equations (18) is of the form

$$
\begin{cases}
c^i[T_1(\xi^i,t^F)-T_0]+\dfrac{1}{\rho_1 c_1}\displaystyle\sum_{j=1}^{N_1}\int_{\Gamma_{1j}}\left(\int_{t^{F-1}}^{t^F} T_1^*(\xi^i,x^j,t^F,t)dt\right)d\Gamma_{1j}q_1(x^j,t^F)= \\
=\dfrac{1}{\rho_1 c_1}\displaystyle\sum_{j=1}^{N_1}\int_{\Gamma_{1j}}\left(\int_{t^{F-1}}^{t^F} q_1^*(\xi^i,x^j,t^F,t)dt\right)d\Gamma_{1j}[T_1(x^j,t^F)-T_0] + \\
+\displaystyle\sum_{l=1}^{L}\int_{D_l}T_1^*(\xi^i,x^l,t^F,t^{F-1})dD_l\ [T_1(x^l,t^{F-1})-T_0]\ ,\ \xi^i\in D, \\
c^i[T_4(\xi^i,t^F)-T_0]+\dfrac{1}{\rho_4 c_4}\displaystyle\sum_{f=1}^{F}\sum_{j=1}^{N_4}\int_{\Gamma_{4j}}\left(\int_{t^{f-1}}^{t^f} T_4^*(\xi^i,x^j,t^F,t)dt\right)d\Gamma_{4j}q_4(x^j,t^f) \\
=\dfrac{1}{\rho_4 c_4}\displaystyle\sum_{f=1}^{F}\sum_{j=1}^{N_4}\int_{\Gamma_{4j}}\left(\int_{t^{f-1}}^{t^f} q_4^*(\xi^i,x^j,t^F,t)dt\right)d\Gamma_{4j}[T_4(x^j,t^f)-T_0],\xi^i\in D_4
\end{cases}
\tag{20}
$$

where $c^i=c(\xi^i)=1$ for internal points, $c^i=0.5$ for boundary points (constant boundary elements).

The above boundary integral equations can be rewritten (compare [2,3])

$$\begin{cases} \sum_{j=1}^{N_1} G^1_{ij}\, q^F_{1j} = \sum_{j=1}^{N_1} H^1_{ij} [T^F_{1j} - T_0] + \sum_{l=1}^{L} P^1_{il} [T^{F-1}_{1l} - T_0], \quad i=1,\dots,N_1 \\ \sum_{f=0}^{F-1} \sum_{j=1}^{N_4} G^{f,4}_{ij}\, q^{F-f}_{4j} = \sum_{f=0}^{F-1} \sum_{j=1}^{N_4} H^{f,4}_{ij} [T^{F-f}_{4j} - T_0] \quad , \quad i=1,\dots,N_4 \end{cases} \tag{21}$$

The following denotations will be used:

- T^F_1, q^F_1 - the values of temperatures and heat fluxes at the nodes of external boundary $\Gamma_1 - \Gamma_c$ at the moment t^F,
- T^F_{1c}, q^F_{1c} -the values of temperatures and heat fluxes at the nodes of common boundary Γ_c between casting and mould on the casting side at the moment t^F,
- T^F_4, q^F_4 - the values of temperatures and heat fluxes at the nodes of external boundary $\Gamma_4 - \Gamma_c$ at the moment t^F,
- T^F_{4c}, q^F_{4c} -the values of temperatures and heat fluxes at the nodes of common boundary Γ_c between casting and mould on the mould side at the moment t^F.

Then the integral equations (21) can be written in the matrix form

$$\begin{cases} [\mathbb{G}^1 \quad \mathbb{G}^1_c] \begin{bmatrix} q^F_1 \\ q^F_{1c} \end{bmatrix} = [\, \mathbb{H}^1 \quad \mathbb{H}^1_c \,] \begin{bmatrix} T^F_1 \\ T^F_{1c} \end{bmatrix} + [\mathbb{P}^1][T^{F-1}_1 - T_0] \\ [\mathbb{G}^4 \quad \mathbb{G}^4_c] \begin{bmatrix} q^F_4 \\ q^F_{4c} \end{bmatrix} = [\, \mathbb{H}^4 \quad \mathbb{H}^4_c \,] \begin{bmatrix} T^F_4 \\ T^F_{4c} \end{bmatrix} + \sum_{f=1}^{F-1} [\mathbb{H}^{f,4} \quad -\mathbb{G}^{f,4}] \begin{bmatrix} T^f_4 - T_0 \\ q^f_4 \end{bmatrix} \end{cases} \tag{22}$$

Assuming the 4th kind condition on the common boundary Γ_c:

$$x \in \Gamma_c : \quad \begin{cases} T^F_{1c} = T^F_{4c} = T^F_c \\ q^F_{1c} = -q^F_{4c} = q^F_c \end{cases} \tag{23}$$

one obtains the following system of equations

$$\begin{bmatrix} \mathbb{G}^1 & -\mathbb{H}^1_c & \mathbb{G}^1_c & 0 \\ 0 & -\mathbb{H}^4_c & -\mathbb{G}^4_c & \mathbb{G}^4 \end{bmatrix} \begin{bmatrix} q^F_1 \\ T^F_c - T_0 \\ q^F_c \\ q^F_4 \end{bmatrix} = \begin{bmatrix} \mathbb{H}^1 & 0 \\ 0 & \mathbb{H}^4 \end{bmatrix} \begin{bmatrix} T^F_1 - T_0 \\ T^F_4 - T_0 \end{bmatrix} + \tag{24}$$

$$+ \begin{bmatrix} [\mathbb{P}^1][T^{F-1}_1 - T_0] \\ 0 \end{bmatrix} + \begin{bmatrix} 0 \\ \sum_{f=1}^{F-1} [\mathbb{H}^{f,4} \quad -\mathbb{G}^{f,4}] \begin{bmatrix} T^f_4 - T_0 \\ q^f_4 \end{bmatrix} \end{bmatrix},$$

from which the unknown values of temperature and heat fluxes in the nodes on Γ_c boundary and missing values of temperature and heat fluxes in the remaining boundary nodes are calculated. The temperature field in internal nodes ξ^i results from the formulas

$$\begin{cases} \xi^i \in D: \; T_1(\xi^i, t^F) = \sum_{j=1}^{N_1} H^1_{ij} [T^F_{1j} - T_0] - \sum_{j=1}^{N_1} G^1_{ij} \; q^F_{1j} + \sum_{l=1}^{L} P_{il} [T^{F-1}_{1l} - T_0], \\ \xi^i \in D_4: \; T_4(\xi^i, t^F) = \sum_{f=0}^{F-1} \sum_{j=1}^{N_4} H^{f,4}_{ij} \; [T^{F-f}_{4j} - T_0] - \sum_{f=0}^{F-1} \sum_{j=1}^{N_4} G^{f,4}_{ij} \; q^{F-f}_{4j}. \end{cases} \tag{25}$$

The obtained temperature field is corrected by adding to the solution the previously substracted temperature T_0.

Numerical experiments show that the accuracy of solution improves significantly when the analysed object D is a convex one. This fact makes it neccessary to divide the considered area into the convex sub-areas (for example line A-A in Figure 7). The numerical procedure of sub-areas coupling is identical as in the case of 4th kind boundary condition with ideal contact, at the same time the thermophysical parameters of considered parts are the same.

EXAMPLES OF NUMERICAL REALIZATION

In order to verify the idea of coupling the 1st and 2nd scheme of BEM for sub-areas with essentially different thermophysical parameters the following numerical experiment has been executed. Two half-spaces separated by a plane (compare Figure 5) have been considered.

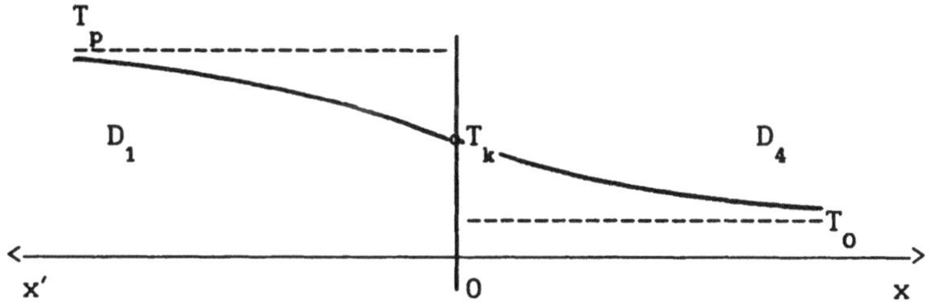

Fig.5. Temperature field in D_1 and D_4

The heat exchange in this system is described by the equations

$$\frac{\partial T_1(x',t)}{\partial t} = a_1 \frac{\partial^2 T_1(x',t)}{\partial x'^2} \quad , \quad \frac{\partial T_4(x,t)}{\partial t} = a_4 \frac{\partial^2 T_4(x,t)}{\partial x^2}, \tag{27}$$

with boundary conditions

$$x = \infty : \; T_4(x,t) = T_0 \quad , \qquad x' = \infty : \; T_1(x',t) = T_p \; , \tag{28}$$

$$x = x' = 0 : \; -\lambda_1 \frac{\partial T_1(x',t)}{\partial x'} = \lambda_4 \frac{\partial T_4(x,t)}{\partial x}, \; T_1(x',t) = T_4(x,t) \tag{29}$$

and initial condition

$$t = 0 : \quad T_1(x',t) = T_p \quad , \qquad T_4(x,t) = T_0 \; . \tag{29}$$

It should be pointed out that this model can be solved by analytical methods [5].
The same example has been solved numerically, in particular two sub-areas (metal and sand mould with thicknesses 0.4 m) have been divided into 20 internal linear elements, at the same time it was assumed that for x=0.4: $-\lambda_1 \frac{\partial T}{\partial x}1 = 0$ and $x' = 0.4$: $-\lambda_4 \frac{\partial T}{\partial x}4 = 0$.
The comparision between analytical and numerical results is shown in Figure 6 and this verification was in all respects

favourable.

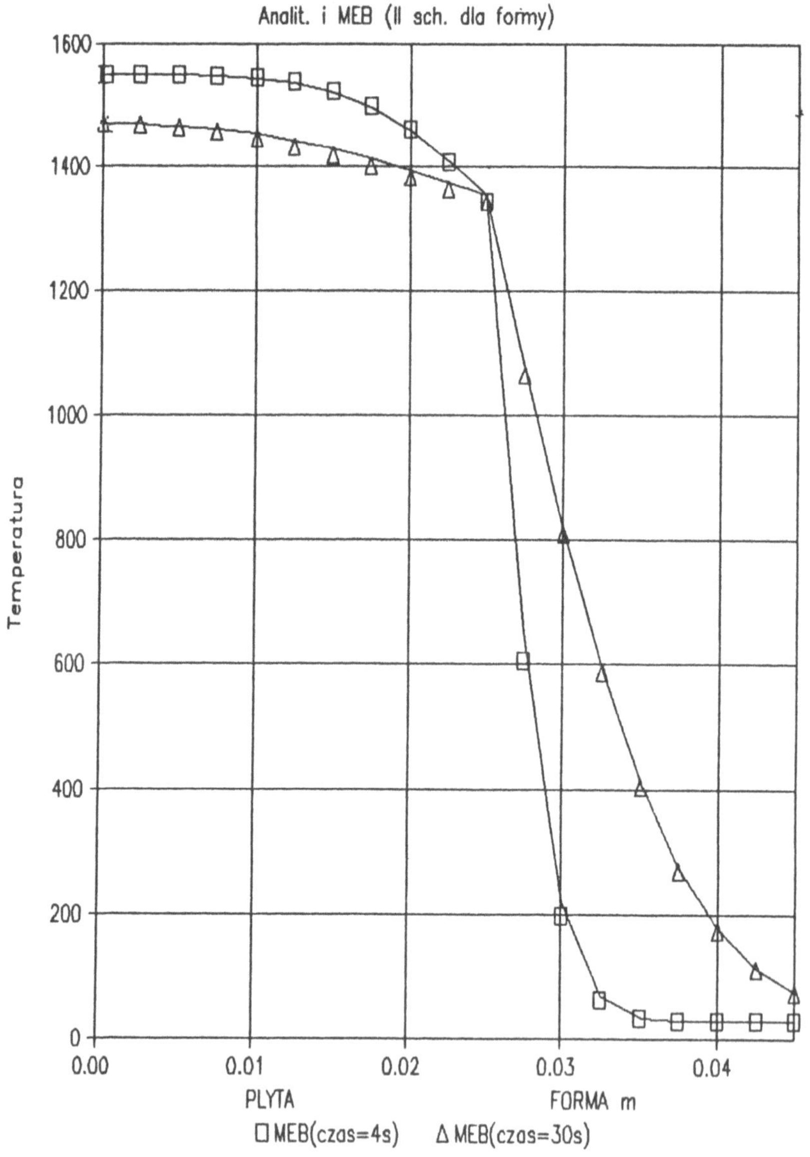

Fig. 6. Temperature field in D and D_4 sub-areas

In the second stage of investigations the examples of the method utilization in the typical practical foundry problems have been worked out [2]. One of them will be below presented.

The steel thermal center of T type shown in Figure 7 has been considered.

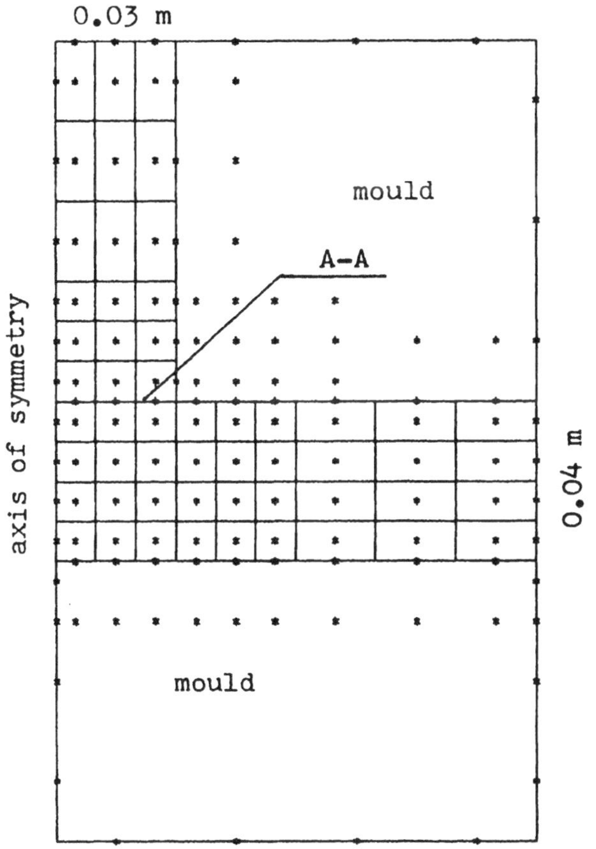

Fig 7. The thermal center of T type

The thermal processes in the casting and mould sub-areas are described by a system of parabolic equations (2),(5) with boundary conditions (6) and condition in the form $-\lambda_m \frac{\partial T_m}{\partial x}=0$ on the arbitrary distinguished rectangle bordering the considered object. The initial condition in the form t=0: $T_1=T_p$, $T_4=T_0$ has been assumed.

Because the casting area is not a convex one, therefore the thermal center was formally divided into two convex parts (line A-A in Figure 7). As it was mentioned the 1st scheme of the BEM supplemented by correction procedures for casting area and 2nd scheme for the mould sub-areas have been used. In the Figures 8 and 9 the calculated temperature fields for 60s, 120s are shown.

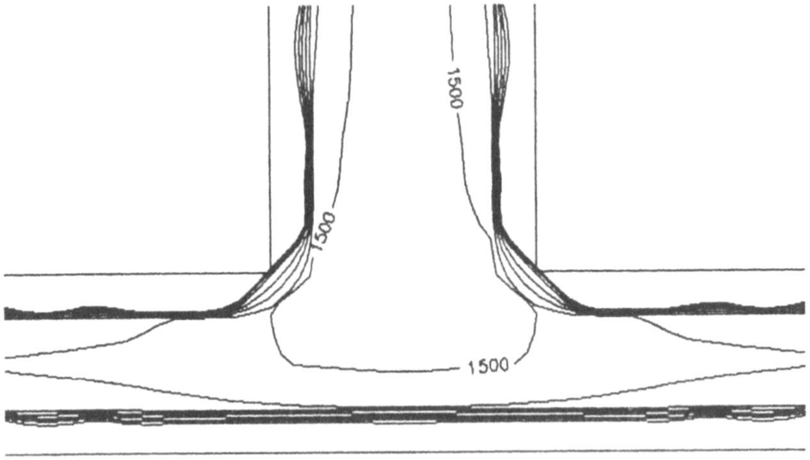

Fig. 8. Temperature field for t=60s

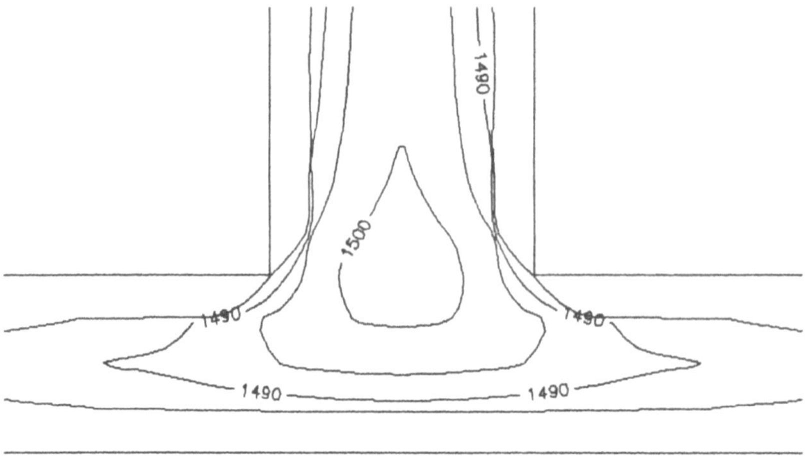

Fig. 9. Temperature field for t=120s

References

1. Mochnacki, B., Substitute Thermal Capacity of Metal Solidifying in an Interval of Temperature, Bull. of the Pol. Ac. of Sc., Techn. Sc., Vol. 32, No 3-4, pp. 127-143, 1984

2. Majchrzak, E., Utilization of the Boundary Element Method in the Thermal Theory of Foundry, qualifying work, Publications of Silesian Techn. Univ., Mechanics No 102, Gliwice, 1991

3. Brebbia, C. A., Telles, J. C. F. and Wrobel, L. C., Boundary Element Techniques, Springer-Verlag, Berlin, Hildeberg, New York and Tokyo, 1984

4. Hong, C. P., Umeda, T. and Kimura, Y., Numerical Models for Casting Solidification: Part I. The Coupling of the BEM and FDM for Solidification Problems, Metall. Trans. B, Vol. 15B, pp. 91-99, 1984

5. Rubinstein, L. I., Stefan's Problem, Zwajgzne, Riga, 1967

Free Surfaces Computation in a Steady-State Finite Element Model for Tube Rolling

M. Vacance, E. Massoni, J.L. Chenot

Centre de Mise en Forme des Matériaux (CEMEF), Ecole des Mines de Paris, Sophia Antipolis, 06560 Valbonne, France

ABSTRACT

A three dimensional steady state finite element model is developed to investigate seamless tube hot rolling. The steady state is modelled, so that the equilibrium resolution (flow formulation) must be supplemented by a technique determining the bounds of the volume occupied by the material. In tube rolling several free surfaces (internal and external) exist. Their determination is essential for a good simulation. Streamline method and an optimal control method are tested for this rolling simulation. Comparison of the two methods is made, according to different kinds of free surfaces to be computed.

INTRODUCTION

Tube rolling operation generates complex deformation of the material. The use of a Finite Element model is relevant for stress, strain, temperature and rolling forces determination. A steady state approach is carried out in order to shorten computing time. However free surfaces determination must be done.

The objectives of our code are the external shape determination in the stationnary phase, as a function of process parameters. Several free surfaces are determined. Internal and external surfaces of the tube are computed before the roll bite where pre-deformation occurs, and along the roll gap where bulging occurs. Contact zones are deduced from this determination.

Although equilibrium and shape determination interfer, resolution is made iteratively and independently on the two points up to global convergence. Two free boundary algorithms have been tested for this purpose. The streamline method and an optimal control method. Comparison of the two methods is made. This shape detemination on one stand is used to investigate rolling in the following stands.

THE PROCESS

The Multi Stand Pipe Mill is part of an efficient seamless tube hot rolling process used in DALMINE plant. A thick cylindrical preform is mounted on a mandrel and is successively rolled by eight pairs of rolls (fig. 1). In the first stands the material is pressed on the mandrel and lateral bulging occurs. Then the shape is regularised to the tube final dimensions.

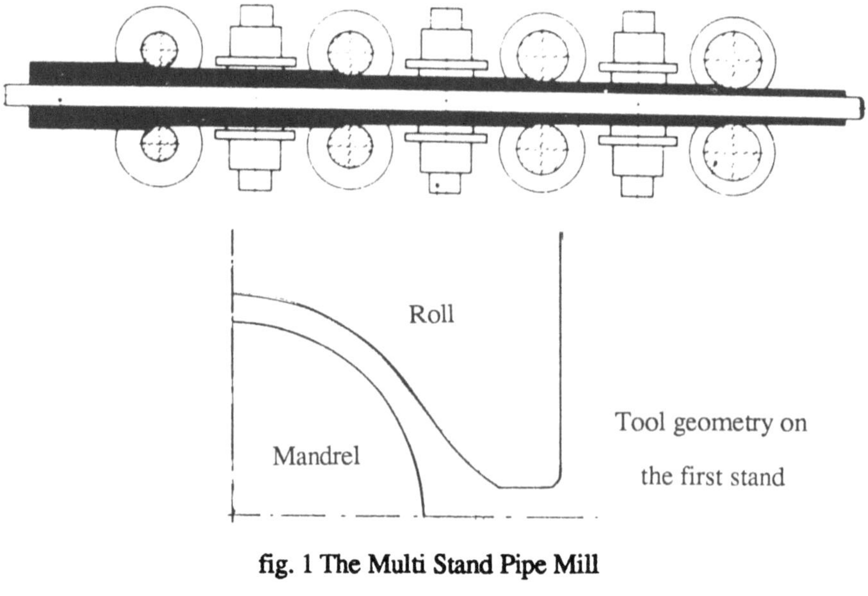

fig. 1 The Multi Stand Pipe Mill

The tube undergoes complex deformation on each stand. A Finite Element model has been developed in order to investigate this process. Rolling forces and torques are determined, as well as the final shape of the tube.

THE MODEL

The MPM finite element model is based on a steady state approach. Computation is made stand by stand, taking into account the presence of the grooved rolls and the mandrel. When computing one stand, the neighbouring stands are represented by front and back tensions. (fig.2)

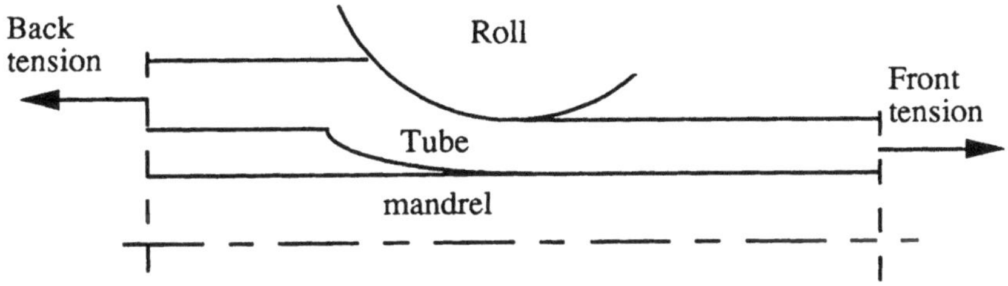

Fig. 2 Front and back tension on the computed stand

For each stand, we only simulate steady state, in order to avoid heavy and long incremental simulation of the transient phase,which have no industrial interest. The position of the material is unknown a priori when steady state is reached, and must be determined by the model. This location may also depend on operating conditions (frictional conditions, speed of the tools...) .

We must consider two kinds of boundary surfaces:

- Surfaces in contact with the tools, the geometry of which is known, but the limits of the contact zones are unknown.
- Surfaces free from any contact the geometry of which is unknown and must be determined.

We then have to solve simultaneously equations permitting the determination of the domain occupied by the material, and equilibrium of the material on this domain.

Although these two points are interdependent we chose an iterative decoupled resolution. The algorithm is presented figure 3.

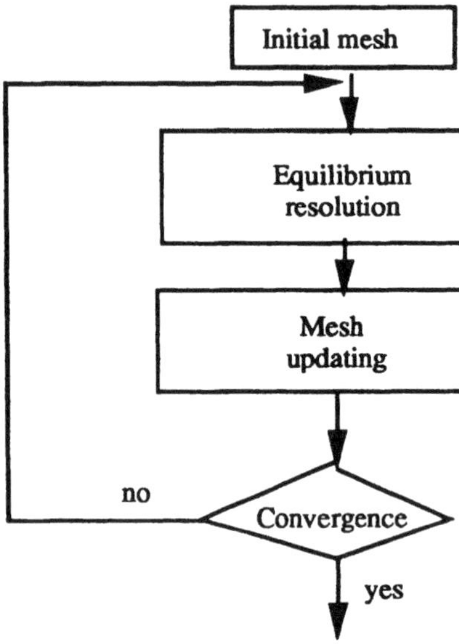

Fig. 3 Resolution algorithm

The flow formulation [1] is used to find the equilibrium velocity field.

<u>Rheology:</u>

Rheology is set as a Norton Hoff incompressible viscoplastic law:

$$\sigma = -p\,\mathbf{I} + 2\,K\,(\sqrt{3}\,\dot{\bar{\varepsilon}})^{\,m-1}\,\dot{\varepsilon} \qquad (1)$$

$$\mathrm{Tr}(\dot{\varepsilon}) = \mathrm{div}\ v = 0 \qquad (2)$$

where σ is the stress tensor, $p = -\mathrm{Tr}(\sigma) / 3$ is the hydrostatic pressure, $\dot{\varepsilon}$ the strain rate tensor associated with the velocity field V, $\dot{\bar{\varepsilon}}$ is the equivalent strain rate.

K, the material consistency, may depend on strain $\bar{\varepsilon}$ and temperature T according to the law:

$$K = A(\bar{\varepsilon} + \bar{\varepsilon}_0)^n \exp(Q/RT)\,.(\sqrt{3})^{-m-1} \qquad (3)$$

where parameters A, n, m, Q and $\bar{\varepsilon}_0$ have been determined experimentally for the steel used.

Tribology:

Friction conditions at the interface is computed by the Norton friction law which links the friction stress τ with the tool-workpiece sliding velocity ΔV by the relation:

$$\tau = -\alpha K \|\Delta V\|^{(p-1)} \Delta V \tag{4}$$

Different coefficients α are used to simulate tube-roll (α_1) and tube-mandrel (α_2) contact.

Resolution

Equilibrium is written in terms of the principle of virtual work where only velocity remains as an unknown:

$$\forall V^* \quad \int_\Omega 2K(\bar{\varepsilon},T)\,(\sqrt{3}\,\dot{\bar{\varepsilon}})^{m-1}\,\dot{\varepsilon} : \dot{\varepsilon}^*\, d\Omega - \int_{S^d} T^d . V^*\, ds$$

$$+ \int_S \alpha\ K(\bar{\varepsilon},T)\,\|\Delta V\|^{p-1}\,\Delta V^*\, ds + \int_\Omega \rho K(\bar{\varepsilon},T)\,\{div(V).div(V^*)\}\, d\Omega = 0 \tag{5}$$

where $\dot{\varepsilon}^* = 1/2\{gradV^* + {}^t gradV^*\}$, S^d is the surface where external stresses T^d are prescribed. A penalty factor $\rho = 10^7$ has been used to enforce incompressibility [2].

Classical discretization by the Finite Element Method leads to a set of non linear equations in the nodal velocities, solved by the Newton Raphson method.

MESH UPDATING: FREE SURFACES COMPUTATION

Several free surfaces must be computed for tube rolling. In the first stand we observe two main zones of free deformation, for both internal and external surface of the tube.

A pre-deformation zone is located before contact with the rolls and the mandrel. However the most important deformation is a lateral bulging between the rolls (fig. 4).

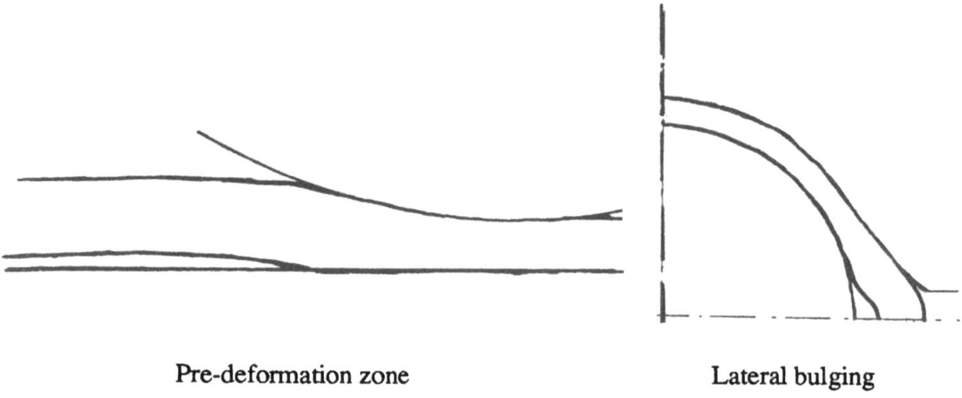

Fig. 4. Deformation zones on the first stand

Determination of the contact zones

Geometry of the contact zone are deduced from the geometry of the tools, assumed perfectly rigid. However we must determine whether a node can be considered in contact or not. In our case we chose a geometrical and force criterion. A node is considered in contact if it touches a tool and if the force it undergoes tends to push it against this tool. These forces are directly deduced from the equilibrium resolution.

The remaining nodes of the boundary are free to move and their correct position must be determined.

Two methods are used in our code in order to update the domain used by the material.

Streamline metod

- The first method derives from streamline method. Its aim is to calculate the position of the nodes step by step by integrating the velocity of a material point along its path along the stand. Relative displacement is given by the set of equations [3,4] :

$$\Delta y = \int (V_y/V_x)\, dx \quad \text{and} \quad \Delta z = \int (V_z/V_x)\, dx \tag{6}$$

where x is the rolling direction. Integration is made line by line,with two Gauss points by linear element. The starting position is given by the tube geometry, far enough before the roll bite.

This simple method gives a good prevision of the lateral bulging on the first stand. The computed surfaces are smooth. Comparison of the computed bulging with experimental measurements is shown in figure 5. Notice the influence of friction coefficients on the bulging.

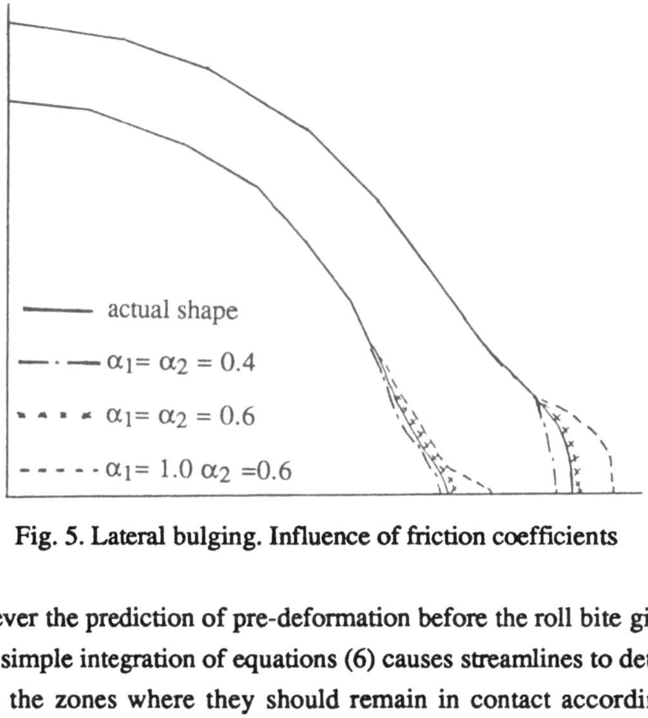

Fig. 5. Lateral bulging. Influence of friction coefficients

However the prediction of pre-deformation before the roll bite gives more trouble. A simple integration of equations (6) causes streamlines to detach from the roll in the zones where they should remain in contact according to the criterion described above. Since it is a local method, the contact zone with the tools is not taken into account for the determination of the free boundary. This problem has already been found [4] , and solutions to avoid this loss of contact depend on the geometry of the problem. They generally consist of operating a post smoothing algorithm in order to get a smooth connexion between the free surface and the contact zone.

The amplitude of loss of contact increases with friction coefficients and the roll speed. We suspect that a part of the problems comes from a poor determination of the velocity field before the roll bite.

Figure 6 shows the deformation of the tube with this method. All external surfaces have been computed. No specific smoothing method has been used to keep contact zone in order to view the importance of the detachment.

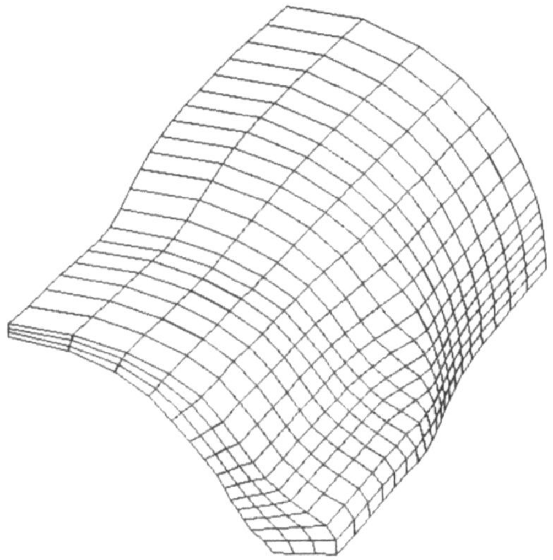

Fig. 6. Global computation with streamline method. Observe detachment

In order to solve this problem we must take into account the presence of contact zones as boundary conditions when determining the free surfaces. A global method has also been tested in our case.

OPTIMAL CONTROL METHOD

This second method is an optimal control method. Since, in a steady state process, no material should cross the external surfaces of the product, this method tries to find the boundary Γ which minimizes the flux of material across it, for a given velocity field.

For this we introduce the functional:

$$j = \int_{\Gamma(X_f)} (V(X_f) . n(X_f))^2 \, ds \qquad (7)$$

which is minimized with respect to the position of the nodes (X_f) by Newton Raphson method. We assume that the nodes are able to move along the normal direction of the previous surface, and we determine the amplitude of the motion.

This method has already been successfully tested in stationnary shape rolling simulation [5, 6]. A more detailed description of this algorithm is given in this proceeding [7]. Although more computational time and more memory is needed, this method presents a greater flexibility to take into account boundary conditions, such as contact zones.

This method has been tested for the prediction of tube bulging only. The results are similar than with the streamline method. Figure 7 compares the results of the two methods.

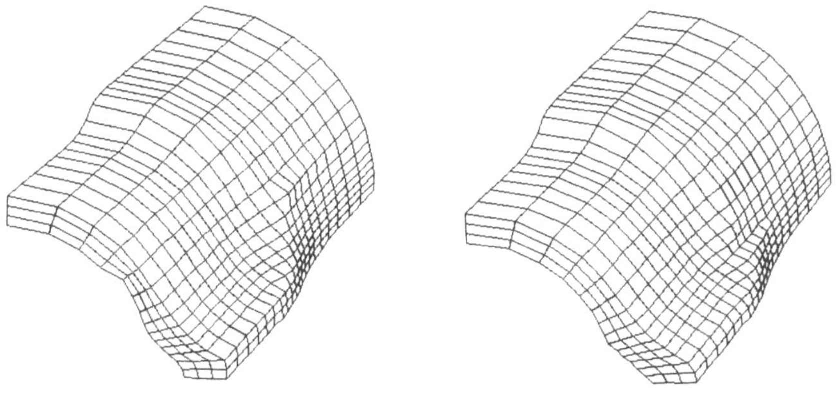

Optimal control method Streamline method

Fig. 7 Comparison of lateral bulging with the two methods

Tests have been made to compute all the free surfaces of the tube with this method. If no particular boundary condition is taken in order to prevent contact zones from motion, we observe a computed shape very close to the one we computed with streamline method. (see figure 6 and 8).

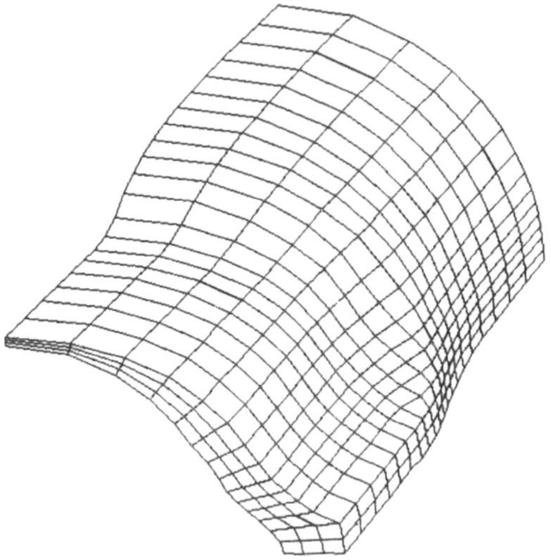

fig. 8. Global computation with optimal control method. Observe detachment.

Now let us prevent any no motion of nodes in contact. The surface we obtain presents a smooth connexion between the free surface and the contact zone, as it is shown on figure 9.

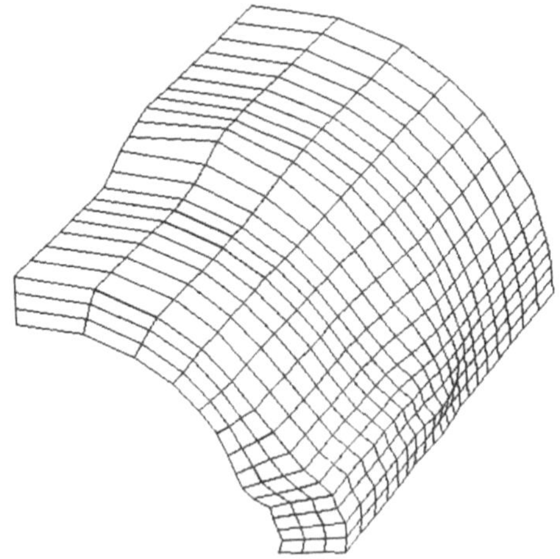

Fig. 9. Global computation . Contact zones are considered.

However the lateral surfaces are irregular. Here indeed the main contribution to the value of the functional in equation (7) (95 % of j) comes from the zone where computation has given a regular surface. Imposing boundary conditions seems to increase the difference of value between the two zones. With these conditions it can be understood that minimization is imperfect in zones where the functional is already very small.

In order to avoid this problem we envisaged an hybrid method. We integrate steamline equations only when the streamlines don't reach any contact zone, and use the optimal control method elsewhere, in order to take into account contact zones as boundary conditions. A computation with this hybrid method is shown on figure 10.

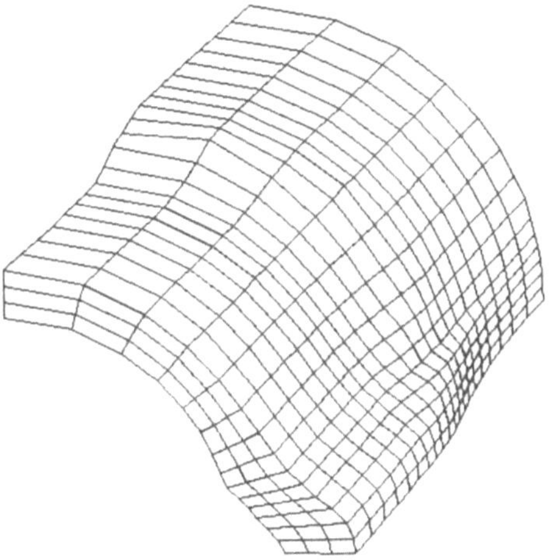

Fig. 10. Hybrid computation of the whole surface.

APPLICATION : COMPUTATION OF THE FOLLOWING STANDS.

The objective of the code is to predict deformation in the eight stands of the mill. Therefore one of the main scope of the free surface computation is the determination of the exact shape of the tube after rolling on one stand, in order to be able to compute rolling in the following stand with accurate initial data. The deformed shape of the tube at the exit of one stand is then used to create the initial mesh for the following stand. Material parameters such as temperature and deformation, determined by our code, are also transmitted to the next stand as initial data.

Figure 11 shows the initialisation of computation of the second stand. The entry section of the mesh used for the second stand correspond to the exit section of the first stand. Deformation on the second stand is shown. Lateral bulging is much smaller than in the first stand.

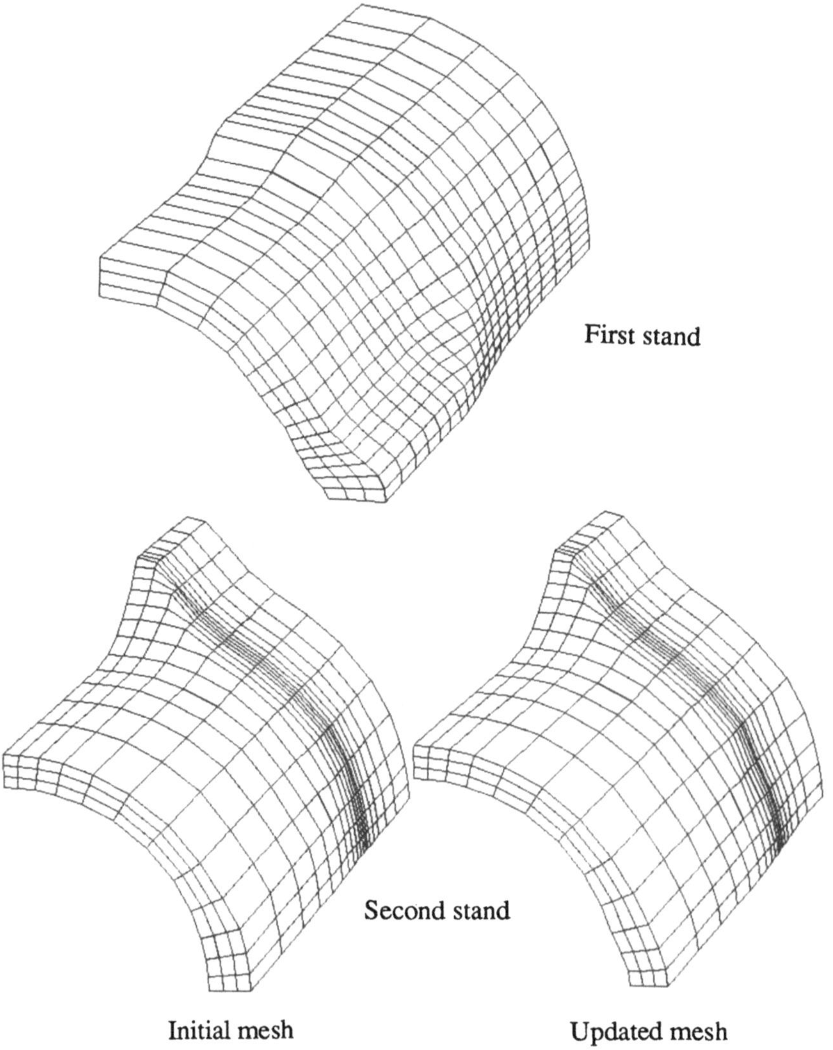

Fig. 11. Use of the free surface determination in the first stand to compute the second one.

CONCLUSION

This paper describes two different free surfaces algorithms used in the field of tube rolling modelling with a steady state approach.

Streamline method is fast and simple to implement. It gives very good results when the streamlines remain free from any contact. However, the local formulation prevents us from taking into account the presence of contact zones.

The optimal control method makes it possible to take into account boundary conditions for the determination of the free surfaces. However, since it is a minimization method, inhomogeneity of flux may causes trouble, and generates irregular surfaces.

A combination of both methods according to the location of the surfaces is shown to be a good compromise.

ACKNOWLEDGEMENT

The code used in this study is developed with financial support from Centro Sviluppo Materiali -Rome, INNSE - Milan, and DALMINE -Dalmine.

REFERENCES

[1] M. Vacance, E. Massoni, J.L. Chenot, Rovelli, C. Cumino, M. Lubrano, Multistand pipe mill finite element model, Journal of Materials Processing Technology, 24 (1990) 421-430 .

[2] O.C. Zienkiewicz , Flow formulation for numerical solution of forming processes. In Numerical Analysis of Forming Processes, 1984, J. Wiley.

[3] R.I.Tanner , R.E. Nickell , R.W. Bilger , Finite Element methods for the solution of some incompressible Non-Newtonian fluid mechanics problems with free surface. Comp. Meth. in app. mech. and Eng. 6 (1975) pp 155-174.

[4] K.Yamada , S. Ogawa ,S. Hamauzu, T. Kikuma : Three-dimensional analysis of mandrel rolling using rigid-plastic Finite Element method.Proc. Numiform 89 (Fort Collins, July 1989), E.G. Thomson et al eds, A.A. Balkema, Rotterdam,1989

[5] C. Bertrand-Corsini , P. Montmitonnet , J.L. Chenot , F.Fau , P.Buessler : A three dimensional analysis of hot rolling with a steady state thermomechanical coupling.Numiform 89 (Fort Collins,July 1989), Thomson et al. (eds) .1989 Balkema, Rotterdam.

[6] P. Montmitonnet, A. Antunes, M. Vacance,P. Buessler, T. Iung, J.L. Chenot , Use of a Finite Element code for optimisation of rolling operations. Proc. ICTP (Kyoto 1990) .

[7] P. Montmitonnet, A. Bern, J.L. Chenot, Spread in hot rolling: an optimal control approach. Proc. present conference.(Southampton july 1991).

Prediction of the Free Surface in a Continuous Casting Mould and its Influence on the Flow Pattern and the Temperature Distribution

A. Preuer

VOEST Alpine Stahl Linz GmbH, A-4010 Linz, Austria

INTRODUCTION

Fluid flow is one of the main physical phenomena in the continuous casting process. Because of the high temperature of liquid steel, it is very hard or impossible to measure the flow pattern. Numerical computations can be used to reduce the experimental effort and are a helpful tool for investigating these phenomena. In the continuous casting process the liquid steel flows through an entry nozzle into the slab caster, Fig.1. The flow pattern has a great influence on the surface quality of the slab as it controls the temperature distribution and therefore the solidification process.

The aim of this investigation is the prediction of the flow pattern and the temperature distribution in the continuous casting mould especially the shape of the free surface which depends on the flow pattern, pressure distribution and, as a special feature in this case, on the casting powder swimming on the steel surface. Free surfaces with swimming casting powder or slag on it play an important role in metallurgy (continuous casting, ladle metallurgy) and therefore the computation of this phenomena will be performed in this analysis. The numerical computation was done with the fluid flow package Phoenics. This code is a general-purpose code for fluid flow and heat transfer. The code is based on the control volume finite difference method. For the prediction of the free surface an additional routine was written and linked to the Phoenics code.

Usually the free surface in continuous casting computations is modeled as a frictionless plane of symmetry, see Flint [1],Yao et al. [2], Thomas [3], Thomas and Mika [4], Flood et al. [5], Matys et al. [6]. As can be seen from the results below an essential difference consists between the flow pattern and therefore between the temperature distribution of the plane and the computed free surface.

PHYSICAL MODEL

The basic equations for the physical model which have to be modeled are the continuity equation, the momentum equations and a transport equation for the temperature (enthalpy), Szekely [7]. The flow behaviour is assumed to be a steady state and the flow domain is three dimensional, Fig.1. The Navier Stokes equations are time averaged because of the "high Reynolds number" flow of liquid steel even at low velocities. The closure of the equations was performed with the k-ε turbulence model, see Rodi [8]. After time averaging and under the assumptions that density and laminar viscosity are constant the governing equations read.

continuity

$$\frac{\partial u_i}{\partial x_i} = 0 \tag{1}$$

momentum

$$u_j \frac{\partial u_i}{\partial x_j} = -\frac{1}{\rho_{st}} \frac{\partial p}{\partial x_i} + \frac{\partial}{\partial x_j}\left(\nu_l \frac{\partial u_i}{\partial x_j} + \nu_t\left(\frac{\partial u_i}{\partial x_j} + \frac{\partial u_j}{\partial x_i}\right) - \frac{2}{3} K \delta_{ij}\right) + \rho_{st} g_i \tag{2}$$

turbulent kinetic energy K

$$u_i \frac{\partial K}{\partial x_i} = \frac{\partial}{\partial x_i}\left(\frac{\nu_t}{\sigma_K} \frac{\partial K}{\partial x_i}\right) + \nu_t\left(\frac{\partial u_i}{\partial x_j} + \frac{\partial u_j}{\partial x_i}\right)\frac{\partial u_i}{\partial x_j} - \varepsilon \tag{3}$$

dissipation rate ε

$$u_i \frac{\partial \varepsilon}{\partial x_i} = \frac{\partial}{\partial x_i}\left(\frac{\nu_t}{\sigma_\varepsilon} \frac{\partial \varepsilon}{\partial x_i}\right) + c_{1\varepsilon}\frac{\varepsilon}{K}\left(\frac{\partial u_i}{\partial x_j} + \frac{\partial u_j}{\partial x_i}\right)\frac{\partial u_i}{\partial x_j} - c_{2\varepsilon}\frac{\varepsilon^2}{K} \tag{4}$$

The turbulent viscosity ν_t is calculated from

$$\nu_t = c_\mu \frac{K^2}{\varepsilon} \tag{5}$$

u_i ... velocity component in direction i m/sec
p ... pressure N/m^2
ρ_{st} ... density of steel kg/m^3
ν_l ... laminar viscosity m^2/sec
ν_t ... turbulent viscosity m^2/sec
g_i ... gravitational acceleration in direction i m/sec

c_μ, $c_{1\varepsilon}$, $c_{2\varepsilon}$, σ_ε, σ_K ... model constants, see Rodi [8]

The natural convection can be neglected, hence the temperature distribution can be predicted with the known velocity field and turbulent viscosity computed before. Therefore the transport equation for the temperature is linear for constant specific heat and constant thermal conductivity.

transport equation for temperature ($T = \frac{h}{c_p}$)

$$\rho_{st} u_i \frac{\partial T}{\partial x_i} = \frac{\partial}{\partial x_i}\left(\left(\frac{\rho_{st}\nu_l}{Pr_l} + \frac{\rho_{st}\nu_t}{Pr_t}\right)\frac{\partial T}{\partial x_i}\right) \tag{6}$$

T ... temperature °C
h ... sensible enthalpy J/kg
Pr_l ... laminar Prandtl Number = $\frac{\rho_{st} \nu_t c_p}{\lambda}$
c_p ... specific heat J/kg K
λ ... thermal conductivity W/m K
Pr_t ... turbulent Prandtl Number = 1, see [8]

Further assumptions
- For symmetric outflow conditions through the entry nozzle the model is double symmetric, only one quarter of the mould has to be modeled, Fig.1.
- For asymmetric outflow conditions the model is symmetric, the half mould has to be modeled.
- in casting direction the model extension is three times the mould length = 2.4 m.
- casting size (cross section) = 1.35/0.21 m
- viscosity does not depend on temperature
- the fluid flow is modeled only within the solidified shell, Fig.1

Material properties

density ρ_{st}	= 7000	kg/m^3
laminar viscosity ν_l	= 0.000001	m^2/sec
specific heat c_p	= 750	J/kg K
thermal cond. λ	= 30	W/m K

For reasons of security, no detailed information about the entry nozzle geometry nor the boundary conditions can be given.

CALCULATION OF THE FREE STEEL SURFACE

See Fig.1
As already mentioned above, the free surface is usually treated as a frictionless plane of symmetry. In this analysis the shape of the free surface will be computed. For the casting process a casting powder addition on the steel surface is necessary. Because of the high temperature and the lower density of the powder it will be melt and float on the steel surface.

The general condition on the interface between two fluids (fluid 1 and fluid 2) can be found in Landau and Lifschitz [9].

$$\left[p^{(1)} - p^{(2)}\right] n_i = \left[\sigma'^{(1)}_{ij} - \sigma'^{(2)}_{ij}\right] n_j + \gamma \left(\frac{1}{R_1} + \frac{1}{R_2}\right) n_i + \frac{\partial \gamma}{\partial x_i} \quad (7)$$

pressure viscous force surface tension

where

p ... pressure in the fluid N/m^2
σ'_{ji} ... viscous part of the stress tensor N/m^2
γ ... surface tension N/m
$1/R_1 + 1/R_2$ ... mean Gaussian curvature of the interface 1/m
n_i ... components of the unit normal to the interface

In the first stage some simplifications can be made for the special case in this analysis. The slope of the free surface is small compared with the extension of the mould, so the unit normal to the interface can be fixed parallel to the direction of gravitation. Further, the Gaussian curvature and also the surface tension (=1.7 N/m) are negligible in comparison with the pressure difference at the free surface (=500-1000 N/m^2, depending on the inlet velocity). This can be done also for the viscous force because of the low viscosity. Hence equation (7) in the gravitational direction reduces to

$$p_1 - p_2 = 0 \tag{8}$$

This means, in this case the pressure in the steel bath and the casting powder pressure are equal at the interface. Further simplifications at this point are to assume that no interface friction consists between steel and powder, the surface of the powder is plane and a minimum thickness of the powder layer is achieved (=2 cm), see Fig.1. This defines the following boundary condition at the interface = steel surface.

$$P(x) = \rho_G \, g \, h_G(x) \tag{9}$$

where

p(x) ... internal pressure of steel at point x at the surface
ρ_G ... density casting powder = 2000 kg/m^3
g ... gravitational acceleration
$h_G(x)$ ... casting powder height at point x

The normal velocity is zero and all other variables, except temperature, have a zero normal gradient. For the t emperature boundary condition see below.

Computation of the free surface

First the velocity field and the pressure distribution are computed with the plane surface. They do not satisfy the pressure boundary condition. This defect of the boundary condition can be interpreted as a necessary correction of the steel surface level.

$$p(x) - \rho_G \, g \, h_G(x) = \rho_{St} \, g \, \Delta h_{St}(x) \tag{10}$$

$$\Rightarrow \qquad \Delta h_{St}(x) = \frac{P(x) - \rho_G \, g \, h_G(x)}{\rho_{St} \, g}$$

$$h^n_{St}(x) = h^o_{St}(x) + \Delta h_{St}(x)$$

with

$h^o_{St}(x)$ ... old (given) steel surface level
$\Delta h_{St}(x)$ ... correction of steel surface level
$h^n_{St}(x)$ ... new steel surface level

Notice here that $h_G(x)$ is, because of the above assumptions, a function of $h_{St}(x)$ and therefore a change of h_{St} causes a change of h_G. The computation must provide both the shape of the free surface and the pressure condition for the steel bath on it (free boundary value problem).

The following algorithm is used to solve this problem:

1) compute p and u_i for the given free surface shape = h^o_{St}, (plane surface at the first time).
2) if $p = \rho_G\ g\ h_G(h_{st})$ -> stop, solution is reached
   if $p \neq \rho_G\ g\ h_G(h_{st})$ -> compute the new steel surface h^n_{st} that

$$p = \rho_G\ g\ h_G(h^n_{st})$$

is satisfied (an equilibrium of pressure, steel surface and casting powder layer must be found, in general via an iterative procedure).

$$h^o_{st} = h^n_{st}$$

goto 1)

With the flow package Phoenics the computation of p and u_i was done and for the realization of the free surface algorithm an additional routine was written and linked to the Phoenics code. During the computation the numerical grid will be continually approximated to the predicted shape of the free surface starting with a cartesian grid (boundary fitted coordinates).

BOUNDARY CONDITIONS

Symmetry planes

On symmetry planes all variables have a zero normal gradient. The velocity component normal to the symmetry plane is zero.

Inlet (entry nozzle)

The throughput rate $\dot{m}$ and the inlet velocity V_{in} are fixed and are functions of the casting speed (=downward speed of the slab =CSP).

$$\dot{m} = \rho_{st} * \text{inlet area} * v_{in}$$

Inlet parameters for K and ε, K_{in}, ε_{in} proposed by Spalding [10] without the constant C

$$K_{in} = 0.0045\ v_{in}\ , \qquad \varepsilon_{in} = C\ 0.1643\ K\ /0.1/\text{entry nozzle diameter}$$

C is a constant to adapt the numerical results to the water model results.

$$T_{in} = 1550\ \ C$$

Walls

Normal velocity is set to zero, tangential velocity = CSP (=casting speed) using turbulent wall functions. Temperature boundary conditions are always transformed to an equivalent heat flux q given by

$$q = \alpha_e\ (T_W - T)$$

α_e ... effectiv heat transfer coefficient W/m^2K is a function of cooling water temperature, copper mould, shell thickness and also of the distance from the top

T_W ... cooling water temperature = 40°C
T ... temperature in the first computational cell close to the boundary

An additional temperature source S_T appears through the solidification of the liquid steel on the mould walls (=boundary). It is the latent heat. From the measurements the shape of the solidified shell is known and therefore the necessary shell growth depends on the casting speed.

$$S_T = H/c_p * (\text{shell growth} \quad kg/sec/m^3)$$

where H ... latent heat (= 270000) J/kg

The computed heat flux q was compared and adjusted to the measured heat flux through the mould.

Free Surface
In the same manner as at the wall the temperature boundary condition is transformed to an equivalent heat flux condition, with α_e=50 and the ambient temperature T_a=50. Actually α_e is a function of the casting powder height, but the heat loss at the free surface is quite small compared with the heat loss through the mould wall and a α_e which depends on powder height has no real influence on the temperature distribution. All other boundary conditions are described in the above chapter "Calculation of the Free Steel Surface".

RESULTS

As already mentioned above no detailed results about the influences of the casting parameters like entry nozzle geometry and submergence depth of the nozzle to the computed quantities may be given here. Only the general behaviour and the major differences between the results obtained with a fixed plane and a computed free surface are pointed out.

Fig.3 shows the flow pattern in the left side of the mould for symmetric inlet conditions. The development of two characteristic eddies is observed: one eddy between the impinging jet of the entry nozzle and free surface, another large eddy between the jet and the bottom. This general feature was also achieved with water model experiments and other numerical analysis, see [1] - [4], [6]. The free surface computation has no significant influence to the flow pattern below the impinging jet. So, only the upper part of the mould is plotted in the following Figures.

The comparison of Fig.4 and Fig.5 demonstrates the essential difference of the flow pattern computed with a fixed plane surface and the free surface algorithm for the same casting speed. In the case of the plane surface, Fig.4, the flow accelerates along the free surface. This general behaviour is quite similar for different CSP, only the absolute velocity is changing. On the other hand, in the case of the computed free surface, the higher steel level on the wall due to the higher pressure in this part causes the development of a dead zone in the symmetry plane, Fig.5, an additional eddy at the broad wall, Fig.6. Further the acceleration parallel to the surface is reduced, a feature which was pointed out in Abdullah and Salcudean [11] too. This follows from the pressure distribution in the steel bath with the maximum value between the inlet nozzle and the wall, where the casting powder height is maximum. This is in agreement with the boundary condition, equation (9). The pressure profile on a plane surface has an inverted feature, minimum in the middle. Water models but also light notches ("oscillation marks") on the slab surface, see Fig.2, due to the solidification at the steel surface level at the wall, confirm the computed smooth wave shaped surface, Fig.5 and Fig. 6. Whether an eddy is formed or not depends on the CSP, compare Fig.3 with CSP=1.2 m/min and Fig.5 with CSP=1.8 m/min. As mentioned above, this is a feature that could not be observed with a fixed plane surface. Some other results can be found in Zuba et al. [12].

Fig.6 shows a 3D view of the broad and narrow wall and the free surface with the boundary fitted grid approximated during the computation to the predicted free surface. The slope of the surface normal to the broad mould wall is very small. The flow pattern dominates parallel to the broad wall.

Fig.7 compares the temperature profiles of the flow situations represented in Fig. 3,4,5. The effect of the additional eddy in Fig.5 can be seen very clearly. A consequence of this eddy is a temperature reduction in the upper part of the mould. It is surprising that a small change of the surface is able to drop the temperature level under the level of a lower CSP. This temperature reduction accelerates the solidification, but because of slab surface quality this effect should be avoided. The task is now to improve the process parameter to avoid such effects.

CONCLUSION

A routine for the treatment of special free surface boundary conditions in metallurgy (casting powder in continuous casting, slag in a ladle) was written. But also the case of a constant pressure boundary condition (free surface in atmosphere) can be treated by setting the casting powder density equal to zero. The results demonstrate very well, that a small deformed surface is able to cause an essential change in flow pattern and temperature distribution. Therefore the computation of the free surface is very important for a more realistic prediction of the fluid flow and temperature in a continuous casting mould. A first step was done by this analysis.

In the next stage some improvements will be performed because two assumptions cannot be held: the fixing of the unit normal for the free surface in a ladle and a frictionless interface.

REFERENCES

1. Flint, P.J. A tree dimensional finite difference model of heat transfer, fluid flow and solidification in the continuous slab caster, 9th Process Technology, 73rd Steelmaking Conf., Detroit, ISS, 1990

2. Yao, M., Ichimiya M., Kiyohara S., Suzuuki K-I. and Sugiyama K. Three dimensional analysis of molton metal flow in continuous casting mold (no more informations known)

3. Thomas, B.G. Application of Mathematical Models to the Continuous Slab Casting Mold, Proceedings of the 72nd Steelmaking Conference, Chicago, IL, April 2-5, 1989

4. Thomas, B.G. and Mika, L.M. Simulation of fluid flow and heat transfer inside a continuous slab casting machine, Proceedings of the Second FIDAP Users' Conf., Evanston, Illinois, Oct 2-4, 1988

5. Flood, S.C., Kasai, K. and Katgerman, L. The Modelling of heat and fluid flow in the DC casting of aluminium alloys, pp. 155-170, Proceedings of the Third Int. Phoenics User Conf., Dubrovnik, Yugoslavia, Aug 28-Sep 1, 1989

6. Matys, P., Salcudean, M. and Abdullah, Z. Fluid Flow and Heat Transfer in Continuous Casting Processes, Department of Mechanical Engineering, The University of British Columbia, Vancouver, B.C., Canada

7. Szekely, J. Fluid Flow Phenomena in Metals Processing, Academic Press, New York, 1979

8. Rodi, W. Turbulence Models and their application in hydraulics - A state of the art review, University of Karlsruhe, Federal Republic of Germany, 1980

9. Landau, L.D. and Lifschitz, E.M. Lehrbuch der theoretischen Physik, Hydrodynamik, 2. edition, Chapter II, Zaehe Fluessigkeiten, pp. 52-60, Chapter VII, Oberflaechenerscheinungen, pp. 266-282, Akademie Verlag, Berlin, 1971

10 Spalding, D.B. The PHOENICS Beginner's Guide, CHAM TR/100,1989

11. Abdullah, Z. and Salcudean, M. Free surface flow during the filling of a cylinder, Int. J. for Num. Meth. in Fluids, Vol 11, pp. 151-168, 1990

12. Zuba, G., Preuer, A., Sattler, H. and Vlcek, J. Complex fluid flow analysis in continuous casting, pp. 21/1-21/13, Proceedings of the 5th international continuous casting conference, Linz, Austria, June 18-20, 1990

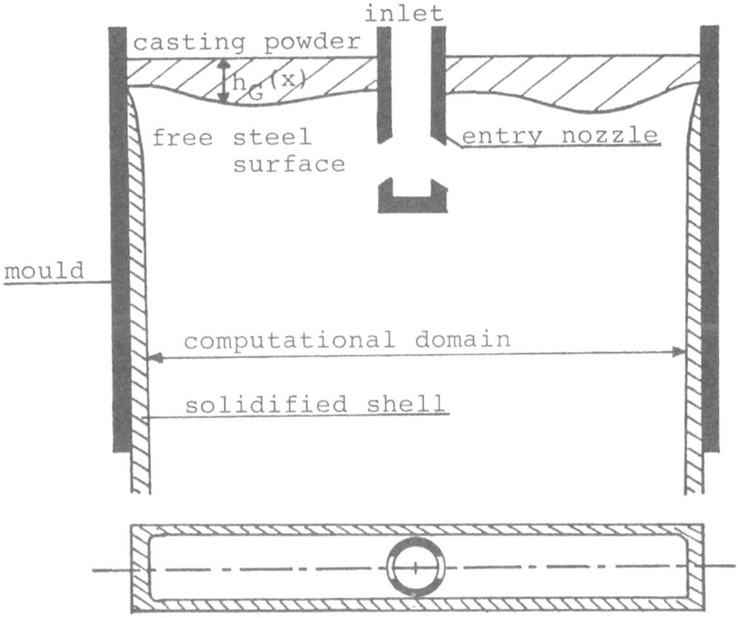

Figure 1: Schematic problem description

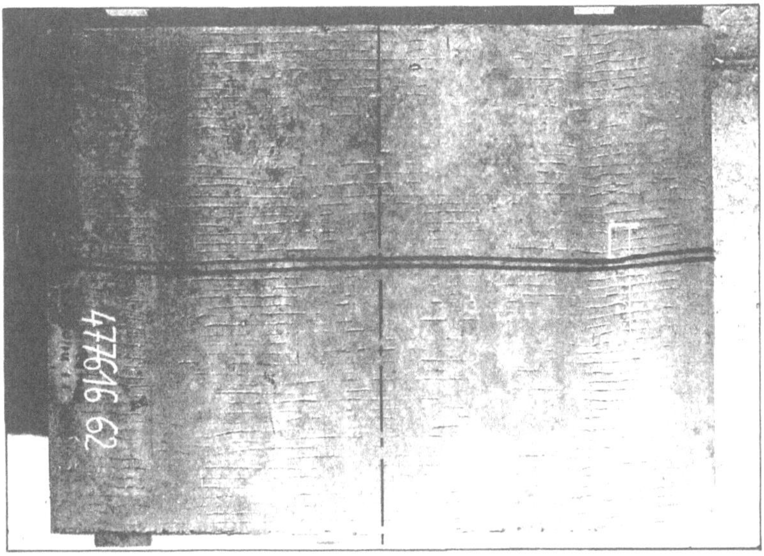

Figure 2: Thin notches ("oscillation marks") at the slab surface, they reflect the liquid steel surface in the mould during the casting process

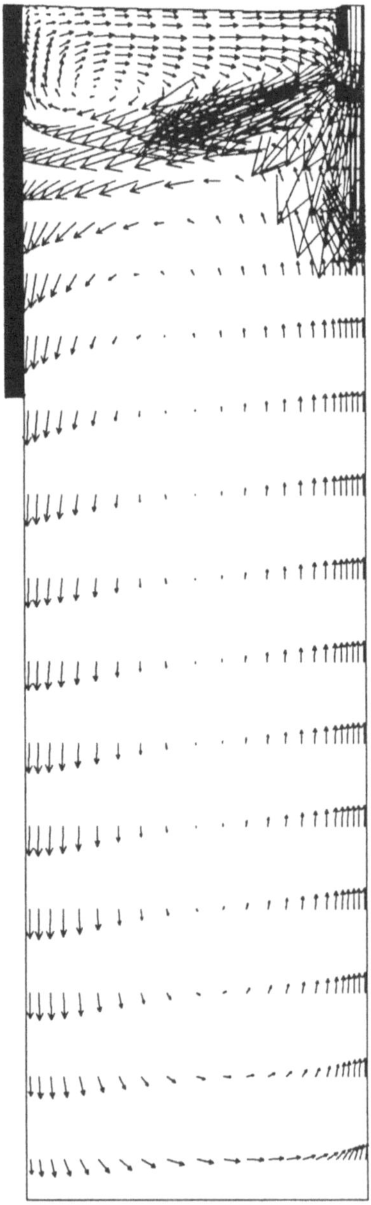

Figure 3: Flow pattern in the symmetry plane by a symmetric entry nozzle, with free surface computation, casting speed = CSP = 1.2 m/min

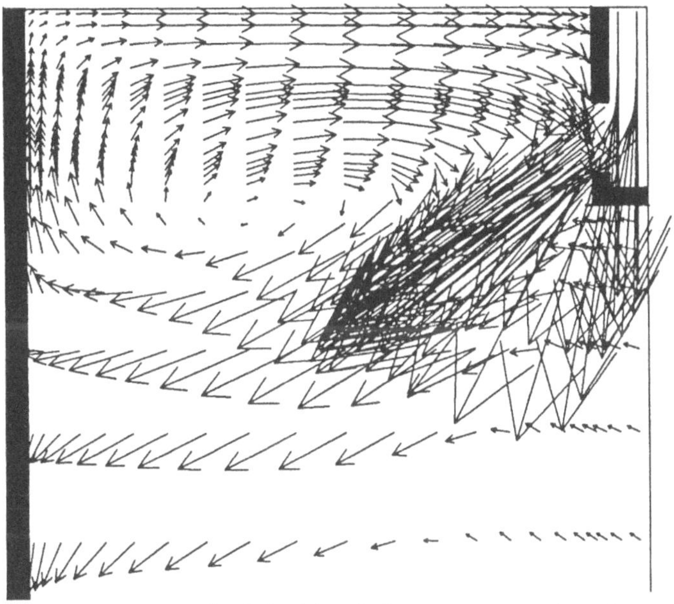

Figure 4: Flow pattern at the left side of the symmetry plane by an asymmetric entry nozzle, with a fixed plane surface, casting speed = CSP = 1.8 m/min

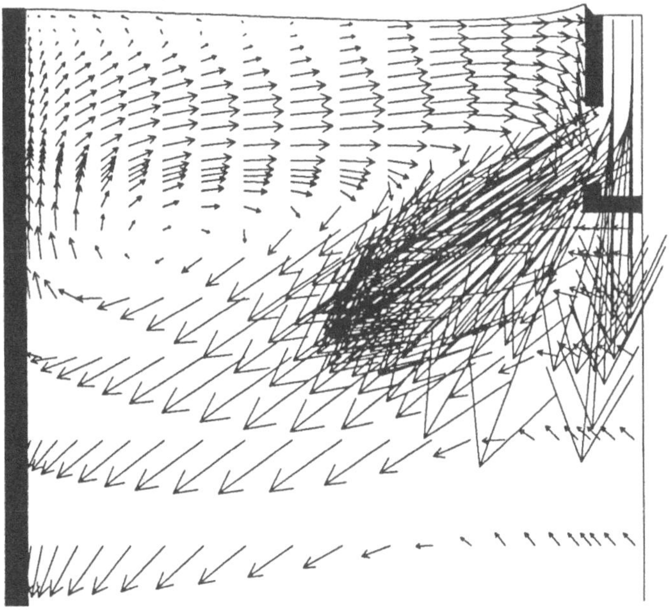

Figure 5: Flow pattern at the left side of the symmetry plane by an asymmetric entry nozzle, with the computed free surface, casting speed = CSP = 1.8 m/min

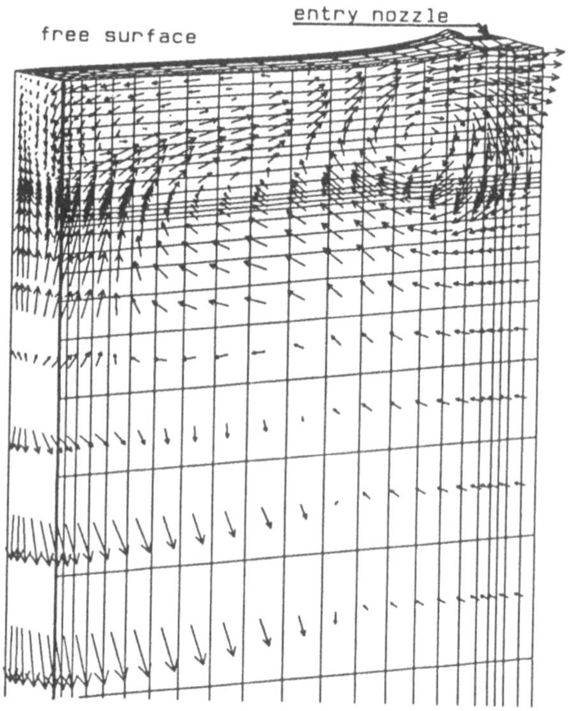

Figure 6: 3D view of one quarter of the mould, flow pattern at the broad and narrow wall, free surface, boundary fitted coordinates

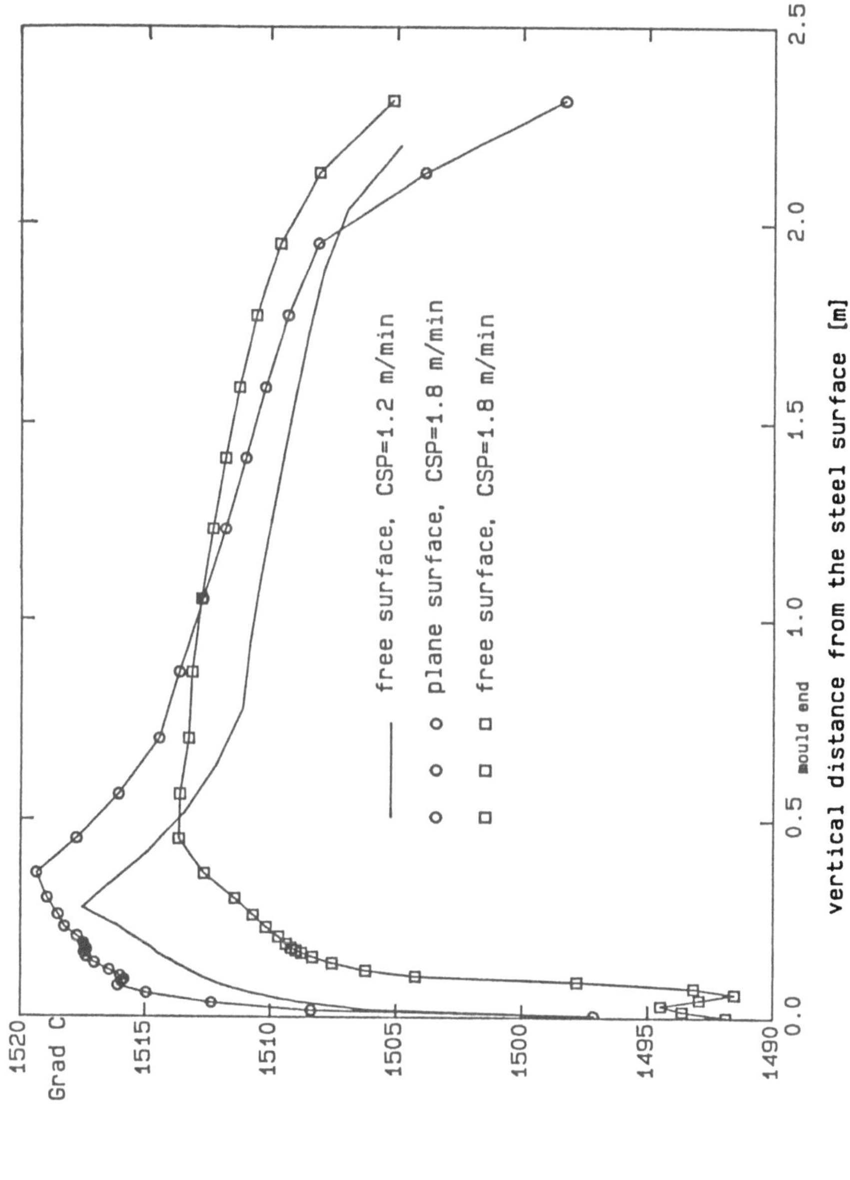

Figure 7: Comparison of vertical temperature profiles along the left vertical boundary in Figure 3,4 and 5

SECTION 3: ELECTRICAL/ELECTROMAGNETIC PROBLEMS

Optimal Shape Synthesis of Electric Devices using the Boundary Element Method

M. Rudnicki (*), A. Savini (**), P. Di Barba (**), K. Pawluk (*)
(*) Department of Fundamental Research, Institute of Electrical Engineering, Warsaw, Poland
(**) Department of Electrical Engineering, University of Pavia, Pavia, Italy

ABSTRACT

The boundary element method (BEM) is applied to the shape synthesis in static problems of electromagnetism. A methodology developed by the authors, based on the idea of the controlled synthesis subregion, is employed. Using BEM in conjunction with the sequential linear programming technique, a part of the boundary is allowed to move so as to ensure the given field distribution in the controlled subregion. A numerical example of synthesis of electric field is given.

INTRODUCTION

Synthesis, as opposed to analysis, brings about the solution of an inverse problem. That is, given the field, we search for a device that produces it. In particular, we can search for the shape to achieve a prescribed field distribution in a part of the device. In practice, an electric device is described by a big number of parameters determining its features. Therefore, practical synthesis problems are limited to some class of solutions since we are not able to solve the full synthesis problem. Other factors that can be extracted from the field include: boundary conditions, internal excitations and environmental parameters. Nevertheless, shape synthesis seems to be the most important from the practical point of view.

The finite element method has been already used to solve problems of shape optimal design in electromagnetic problems [1]. Pironneau [2] in his book gives a deep mathematical background with some practical aspects also considered. Arumugam at al. [3] investigate the problem of optimization of the shape of the poles of an electromagnet. With their approach some boundary nodes of the finite element mesh are assumed to be movable in the given direction. When exact formulas for the field are available, it is possible to find the optimal shape of a device using nonlinear least squares and constrained optimization [4].

There are very few papers, however, on shape optimization using the boundary element method. This fact is surprising since the boundary element approach seems to be ideal for any sort of boundary manipulations. In the author opinion this may be partially explained by the well-known numerical drawbacks of BEM: full, nonsymmetric matrices and difficulties in handling nonlinear problems, contrary to the finite element method. In optimal shape design, employing synthesis controlled subregions, the vast majority of computer time is spent for numerous repeated analyses of the problem in order to minimize the objective function expressed as a mean square root deviation between prescribed and actual field. Hence, the use of the boundary element approach seems reasonable for relatively small problems. In the present paper we show a solution of the problem of optimal shape synthesis of a capacitor. Numerical results are given and a comparison is made with the results of finite element approach.

BOUNDARY INTEGRAL APPROACH

Let us consider the boundary value problem for Poisson's equation. If u is the potential function to be determined and if ε denotes the environmental constant associated with the plane region Ω, then u obeys the following equation:

$$\varepsilon \Delta u(\mathbf{x}) + f(\mathbf{x}) = 0 \qquad \text{in } \Omega \tag{1}$$

where $f(\mathbf{x})$ is a source function and $\mathbf{x}=(x_1,x_2)$ stands for an internal field point of Ω. For an arbitrary observation point $\mathbf{y}=(y_1,y_2)$ the well-known boundary integral formula holds:

$$\alpha(\mathbf{y})u(\mathbf{y}) = \int_S [F(\mathbf{x},\mathbf{y})q(\mathbf{x}) - G(\mathbf{x},\mathbf{y})u(\mathbf{x})]dS(\mathbf{x}) + \int_\Omega F(\mathbf{x},\mathbf{y})f(\mathbf{x})d\Omega(\mathbf{x}) \tag{2}$$

where S is the boundary of Ω, $q(\mathbf{x})=\varepsilon(\partial u/\partial n)$ is a flux across the boundary and F denotes a fundamental solution to equation (1). The fundamental solution is obtained by assuming Ω to be infinite and taking as a source the Dirac delta function. The kernel G is the flux produced by F, i.e. $G=\varepsilon(\partial F/\partial n)$. The coefficient α is 1 for $\mathbf{y}\in\Omega$, $\gamma/2\pi$ for $\mathbf{y}\in S=\partial\Omega$ and 0 for $\mathbf{y}\notin\Omega$ respectively. Here γ denotes the interior angle of S at $\mathbf{y}$. In two dimensions a fundamental solution F and flux G are

$$
\begin{aligned}
F(\mathbf{x},\mathbf{y}) &= -\frac{1}{2\pi\varepsilon}\ln(\mathbf{r}) \\
G(\mathbf{x},\mathbf{y}) &= \varepsilon\left(\frac{\partial F}{\partial n}\right) = -\frac{1}{2\pi}(\mathbf{r}\cdot\mathbf{n})/r^2
\end{aligned}
\tag{3}
$$

where $\mathbf{r}$ links observation and source point and $r=|\mathbf{r}|$. In order to discretize the boundary integral equation (2) the boundary S is divided into a number of straight boundary segments and Ω is partitioned into internal domain elements or cells. Using linear interpolation functions for $\mathbf{u}$ and $\mathbf{q}$ on each boundary segment and linear interpolation functions on cells yields

$$
\begin{aligned}
&u = \mathbf{N}\mathbf{u}^j \quad , \quad q = \mathbf{N}\mathbf{q}^j && \text{for jth segment } S^j \\
&f = \mathbf{H}\mathbf{f}^k && \text{for kth cell } \Omega^k
\end{aligned}
\tag{4}
$$

where $\mathbf{N}=[N_1,N_2]$, $\mathbf{H}=[H_1,H_2,H_3]$ represent linear interpolatory functions and $\mathbf{u}^j$, $\mathbf{q}^j$, $\mathbf{f}^k$ are the nodal values u, q and f respectively. Using the above approximation the following formula is obtained

$$
\alpha(\mathbf{y})u(\mathbf{y}) = \sum_{j=1}^{N}\int_{S^j}F\mathbf{N}dS\mathbf{q}^j - \sum_{j=1}^{N}\int_{S^j}G\mathbf{N}dS\mathbf{u}^j + \sum_{k=1}^{K}\int_{\Omega^k}F\mathbf{H}d\Omega\mathbf{f}^k \tag{5}
$$

In the above formula N denotes the number of boundary segments and K stands for the number of domain elements. After some algebra we arrive at the system of N algebraic equations:

$$
\sum_{j=1}^{N}\mathbf{A}^{ij}\mathbf{u}^i + \varepsilon\gamma^i\mathbf{u}^i = \sum_{j=1}^{N}\mathbf{B}^{ij}\mathbf{q}^j + \sum_{k=1}^{N}\mathbf{D}^{ik}\mathbf{f}^k \tag{6}
$$

where the coefficient matrices are given by:

$$\mathbf{A}^{ij} = -2\pi\varepsilon \int_{S^j} G(\mathbf{x},\mathbf{y}^i)\mathbf{N}dS$$

$$\mathbf{B}^{ij} = -2\pi\varepsilon \int_{S^j} F(\mathbf{x},\mathbf{y}^i)\mathbf{N}dS \quad , \qquad \mathbf{u}^i = \mathbf{u}(\mathbf{y}^i) \quad \mathbf{y}^i \text{ - ith boundary node} \tag{7}$$

$$\mathbf{D}^{ik} = -2\pi\varepsilon \int_{\Omega^k} F(\mathbf{x},\mathbf{y}^i)\mathbf{H}d\Omega$$

When Ω is a bounded domain then from equations (5) and (6) it follows that

$$\gamma^i = -2\pi \sum_{j=1}^{N} \int_{S^j} G(\mathbf{x},\mathbf{y}^i)ds \tag{8}$$

The coefficients $\mathbf{A}^{ij}$, $\mathbf{B}^{ij}$ are obtained with the use of exact integration process while the calculation of $\mathbf{D}^{ik}$ requires numerical quadrature to be used. For the nonsingular case a seven-point symmetric quadrature formula with uniform sampling points is used. Similarly, when the collocating point is a node of internal cell, then singularity is removed through integration on the normalized cell. Details can be found elsewhere [5], [6].

MODEL PROBLEM OF SHAPE OPTIMAL DESIGN

Generally, for mixed boundary conditions one can abstractly formulate an optimal shape design problem as follows [2]:

$$\min_{\Omega\subset\mathcal{A}} \mathbb{E}(\Omega) = \int_{\Omega_c} |\Delta(u(\Omega)) - u_d| d\Omega(\mathbf{x}) \tag{9}$$

where $u(\Omega)$ is a solution of a state equation (1) along with mixed boundary conditions and

$$\mathcal{A} = \left\{ \Omega\colon\ C\supset\Omega\supset D,\ \Omega \text{ - open},\ \Gamma \text{ - Lipschitz} \atop D,\ C \text{ - bounded domains} \right\} \tag{10}$$

denotes a set of allowable domains. $\mathbb{E}$ is our objective function to be minimized and u_d is the prescribed field in a controlled subregion Ω_c.

Generally, shape optimization problems might be classified as follows:
1) Searching for the best shape among an admissible set of shapes
2) Searching for the best location of an object given its shape
3) Searching for the part of the boundary that is not accessible for physical measurement given the rest of the boundary and the state equation (shape identification).
4) Free boundary problems

In the present paper we deal with a problem belonging to the first group. In the discrete representation, the problem (9) corresponds to that of the nonlinear least squares. There are some effective numerical algorithms and computer codes that are best suited to the particular form of the objective function. For unconstrained problems the best of them is, undoubtedly, the Levenberg-Marquardt method. But in our case we are going to investigate a problem that is constrained by its very nature. Hence, in order to avoid excessive distortions and degeneration of the region, some constraints must be imposed on the design variables i.e. the coordinates of the design boundary nodes. Therefore, problem (9) becomes a constrained nonlinear optimization problem that can be solved using any nonlinear programming algorithm. Nevertheless, the derivative free techniques should be preferred because of the objective function complexity. Sequential linear programming (SLP) with move limits has been found a reliable method for the shape optimal design [7]. It is also referred to as Kelley's "cutting plane method". The idea is to linearize nonlinear objective and constraint functions in the neighborhood of a current design vector $\mathbf{X}^p$. The move limits are imposed to limit the design change to the region of applicability of linearization and to prevent an unbounded solution. These move limits are reduced as the optimization progresses in order to effectively deal with the case of non- vertex optimum. They can be considered as the extra constraints that "box in" the current design vector. The range of move during the search may be fixed at each step or varied. The latter is called the "adaptive move limit" and was adopted by the authors in the computer implementation of the algorithm. The procedure uses large move limits at the start of computation to speed progress of design towards the constraint boundary and then gradually decrease the freedom of movement until satisfactory convergence is attained. The method is characterized by many iterations with short steps. The move limits at a given step are set equal to the 80% of the corresponding values at the previous step. On exit move limits magnitude has to be of the order of the given numerical tolerance supplied by the user.

GOVERNING EQUATIONS OF ELECTRIC FIELDS

Under static conditions, Maxwell's law for the electrical field **E** reduces to:

$$\nabla\times\mathbf{E} = 0 \tag{11}$$

Employing the electric scalar potential (or voltage) u we have:

$$\mathbf{E} = -\nabla u \quad , \quad \nabla\times\nabla u = 0 \tag{12}$$

Making use of the constitutive law $\mathbf{D}=\varepsilon\mathbf{E}$ and of the differential form of Coulomb's law: $\nabla\times\mathbf{D}=\rho$, we finally arrive at the Poisson's equation for electric field:

$$-\varepsilon\Delta u = \rho \tag{13}$$

Here **D** is the electric flux density, ε denotes material permittivity: ρ is the charge. When $\rho=0$, the above equation reduces to Laplace's equation. To solve equation (13) some boundary conditions must be specified. In the following, we shall make use of the fact that on equipotential lines the normal derivative $\partial u/\partial\mathbf{n}$ vanishes, e.g. electric flux flows normal to the boundary.

FORMULATION OF THE SYNTHESIS PROBLEM

To demonstrate the outlined strategy the following synthesis problem was formulated. Let us consider the cross section of a parallel plate air capacitor of finite shape (Fig. 1). Our goal is to change the shape of the plates in order to have uniform electric field in the whole capacitor. To this end we define the "controlled subregion" adjacent to the edge of the capacitor and require the normal components of electric field in the controlled subregion to be equal to those in the center of the capacitor (Fig. 2). The governing state equation is:

$$\Delta u = 0 \quad \text{in } \Omega$$

$$u = 1\big|_{S_1} \qquad u = 0\big|_{S_2} \qquad \frac{\partial V}{\partial n}\bigg|_{S_3} = 0 \qquad \frac{\partial V}{\partial n}\bigg|_{S_4} = 0 \tag{14}$$

We define a movable part of the boundary composed of the four nodes. These nodes are allowed to move perpendicularly to the plates of the capacitor. To avoid region distortion side inequality constraints are imposed. For reasons of symmetry extra equality constraints are also introduced.

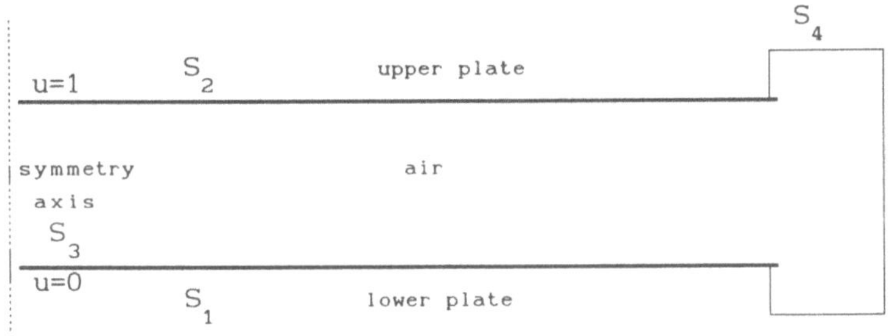

Fig. 1 Cross-section of the air capacitor

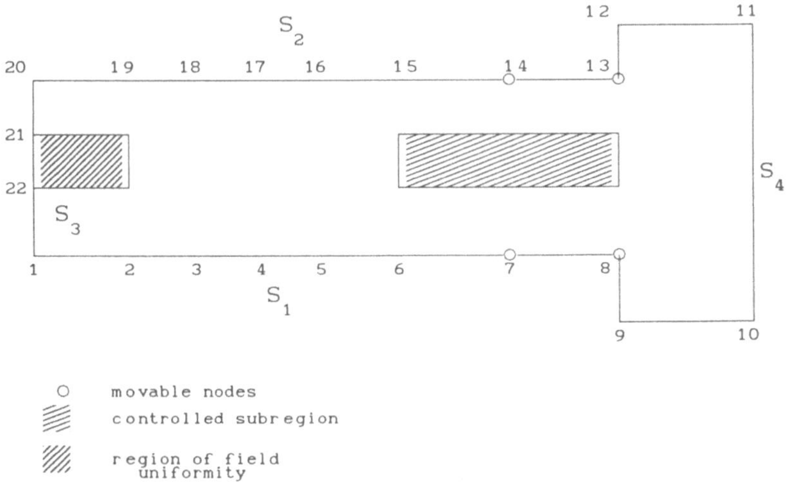

Fig. 2 Boundary element model

The whole boundary is composed of the four parts on which the boundary conditions are prescribed. Both plates of the capacitor are kept at given voltages while on S_3 due to symmetry the homogeneous Neuman condition holds. Moreover, we have to prescribe the boundary condition on S_4 . As mentioned above, on equipotential lines the normal derivative of electrical field is zero. Therefore, we replace the right

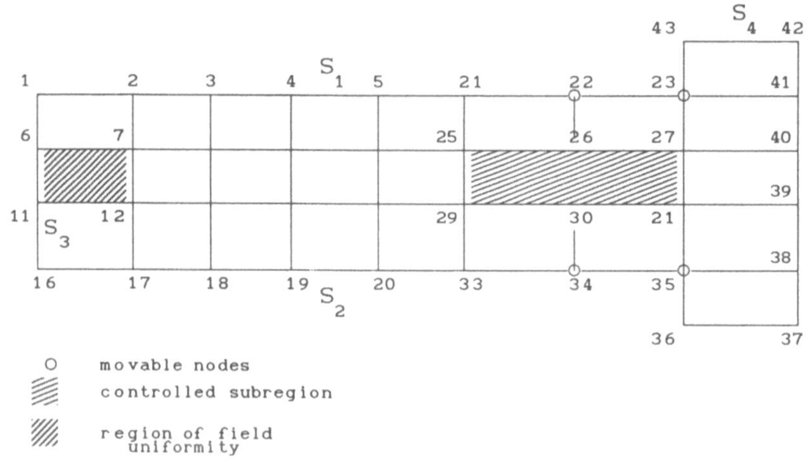

Fig. 3 Finite element model of the problem

edge of the capacitor with an field line to prescribe a homogeneous Neuman condition on it. This assumption is valid as long as the field line is far enough from the edge.

NUMERICAL RESULTS

The boundary element solver using linear boundary segments and linear interior cells has been attached to the SLP optimizer with adaptive move limits. The task of the solver is to perform successive reanalyses of the field problem in order to supply an objective function value to the optimizer. A number of movable boundary nodes is singled-out with their ordinates being the design variables. The optimizer is derivative-free i.e. it does not require the derivatives of an objective function and constraints on input. All derivatives are computed within the program using finite differences. A special subroutine performs the check-out of the computed derivatives and, in case of inconsistencies, a message is printed. The solution progress is monitored and the user can examine the results and take the suggested action.On input the following parameters must be supplied:

- number of inequality constraints
- number of equality constraints
- number of design variables
- maximum number of iterations
- numerical tolerance for terminating the optimization
- increment for computing the finite differences
- initial move limits on design variables

- initial design vector (initial shape)
- objective and constraint functions (user supplied subroutines)

On exit:

- final design vector (final shape)
- objective function value at the optimum
- number of performed iterations
- optimization history
- BEM reanalyses history
- data for plotting
- program run time

To compare our results with those of the finite element simulation we have taken identical data for calculation. In Fig. 3 the finite element mesh for the problem is shown. In Table 1 we give the results for the boundary integral approach described in the present paper. In Table 2 the results of the finite element simulation are given. They are also described in detail elsewhere [9]. Let us mention here that in the finite element simulation we have used triangular elements of the first order so the order of approximation is the same.

Design node number	Initial ordinates of design nodes	Initial field in controlled subregion	Final field in controlled subregion	Final ordinates of design nodes
7	0.0	0.4975	0.5002	0.000
8	0.0	0.4998	0.5000	0.059
13	2.0	0.4668	0.5000	1.940
14	2.0	0.4975	0.5002	2.000

Inequalities: $2. \geq y_{13} \geq 1.35$ $\quad 0 \leq y_7 \leq 0.65$
$2. \geq y_{14} \geq 1.35$ $\quad 0 \leq y_8 \leq 0.65$

Equalities: $y_{13} + y_{14} = 2.$ $\quad y_7 + y_8 = 2.$

Runtime on PS/2 70: 23 min. 58 sec.

Objective function at the optimum: 0.492561D-7

Table 1. Results of BEM simulation

Design node number	Initial ordinates of design nodes	Initial field in controlled subregion	Final field in controlled subregion	Final ordinates of design nodes
34	0.0	0.4772	0.5000	0.001
35	0.0	0.4982	0.5001	0.045
22	2.0	0.4982	0.5001	1.999
23	2.0	0.4998	0.5000	1.995

Inequalities: $2. \geq y_{22} \geq 1.35$ $\quad 0 \leq y_{34} \leq 0.65$

$2. \geq y_{23} \geq 1.35$ $\quad 0 \leq y_{35} \leq 0.65$

Equalities: $y_{22} + y_{34} = 2.$ $\quad y_{23} + y_{35} = 2.$

Runtime on PS/2 70: 7 min. 58 sec.

Objective function at the optimum: 0.167578D-7

Table 2. Results of FEM simulation

CONCLUSION

In the present paper a numerical algorithm for the solution of the optimal shape design problems of electric and magnetic devices is presented. The workability and efficiency of the algorithm is demonstrated on an example of the optimal shape design of plates in the capacitor. The idea of "controlled synthesis region" is employed in the solution. Using the boundary element method it is possible to reduce the dimension of the problem at the cost, however, computational time. Results of both simulations (FEM and BEM) appeared actually the same.

Presented algorithm may be useful for the optimal shape synthesis of any electrical and magnetic device.

REFERENCES

1. Haslinger, J. and Neittaanmäki, P. Theory and Finite Element Approximation of Optimal Shape Design Problems for Partial Differential Equations, John Wiley&Sons, Chichester, 1984.

2. Pironneau, O. Optimum Shape Design for Elliptic Problems, Springer-Verlag, New York, 1984.

3. Arumugam, G., Neittaanmäki, P. and Salmenjoki, K. Sensitivity Analysis for the Design of an Electromagnet, University of Jyväskylä, Finland, 1988 (preprint 82).
4. Mustarelli, P., Rudnicki, M., Savini, A., Savoldi, F. and Villa, M. Synthesis of Magnetic Gradients for NMR Tomography, Magnetic Resonance Imaging, Vol. 8, pp. 101-105, 1990.
5. Brebbia, C.A. The Boundary Element Method for Engineers, Pentech Press, London, 1978.
6. Ligett, J.A. and Liu, P.L.F. The Boundary Integral Equation Method for Porous Media, George Allen and Unwin, Boston, 1983.
7. Best, M. and Ritter, K. Linear Programming: Active Set Analysis and Computer Programs, Prentice Hall, New Jersey, 1985.
8. Pawluk, K. and Rudnicki, M. The State of Art in the Synthesis of Electromagnetic Field, Electromagnetic Fields in Electrical Engineering, (Eds Savini, A. and Turowski, J.), pp. 287-292, Plenum Press, New York, 1988.
9. Rudnicki, M. and Savini, A. Optimal Synthesis Methodologies in Electromagnetism, Proc. AMSE Conf. on Modeling & Simulation, Istanbul, June 29-July 1, 1988, Vol. 2A, pp. 51-58.

Modelization of the Non-Linear Moving Parts of an Electromechanical Relay

F. Delincé (*), P. Dular (*), A. Genon, W. Legros, A. Nicolet, M. Umé

University of Liège, Department of Electrical Engineering, Institut Montefiore - Sart Tilman B 28-4000 Liège, Belgium

(*) These authors are Research Assistants with the Belgian National Fund for Scientific Research

ABSTRACT

The modelization of an electromechanical relay requires a transient modelization since there are no periodic phenomena like in a rotating machine (Juffer[1]). Moreover, the movement plays a fundamental role in the electromagnetic modelization of the closure of the relay.

The paper presents the coupling between the finite element and the boundary element methods as an efficient technique for the two-dimensional modelling of an electromechanical relay. On the one hand, the finite elements are used inside the magnetic and conducting parts where they permit to take into account the induction of current in conductors and the saturation in ferromagnetic materials. On the other hand, the boundary elements are used around the relay where they prevent the deformation of the meshes.

INTRODUCTION

Movement is of crucial importance in electromechanical relays. A correct treatment of the movement involves a coupling between the mechanical model and the electromagnetic model. On the one hand, electromagnetic phenomena produce forces that mechanically act on the devices producing, among others, movement.

On the other hand, mechanical movement acts on the electromagnetic model : it transforms the geometry of the problem and it produces phenomena such as the motional induced electromotive force.

In this paper, the moving parts of the relay are rigid. The only mechanical degree of freedom is a vertical translation.

The mechanical model consists of the simple movement equation where electromagnetic force is introduced. For the electromagnetic field modelization, we use an existing program called LUCIE, which is based on the coupling of the finite element method and of the boundary element method.

This powerful method allows a rigorous treatment of open boundary problems with BEM together with a correct modelization of the saturable ferromagnetic materials. The modularity of the LUCIE software has allowed an easy and workable introduction of movement in the modelization.

The philosophy behind the use of the coupled method to model the movement is to use the finite element method inside the rigid moving parts that can be the seat of phenomena such as saturation and eddy currents.

The boundary elements are used for linear magnetic and non-conducting media wherein rigid parts can move. The advantages of this formulation are that there is no mesh deformation in the finite element domains and that we need no explicit terms for the motional induced electromotive force, which is automatically taken into account by the variation of the boundary element method coefficients.

NUMERICAL METHODS : BOUNDARY ELEMENT - FINITE ELEMENT COUPLING

The equation for the two-dimensional magnetostatics is :

$$\text{div}\ \nu\ \text{grad}\ A = -\ J \tag{1}$$

- A is the vector potential which has only one component;
- J is the current density;
- n is the magnetic reluctivity.

In linear magnetic non-conducting media, the direct boundary element method (Brebbia[2]) is based on the relation (2) :

$$cA = \int_{\Gamma} A\ \partial G/\partial n\ d\Gamma - \int_{\Gamma} G\ \partial A/\partial n\ d\Gamma \tag{2}$$

where :

- G is the free space Green function of the two-dimensional Laplace operator;
- c = 0.5 on a smooth boundary;
- $\partial./\partial n$ is for the normal derivative.

Integrals are taken on the boundary Γ of the subdomains and the method involves only A and $\partial A/\partial n$ (tangential induction) on the boundaries.

The finite element formulation (Silvester[3]) is based on the Galerkin method (3) and for domain Ω of boundary Γ we have:

$$\int_{\Omega} (\nu \text{ grad } A \text{ grad } w - Jw) d\Omega - \int_{\Gamma} w \, \nu \partial A / \partial n \, d\Gamma = 0 \tag{3}$$

where ν is the magnetic reluctivity, J the current density and w the weighting function.

The boundary term is usually used to apply a Neuman boundary condition. Here it is used at the interface between finite element and boundary element domains to couple the methods.

Equation (3) remains valid for non-linear magnetic materials where ν is a function of the induction $B = |\text{curl } A|$. In this case, the algebraic system obtained is non-linear and the Newton-Raphson method (Silvester[3]) is used to solve it. A ferromagnetic material is characterized by a function $\nu(B)$. The derivative $\partial\nu(B)/\partial B$ is also necessary for the practical implementation of the Newton-Raphson method. The function :

$$\nu(B) = \alpha + \beta B^6 \tag{4}$$

has been used in the numerical model. It is a classical choice in electrotechnics (Rougé[4]) which is also workable in finite element computations.

INDUCTIVE EFFECTS

Using Ohm's law, the inductive effects in the coils can be treated by introducing (Bourmanne[5]) :

$$J = - \sigma(U + \partial A/\partial t) \tag{5}$$

as the expression for current density in equation (3). The conductor is characterized by its electrical conductivity σ. The time derivative of A expresses the inductive effects, since U = grad V is due to an externally imposed electromotive force and can be interpreted as the terminal voltage of the conductor (per unit of length).

If section S of the wire of the coil is very small, it prevents eddy currents and only a global inductive effect is allowed. If section S of the wire is the same everywhere in an N-turn coil, the current density J is uniform. In this case, the total current I in the coil is related to the terminal voltage U by :

$$N I = - \sigma \frac{\partial}{\partial t} \int_{coil} A \, dS + \sigma S U \tag{6}$$

and the current density J is expressed as the ratio I/S of the current to the section of the wire.

The numerical method used to discretize the transient problem is a semi-discrete Galerkin method. The first step is a spatial discretization of the domains leaving a continuous time variation of the degrees of freedom. This spatial discretization is a classical finite element meshing. It leads to a system of ordinary differential equations. The second step is a time discretization of the degrees of freedom. A finite difference scheme is generally used. In our particular case, a backward Euler method is used. It leads to an algebraic system of equations.

MOTIONAL INDUCED ELECTROMOTIVE FORCE

One of the major difficulties related to the numerical modelization of the movement is to take into account the motional induced electromotive force.

The problem can be analyzed as follows.

We have related current density $\vec{J}$ in a conductor to the electric field $\vec{E}$ through Ohm's law :

$$\vec{J} = \sigma\vec{E} \tag{7}$$

However this law is only valid for a conductor at rest. To determine the relation between current density and the electric field in a restless conductor, we have to consider a referential associated to the conductor and to use the non-relativistic field transformation formulae. We obtain (Landau[6]) :

$$\vec{J} = \sigma\left(\vec{E} + \vec{v}X\vec{B}\right) \tag{8}$$

This modified Ohm's law for restless conductors relates the current density in the conductor to the electric field $\vec{E}$ and to the induction $\vec{B}$ defined in a <u>referential at rest</u> with respect to which the conductors are moving.

If the electromotive force acting on a closed linear circuit is considered, it can be shown that the first term of the right-hand member of equation (8) corresponds to a variation of the magnetic flux due to the time variation of the magnetic field itself when the position of the circuit does not change.

The second term of the right-hand member of equation (8) corresponds to the magnetic flux variation due to the displacement of the conductor in a magnetic field constant in time.

Those results can be summed up by Faraday's induction law. The induced electromotive force in a circuit is equal to the total magnetic flux variation through the circuit whatever the origin of the variation : time variation of the field or movement of the circuit.

In the numerical treatment of the eddy currents the flux variation on a time step is computed as the difference between nodal values of the vector potential at two successive moments.

If the node belongs to a moving object, the flux variation is computed with two vector potential nodal values taken at different points in space but corresponding to the same "particle" of the moving body. The time derivative of the potential defined in this way is the "substantial" or "material" derivative.

If objects are displaced during the computation without any other change to the numerical methods, flux variations will be correctly computed. This automatically takes into account movements of conductors or of magnetic parts since, according to Faraday's induction law, it involves the total flux variation at one point of the conductor.

The coupling between the boundary elements and the finite elements naturally solves the problems related to the modelization of the movement.

FORCES COMPUTATION

The movement of the magnetic parts of the relay is a natural movement due to the electromagnetic forces. The determination of the electromagnetic forces is an important point of the movement modelization. The knowledge of the vector potential A and of the tangential field H_t on the boundaries given by the coupling method suggests the use of the Maxwell stress tensor method (Aldefeld[7]).

This method permits to define, at the boundary of objects, an equivalent normal force density $\vec{F}_n$ and an equivalent tangential force density $\vec{F}_t$ as :

$$\vec{F}_n = \left(\frac{B_n^2}{2\mu_0} - \frac{\mu_0}{2} H_t^2 \right) \vec{n} \tag{9}$$

$$\vec{F}_t = B_n H_t \vec{n} \tag{10}$$

The tangential field H_t and the normal induction B_n are well defined even on surfaces of discontinuity. H_t is directly given by the coupling method since B_n is computed from the vector potential by :

$$B_n = \text{curl } A.\vec{n} = \frac{\partial A}{\partial y} . n_x - \frac{\partial A}{\partial x} . n_y$$

$$= \left(- \frac{\partial A}{\partial y}\frac{\partial y}{\partial \xi} - \frac{\partial A}{\partial x}\frac{\partial x}{\partial \xi} \right) / J(\xi)$$

$$= - \frac{\partial A}{\partial \xi} / J(\xi) \tag{11}$$

where ξ is the local coordinate of the form functions used for the boundary element discretization and $J(\xi)$ the corresponding jacobian.

The total force acting on an object of boundary Γ is given by :

$$\vec{F} = \int_{\Gamma} \left(\vec{F}_n + \vec{F}_t\right) d\Gamma \qquad (12)$$

MOVEMENT EQUATION

The displacement of an object is computed by solving the equation of the movement involving the magnetic forces.The following quantities are defined :

- position : x;
- speed : v = dx/dt;
- mass : M;
- elastic constant of a spring : k;
- friction coefficient : f;
- other forces including the magnetic forces : F(t).

The equation of the movement is :

$$M \frac{dv}{dt} = F(t) - fv - kx \qquad (13)$$

The time discretization of this equation by the backward Euler method gives :

$$v_{t+\Delta t} = \frac{Mv_t + F(t)\,\Delta t - kx_t\,\Delta t}{k\,\Delta t^2 + f\,\Delta t + M} \qquad (14)$$

$$x_{t+\Delta t} = \frac{Mx_t + Mv_t\,\Delta t + F(t)\,\Delta t^2 + fx_t\,\Delta t}{k\,\Delta t^2 + f\,\Delta t + M} \qquad (15)$$

Equations (14) and (15) relate the position $x_{t+\Delta t}$ and the speed $v_{t+\Delta t}$ at time $t+\Delta t$ to the position x_t and to the speed v_t at time t.

The movement is determined if an initial position x_0 and an initial speed v_0 are specified. Limits of movement are specified to avoid non-physical interpenetration of moving parts.

INTRODUCTION OF THE MOVEMENT IN THE SOFTWARE

The basis of the modelization of the movement in electromagnetic devices is an existing 2D magnetic field computation software called LUCIE. The central feature of the software is the use of the coupled FEM-BEM. The modular structure of the software is designed to make further developments and modifications as easy as possible (Delincé[8]).

The algorithm used to model the movement is the following :

1. Compute the electromagnetic state of the system.
2. Compute the magnetic forces and torques for this state.
3. Compute the displacements of the moving bodies with equations (14) and (15).
4. Modify the geometry in the database of the software using the computed displacements.
5. Start again at step 1.

This procedure involves some implicit hypotheses. The time-varying electromagnetic force is considered as constant on a time step in equations (14) and (15), the displacements (i.e. the geometry variations) should not be too large on a time step.

Practical considerations show that an adaptative time step is necessary in order to keep the computational time within reasonable limits while remaining as close as possible to the hypotheses.

MODELIZATION OF THE RELAY

Figure 1 shows the geometry of the 2D model of the open relay and the associated meshing.

The magnetic core is composed of an upper moving part and of a lower static part (magnetic characteristic $\alpha = 5.24\ 10^2$ m/H and $\beta = 2.63\ 10^3\ mH^{-1}T^{-6}$). The depth of the structure is 1.6 cm and, for the 2D modelization, all the characteristics of the relay are translated into equivalent characteristics for a standard depth of 1 meter and noted by a prime('). The coil has 1430 turns of thin wire and its resistance R is 18 Ω (R' = 18 Ω / 0.016 m = 1125 Ω/m). According to its dimensions (section of 46 mm x 7 mm) the associated conductivity is $\sigma = 1.13\ 10^{-7}$ S/m. The mass M of the moving part is 112g (M' = 7000 g/m) and the elastic constant of the spring supporting the moving part is k = 137 N/m (k' = 8584 N/m^2). Friction is neglected (f=0).

The relay is fed by a step of voltage U of 12 V (U' = 750 V/m) at the terminals of the coil. The total displacement of the moving part is 7 mm.

The coil and the saturable magnetic parts are modelized with finite elements. The surrounding air is modelized with boundary elements (Figure 1).

Figure 2 shows the field lines in the open relay.

Figure 3 shows the time variation of the position of the moving part.

Figure 4 shows the time variation of the speed of the moving part.

Figure 5 shows the time variation of the force acting on the moving part.

Figure 6 compares the computed (continuous line) and the measured (dashed line) time variation of the current in the coil.

The motional induced electromotive force is responsible for the decrease of the current just before the closure is completed.

CONCLUSIONS

The coupling between the finite elements and the boundary elements appears to be a natural and workable method to model the movement in electromechanical devices.

The rigid saturable moving parts are easily modelled with the finite elements. The media around the relay in which rigid parts move is linear magnetic, non-conducting (air) and open (infinite). Its treatment with boundary elements is therefore natural and avoids all the problems of mesh deformations. The paper shows that the motional induced electromotive force is easily taken into account. The basis of the modelization is the 2D magnetic field computation software called LUCIE using the coupling between the boundary element and the finite element methods. Its modular structure has allowed an efficient and safe modification of the software.

The algorithm of the movement modelization shows that the modifications inside the program have been kept to a minimum. Most of the computations proper to the movement are gathered in an independent module.

The numerical computations show the accuracy of the method. Further developments will be the extension to three-dimensional computations.

REFERENCES

1. Juffer, M., Transducteurs électromécaniques, Traité d'électricité, Vol IX, Presses Polytechniques Romandes, Lausanne, 1985.

2. Brebbia, C.A., The Boundary Element Method for Engineers, Pentech Press, London, 1984.

3. Silvester, P.P., Ferrari, R.L., Finite Elements for Electrical Engineers, Cambridge University Press, Cambridge, 1990.

4. Rougé, M.R., Représentation algébrique des caractéristiques magnétiques, Bulletin de la Société Française des Electriciens, 5ième Série, Tome VI, n° 69, September 1936, pp. 881-892

5. Bourmanne, P., Delincé, F., Genon, A., Legros, W., Nicolet, A., Skin Effect and Proximity Effect in Multiconductor Systems With Applied Currents and Voltages, 35th MMM Conference, San Diego, USA, October 1990.

6. Landau, L., Lifchitz, E., Electrodynamique des milieux continus, Edition Mir, Moscow, 1969.

7. Aldefeld, B., Forces in Electromagnetic Devices, Compumag 1978, Grenoble, pap. 8.1.

8. Delincé, F., Genon, A., Legros, W., Nicolet, A., Umé, M., DDLMU, Degrees of Freedom Management Module for Numerical Modelization, RRES 91, Reliability and Robustness of Engineering Software, Milan, Italy, 22-24 April 1991.

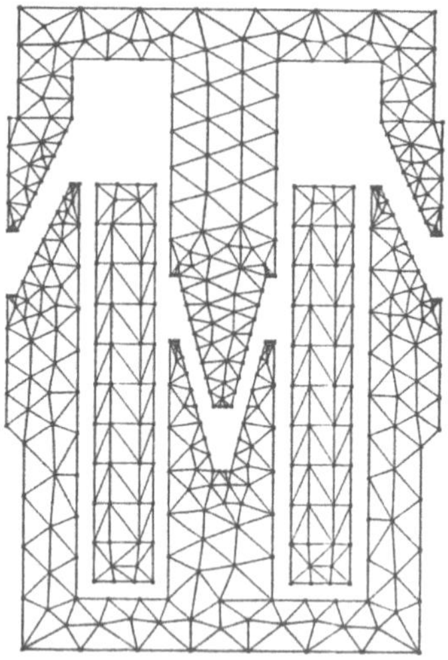

***Figure 1** : Meshing of an electromechanical relay.*

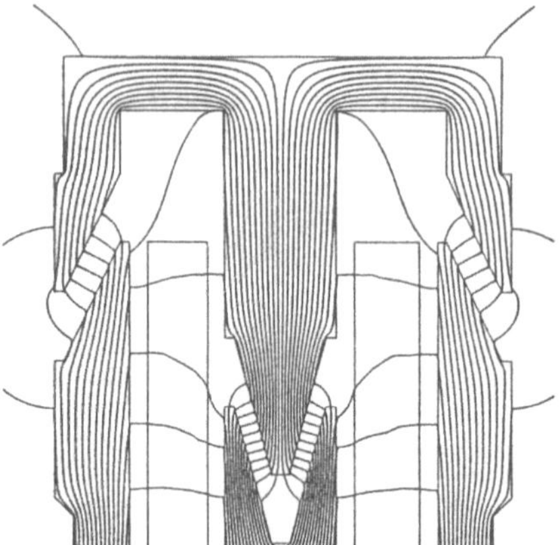

***Figure 2** : Field lines in the relay.*

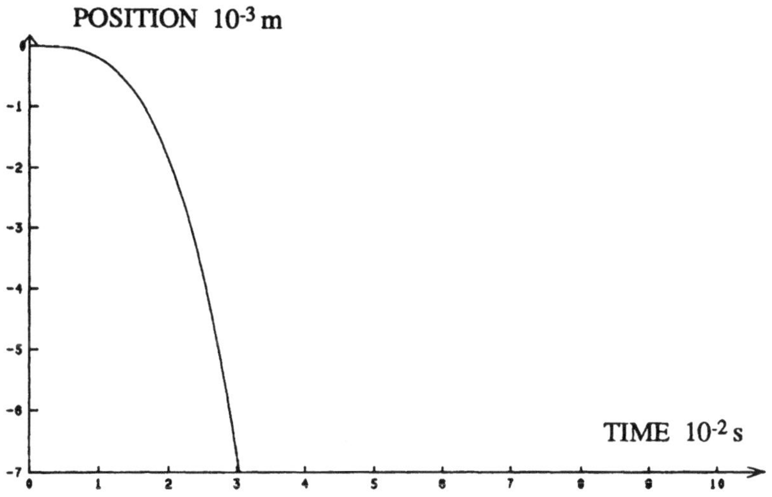

***Figure 3** : Displacement of the moving part.*

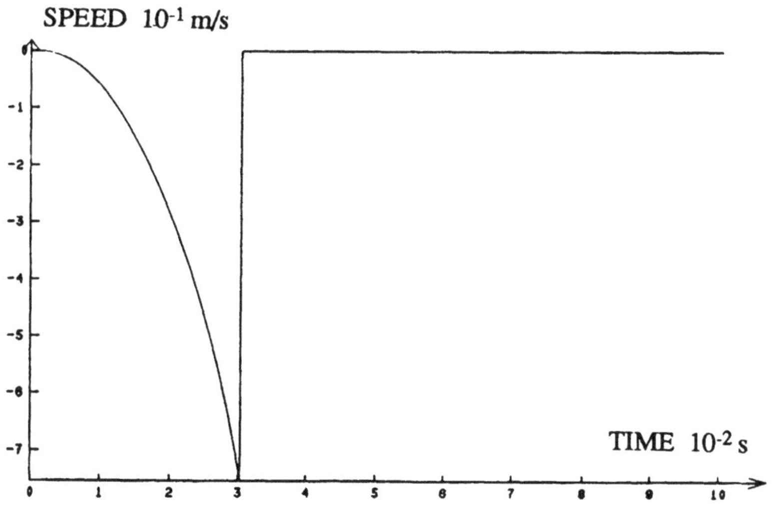

***Figure 4** : Speed of the moving part.*

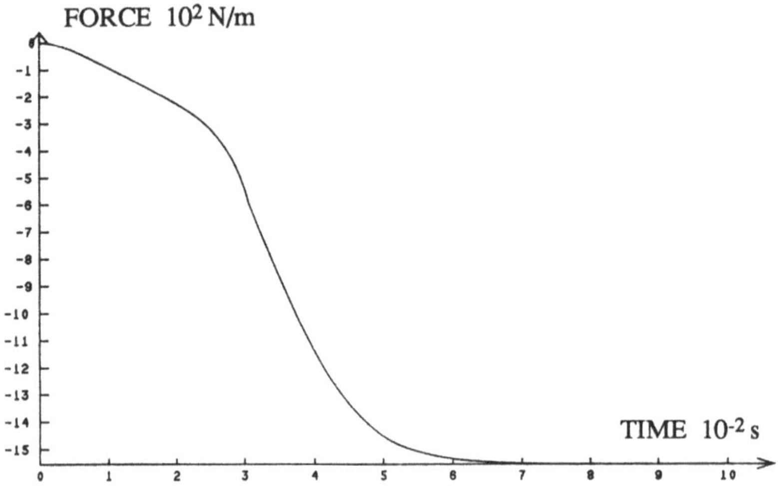

Figure 5 : Force acting on the moving part.

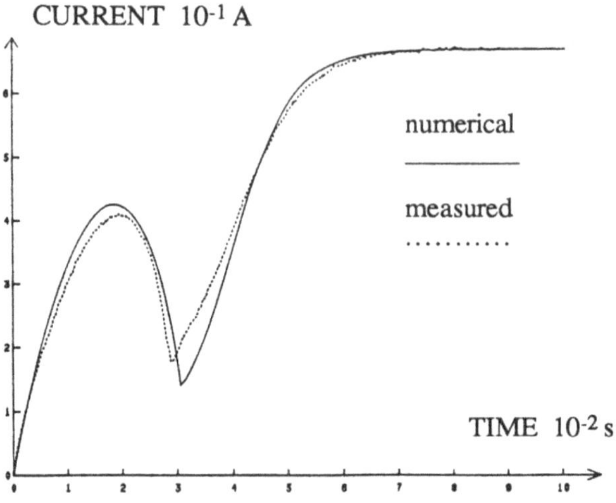

Figure 6 : Current in the coil of the relay.

SECTION 4: SCIENTIFIC APPLICATIONS

Moving Boundary Problems and Solution Strategies in Semiconductor Process Simulation

E. Rank

Numerische Methoden und Informationsverarbeitung, FB Bauwesen, Universität Dortmund, Postfach 500 500, D-4600 Dortmund 50, Germany

Dedicated to Professor W. Wunderlich on the occasion of his 60th birthday

Abstract

One important topic in numerical semiconductor process simulation is the thermal oxidation of silicon. The process step can be described as a transient coupled system of equations for oxidant diffusion, chemical reaction and large mechanical displacement. A moving interface between oxide and silicon renders the whole system a free boundary problem. Moreover, a volume expansion during oxidation by a factor of more than two causes the overall domain of computation to change during simulation. The approach which is presented in this paper, uses the idea of a *distributed reaction zone* between silicon and oxide. A finite element discretization for this formulation shows significant advantages over conventional approaches. In numerical examples an excellent agreement of the simulated results with microscopical photographs of real structures is demonstrated.

Introduction

Fabrication of ultra-high integrated electronic circuits involves up to a few hundred process steps which are used to form microstructures and to set up the electrical properties of individual devices. Structural defining process steps are etching, deposition and oxidation; redistribution of previously implanted impurities controls the electrical characteristics of devices.

An important problem in VLSI-fabrication is the isolation of adjacent electrically active zones. This is generally done by setting up a range of non-conducting silicon dioxide either by socalled box-isolation or by locally oxidizing silicon. In the first method a trench is etched into the silicon wafer which is filled by silicon dioxide afterwards; the second method uses the chemical reaction of silicon and oxygen to oxide. Whereas box-isolation has the advantage of allowing construction of very narrow isolation zones, local oxidation of silicon generally produces oxide of higher quality.

We shall investigate numerical simulation of local oxidation of silicon in this paper. It turns out that the underlying physical phenomena can be described by a moving-boundary- moving-interface-problem of coupled diffusion, chemical reaction and mechanical displacement. A new algorithm is presented, which has first been published in [1] and which shows considerable advantages over other numerical approaches known from the literature. The code is used very successfully in industry [2]. In numerical examples it will be shown that there is an excellent agreement of numerical simulation and physical experiment.

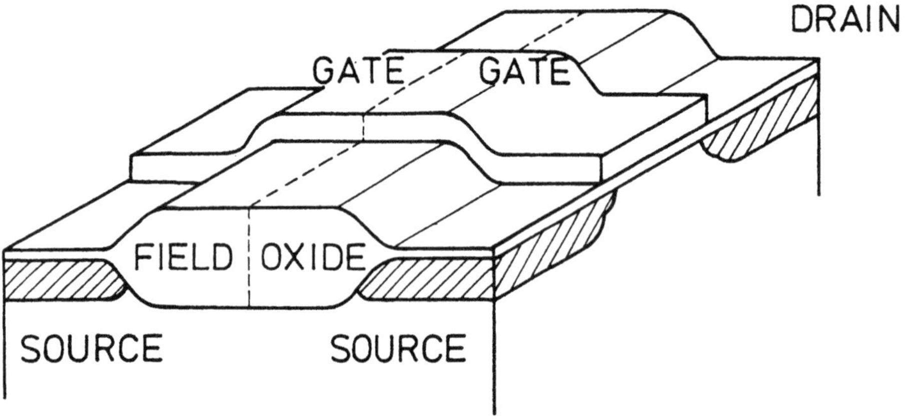

Figure 1 : Three-dimensional view of adjacent electronical devices

Local oxidation of silicon - the physical model

Figure 1 shows a simplified three-dimensional view of a typical structure in highly integrated electronic circuits. Ranges marked as 'source' and 'drain' are electrically active zones of adjacent transistors, the 'field oxide' between the two source regions is set up to isolate the two ranges against each other. This structure grows during local oxidation of silicon. The corresponding process step shall be explained on a two-dimensional cut through the structure as shown in figures 2 and 3. After a thin layer of planar oxide has been deposited onto the silicon, the wafer is masked partly by silicon nitride (figure 2). Then the wafer is oxidized in an atmosphere of oxygen (dry oxidation) or steam (wet oxidation) at temperatures between 900 °C and 1200 °C. During this process step the oxidant penetrates the unmasked surface of SiO_2 and diffuses through the oxide to the interface with silicon. There the chemical reaction of silicon to oxide takes place. This reaction has two consequences : Firstly, silicon is consumed, so the interface moves during oxidation ; secondly, the oxide produced has more than twice the volume of the original silicon - the oxide layer is lifted up (figure 3).

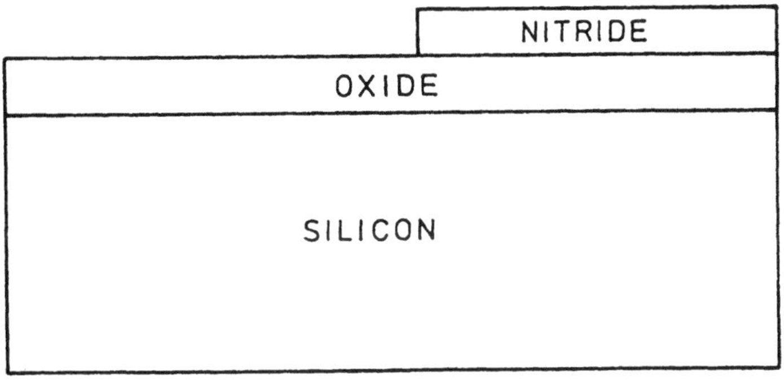

Figure 2 : Initial domain for oxidation

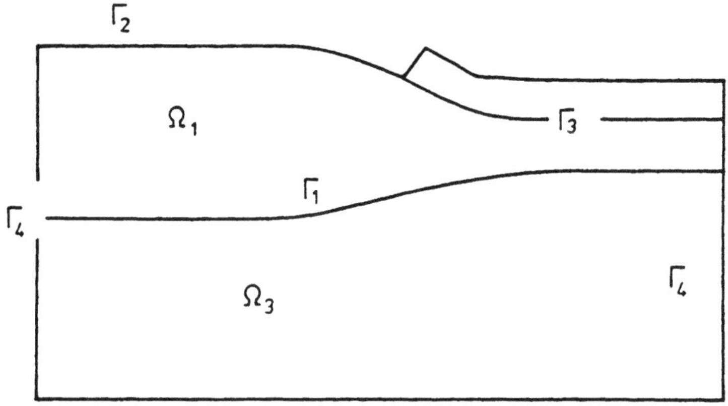

Figure 3 : Domains and boundaries (conventional formulation)

For a mathematical formulation of the problem the silicon dioxide range is denoted by Ω_1 (see figure 3), the exposed boundary by Γ_2, the interface of silicon and oxide by Γ_1, the interface between oxide and nitride by Γ_3 and the symmetry boundary on the left and right of the structure together with the upper boundary of the nitride and the lower boundary by Γ_4. Due to the different timescale of the diffusion process and the structural deformation the transient process can be split into a sequence of timesteps where the structure can be assumed to be stationary. Then the oxidant diffusion during a time interval Δt from t_1 to t_2 is described by the Deal-Grove-Model [3] :

$$\nabla(D_{ox}\nabla C(x,t)) = 0 \quad on \quad \Omega_1 \tag{1}$$

for every fixed time t , with the boundary conditions

$$C = C^* \quad on \quad \Gamma_2 \tag{2}$$

$$\frac{\partial C}{\partial n} = 0 \quad on \quad \Gamma_3 \cup \Gamma_4 \quad . \tag{3}$$

and

$$D_{ox}\frac{\partial C}{\partial n} = k_s C \quad on \quad \Gamma_1. \tag{4}$$

D_{ox} is the diffusion coefficient of the oxidant in oxide, k_s the surface reaction coefficient at the silicon/oxide-interface and C^* the ambient concentration of the oxidant. Consider now a point x at the interface Γ_1, let $N_1[part./nm^3]$ be the number of oxidant molecules incorporated into one unit volume of silicon dioxide, λ the volume expansion factor (=2.2) for the reaction from silicon to oxide. Furthermore it is assumed that the motion of the interface point x is normal to the interface at x. It can easily be seen that, during the time interval Δt the point x moves into the silicon by

$$\Delta x_{Si} = \frac{\Delta t k_s C(x)}{N_1 \lambda} \tag{5}$$

and the oxide layer is lifted at x by

$$\Delta x_{Ox} = \frac{(\lambda - 1)\Delta t k_s C(x)}{N_1 \lambda} \tag{6}$$

The displacement Δx_{Ox} acts as a boundary condition for the coupled mechanical deformation of the oxide layer. This deformation is usually described either elastically (for temperatures below 960 ^oC) or in a viscous flow model for higher temperatures. The solution of the mechanical problem then updates the geometry of the oxide and nitride body for the following time step.

Numerical procedures to simulate this process can be classified into three groups. The first type of method (e.g. [4]) maps the silicon dioxide domain in each time step onto a simple numerical domain (most frequently a rectangle). The second approach uses the boundary element method for diffusion and displacement (e.g. [5]) , the third one (e.g. [6],[7]) models the domain of computation by finite elements. However none of these approaches is able to treat problems of changing oxide topology, for example structures, where two originally seperated oxide ranges grow together during oxidation.

The principles of the new approach presented in this paper have been published in [1]. Basic idea is a regularization/extraction technique, which allows a finite element formulation with no or at most very few mesh-retriangularizations. Moreover it offers the possibility of simulating problems of changing oxide topology without any difficulties.

This new approach uses the idea of a *smooth transition zone* from silicon to oxide. Over the whole domain of simulation a function of *relative density of silicon* η is defined. For this function an ordinary differential equation in every point of the domain can be formulated, coupling the oxidant diffusion via the chemical reaction to the mechanical displacement. No remeshing due to the motion of the interface is necessary, as only the relative silicon density changes its distribution. The shape of the oxide body can, in every time step, be easily extracted from η.

This method is related to the well-established enthalpy methods for Stephan-problems. The regularization with the distributed interface layer can be done so that, in a one-dimensional setting, the regularization/extraction procedure yields precisely the same motion of the interface as the original sharp-interface formulation.

The distributed interface layer

Consider figure 4 which shows a typical 'bird's beak' structure during oxidation similar to figure 3, yet now with a distributed interface zone. Ω_1 is the range of already existing oxide, Ω_3 the range of pure silicon, Ω_4 the 'masking' silicon nitride. The chemical reaction is now not modelled as a reaction *boundary condition* but as a *field reaction* in the interface range Ω_2. The chemical reaction there has three consequences :

- it imposes a distributed sink to the diffusion problem,

- it changes some silicon to oxide, i.e. 'moves' the interface range Ω_2,
- it imposes pre-strains to the mechanical problem, as the produced oxide has 2.2 times the volume of the original silicon.

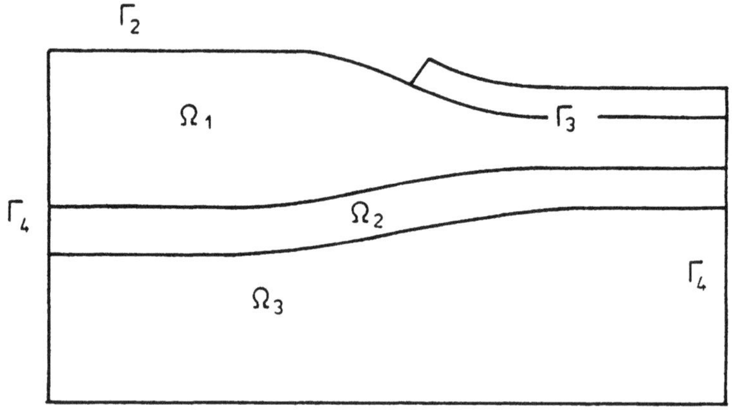

Figure 4 : Domains and boundaries (modified formulation)

To define the interface layer Ω_2, consider on $\Omega_1 \cup \Omega_2 \cup \Omega_3$ a *normalized silicon concentration* η as

$$\eta(x,t) = \frac{C_{Si}(x,t)}{C_{Sio}} \quad , \tag{7}$$

where x is the spatial coordinate, t the time, $C_{Si}(x,t)$ the silicon concentration at time t and point x and C_{Sio} the concentration of silicon atoms in pure crystal. η is 1 in Ω_3, 0 in Ω_1 and between 0 and 1 in the interface Ω_2. (The oxidant diffusion in nitride can be neglected, therefore this part of the domain will only be considered for the structural deformation.)

As the time scale for the diffusion process is much smaller than for the oxide growth, the diffusion steps can again be considered stationary on a sequence of domains.

The oxidant diffusion in $\Omega_1 \cup \Omega_2 \cup \Omega_3$ shall be described by

$$\nabla(D(\eta(t))\nabla C(x,t)) = k(\eta(t))C(x,t) \quad on \quad \Omega_1 \cup \Omega_2 \cup \Omega_3 \tag{8}$$

for every fixed time t , with the boundary conditions

$$C = C^* \quad on \quad \Gamma_2 \tag{9}$$

and

$$\frac{\partial C}{\partial n} = 0 \quad on \quad \Gamma_3 \cup \Gamma_4 \quad . \tag{10}$$

k is the strength of a *spatial sink*. kC defines, how many particles of oxygen *per unit volume* react in a unit time interval to oxide.

Obviously, the chemical reaction in the interface range Ω_2 changes the relative silicon concentration η , i.e. in a time interval $t, t + \Delta t$ and a test volume ΔV (where we assume the oxidant concentration C to be constant), $k(\eta(t))C(x,t)\Delta V \Delta t$ particles of oxygen react with $k(\eta(t))C(x,t)\Delta V \Delta t/(N_1 \lambda)$ unit volumes of silicon to form

$$V_{SiO_2}^{new} = k(\eta(t))C(x,t)\Delta V \Delta t/N_1 \tag{11}$$

unit volumes of silicon dioxide. Letting $\Delta t \to 0$ and noting that η is the *relative* density of silicon it follows that

$$\frac{\partial \eta}{\partial t} = -\frac{1}{\lambda} k(\eta(t))C(x,t)/N_1 \quad . \tag{12}$$

The *additional* volume per unit volume silicon after oxidation is given by

$$V^{new} = \frac{\lambda - 1}{\lambda} \Delta t \Delta V k(x, \eta(t))C(x,t)/N_1 \quad , \tag{13}$$

which produces *strains* in the control volume and thus loads the displacement problem.

A coupling to an *elastic* deformation problem is quite straightforward and has been described in [1]. Strains are used as loads to a displacement problem and the resulting deformed domain defines the geometry for the next time step.

Whereas the assumption of an elastic deformation is realistic for silicon and nitride and also for oxide below 960 $^{\circ}C$, oxide at higher temperatures is better described in a viscous flow model. For an implementation of a mixed elastic/viscous model the standard set of variables for the viscous flow computation (velocities, stress- and strain-rates) have to be transformed to variables of displacement type. This procedure and numerical results for a viscous flow computation have been reported in [9].

Numerical approximation

For the numerical approximation a standard finite element method with linear (in the elastic model) or quadratic (in the viscous flow model) shape functions is used. The quasi stationary diffusion equation (8),the equations for the mechanical deformation and the ordinary differential equation (12) are coupled at integration points of the finite element mesh. The three subproblems can be solved decoupled, i.e. in a cyclic manner in every time step.The algorithm for the oxidation simulation can now be formulated:

Algorithm :

1. Define an initial domain Ω and a silicon distribution η, thus introducing the dioxide,silicon and interface range. (The discretization of nitride is straightforward and need not be described explicitly.)
2. Define a finite element mesh, initialize time $t = 0$
3. Initialize material coefficients D,k,E and ν at all integration points.
4. Solve the quasistationary diffusion problem (8) to (10).
5. Compute the reacting quantity of oxidant and silicon at each integration point from (11).
6. Update the silicon distribution function η at all integration points according to (12).
7. Compute prestrains at all integration points from the reacting volume.
8. Solve the mechanical problem .
9. Update nodal coordinates.
10. Set $t = t + \Delta t$ and goto 3.

The main advantage of this algorithm compared to the conventional approach with a sharp interface can now be seen:

The interface is defined only implicitly via the relative silicon distribution η, therefore topological changes of the oxide body do not cause any algorithmic problems and it is not necessary to remesh due to the interface motion.

The relation of material parameters $D(\eta)$ and $k(\eta)$ to the measurable parameters D_{ox} and k_s is shown in [1]. A sharp interface *line* can be extracted from the smoothed formulation, (this is necessary, as the oxidation simulator is part of a process simulation system where all other programs need a sharp interface as input) by using a contour line algorithm. The *equivalent sharp interface line* is defined as the contour where the relative silicon density is .5 . For a mathematical analysis of this procedure see again [1].

Another advantage of the formulation with a topologically invariant mesh shall be mentioned. It is often important for technology development not only to predict the shape of the oxide body but also to identfy ranges of *high stresses* in oxide and silicon. There is much evidence that a *viscoelastic* flow model predicts stresses in oxide most reliably for a wide temperature range. An implementation of a this model can now make use of the motion of integration points as *material motion*, carrying along the *history* of stresses and strains. No interpolation of data from one mesh to another is necessary. A detailed formulation of a viscoelastic model and numerical examples shall be presented in a forthcoming paper.

Numerical examples

Figure 5 : SEM-photograph and numerical domain : initial structure

The first example shows in figure 5 the SEM-picture of a structure before oxidation on the right and the initial FE-mesh on the left. Figure 6 shows the real and simulated structures after oxidation. The simulation is in excellent agreement with the physical experiment. This can be seen not only on the contour of the extracted interface but also on the deformation of the nitride mask. A part of it has delaminated from the oxide (the 'needle' on top of the oxide layer); this could be simulated by removing the connection of nitride and oxide in nodes when tension occured.

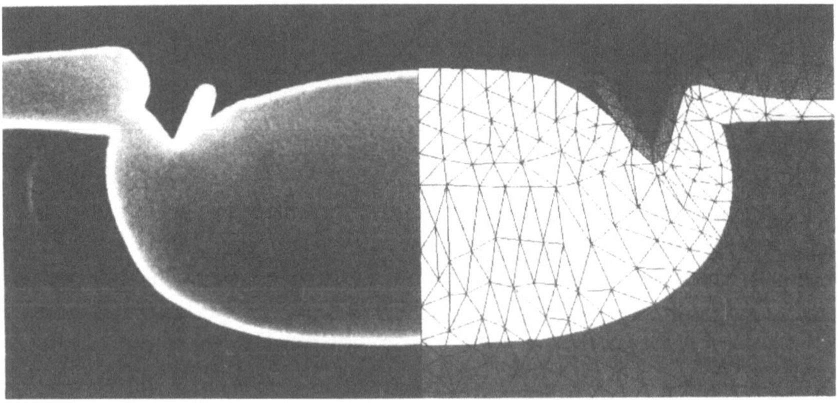

Figure 6 : SEM-photograph and numerical domain : final structure

The second example demonstrates the algorithm's ability to simulate changes in oxide topology. Figure 7 shows the initial structure for a socalled poly-buffered oxidation. In contrast to the first example there is a layer of polysilicon on top of a buffered oxide layer.

In figure 8 an enlarged range near the nitride edge of the initial structure can be seen, showing a shallow initial oxide layer at the free boundary of only 2 nanometer thickness. Oxidation is performed for 100 minutes at 1000 $^{\circ}C$. Figure 9 shows the structure after 5 minutes oxidation. The typical bird's beak has formed in poly silicon, yet the two oxide ranges are still separated.

Figure 10 gives a picture of the structure after 10 minutes. Now the two oxide bodies have joined. Finally the silicon below the buffered oxide range is oxidized, figure 11 showing the result after 100 minutes oxidation time.

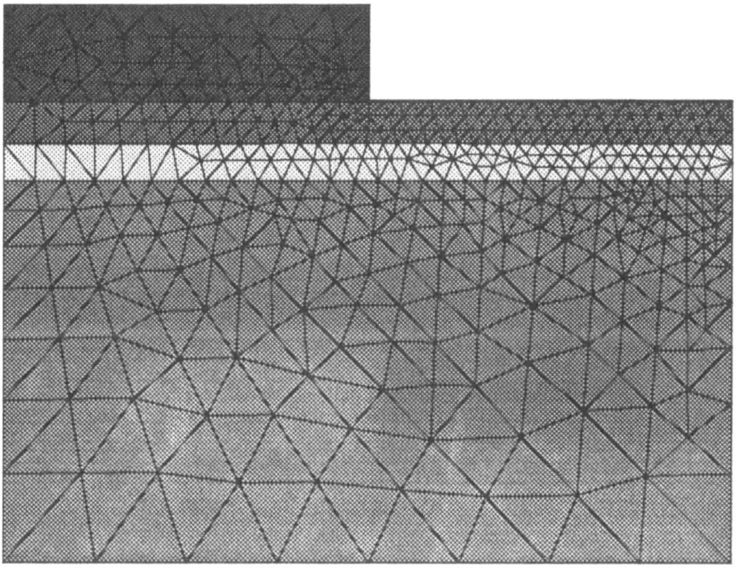

Figure 7 : Initial structure for poly-buffered oxidation

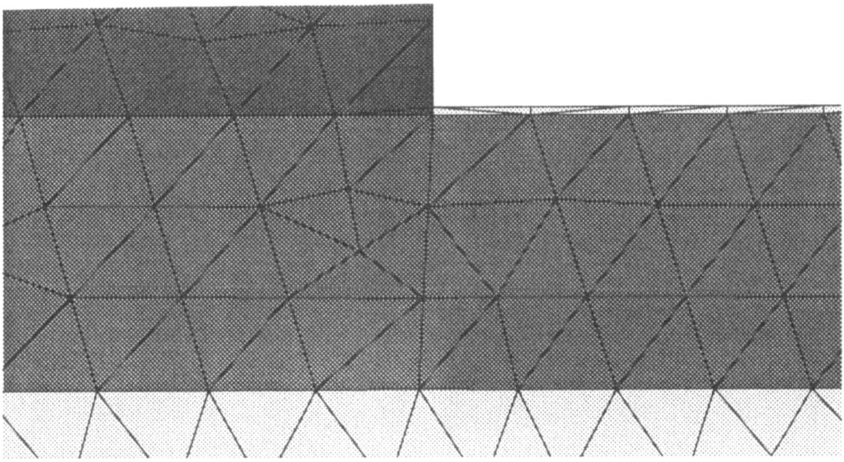

Figure 8 : Initial structure: enlarged range around nitride edge

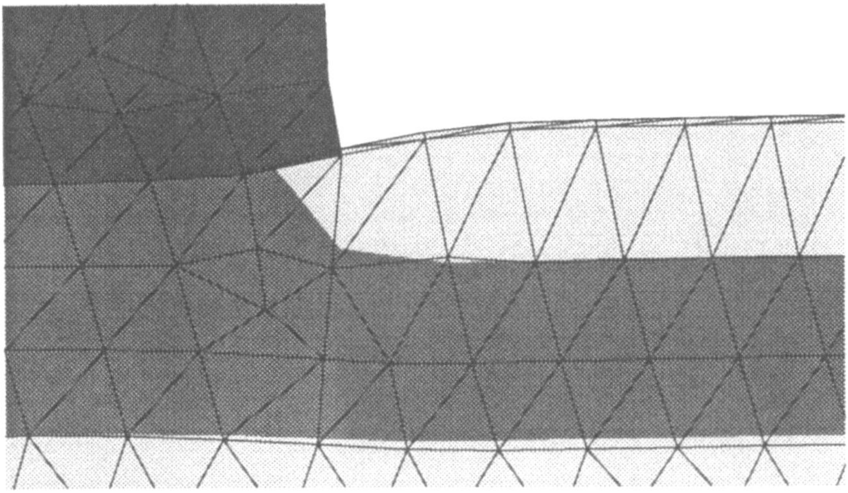

Figure 9 : First bird's beak in poly-silicon

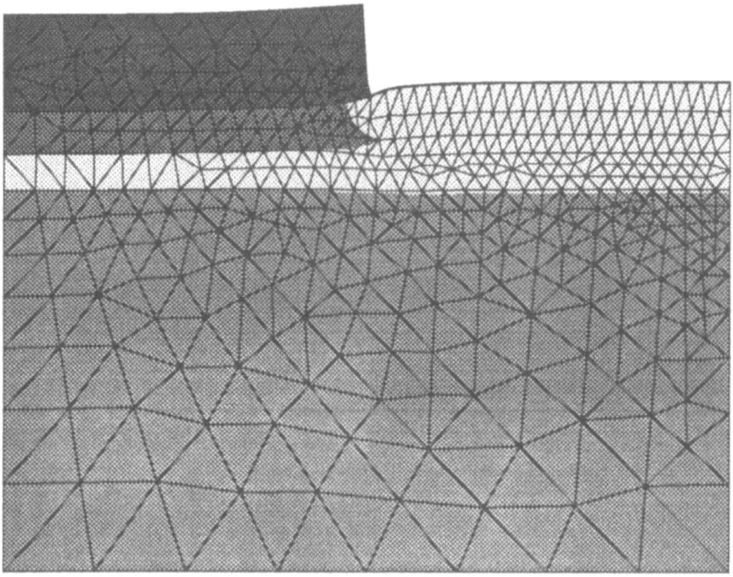

Figure 10 : Oxide ranges grown together

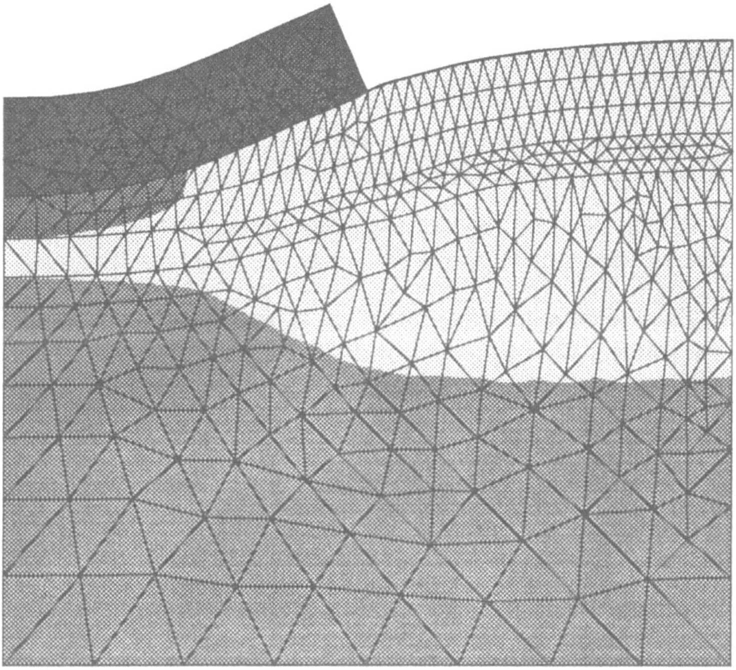

Figure 11 : Structure after oxidation

References

[1] E. Rank, U. Weinert; *A simulation system for diffusive oxidation of silicon. A two-dimensional finite element approach.* , IEEE-CAD , May 1990.

[2] E. Rank:*Structural simulation in the process simulator MIMAS.* Proc. of the 7th VLSI-Process- and Device-Modelling Workshop. Kawasaki, 1990.

[3] B.E. Deal,A.S. Grove.;*General relationship for the thermal oxidation of silicon.* J. Appl. Phys., Vol 36, No 12, 1965

[4] J. Lorenz, J. Pelka, H. Ryssel,A. Sachs,A. Seidl, M. Svoboda; *COMPOSITE - A complete modeling program for silicon technology.* IEEE trans. on computer aided design, Vol CAD-4, 1985

[5] T. Tung,J. Connor,D. Antoniadis; *A boundary element method for modeling viscoelastic flow of thermal oxidation.* IEEE trans. on computer aided design, Vol CAD-7, No 2, pp 215–224, 1988

[6] A. Poncet; *Finite-element simulation of local oxidation of silicon.* IEEE trans. on computer aided design, Vol CAD-4, No 1, 1985

[7] A. Seidl; *Zweidimensionale Simulation der lokalen Oxidation von Silizium.* Doctorial Dissertation, Technische Fakultät der Friedrich-Alexander-Universität Erlangen-Nürnberg, 1988.

[8] O.C. Zienkiewicz; *The Finite Element Method.* 3rd edition. 1977, Mc-Graw-Hill Book Company, Maidenhead , Berkshire, England.

[9] E. Rank; *Finite-element-simulation of local oxidation in semiconductor processing.* Proc. of MAFELAP 1990, Brunel University, Uxbridge; J.R. Whiteman (ed.).

On a Kind of Dynamic Fracture Problem with Moving Boundaries

X.D. Wang
Department of Mechanics, Peking University, Beijing 100871, P.R. China

ABSTRACT

A kind of dynamic fracture problem with moving boundaries which lie on the interface of two bonded dissimilar media are investigated. The motions of the boundaries are not prescribed in the formulation of the problem, so the moving boundaries can be viewed as interface cracks with face contact. By using the method of integral transforms coupled with the technique of transform inversion by contour integration, the governing equations of the problem are formulated in the space — time domain in the form of integral equations. Then the problem is solved by expanding the displacements of the crack faces as the sum of Chebyshev polynomials. Numerical results for several examples are given. The results show that the shear failure has significant effect on the spreading of the interface crack in a tension field.

INTRODUCTION

In studying the fracture initiation in nonhomogeneous structural materials, considerable attention has been directed to interface cracks between bonded dissimilar media[1-8]. The static elasticity solutions for such cracks show that the crack faces make contact near the crack tips [4,5]. For the corresponding dynamic problems, the region in which the crack faces make contact should vary with time and become a moving boundary(or boundaries). So, to investigate the behavior of interface cracks with face contact, moving boundary problem should be dealt with. Unfortunately, there is not a systematic method for such a problem.

The integral transform method has been used to solve dynamic fracture problems with no crack face contact. A large number of papers have been presented. In these papers, the integral transforms of the stress intensity factors are obtained first, then the corresponding results in time — domain are obtained by using the technique of transform inversion. But when a moving boundary(or boundaries) is involved, the problem can not be solved in such a simple way. In such a case, there is interaction between space and time. The solvable governing equations for the problem exist in the space—time domain only. This makes the problem very difficult because derivation of the boundary integral equations requires that basic solution of the problem be given in the space—time domain, which is difficult to obtain in many cases.

Our interest here is to investigate elastodynamic problems with moving boundaries on the interface of two dissimilar materials. A general form of the moving boundaries is considered, which can be reduced to a crack with face contact or a crack propagating with arbitrary speed etc. The governing equations of the problem in the form of integral equations are formulated in the space—time domain. Then the problem is solved by expanding the displacements of the crack faces as the sum of Chebyshev polynomials. Numerical results are given based on the present method, for some special cases.

FORMULATION OF THE PROBLEM

Consider a crack which lies on the interface of two infinite homogeneous, isotropic elastic media, as shown in Fig. 1. The deformation is assumed to be plane strain one and symmetric to y—axis. From the idea of Comninou [4], the crack faces should make frictionless contact near the crack tips. Assume that on the interface there exist two boundaries which vary with time. Denote these boundaries by L_k (k=1,2). For interface crack problem, L_1 and L_2 represent the crack and the region in which the crack faces are open, respectively. Without any loss in generality, it is assumed that all the external loadings act on the moving boundaries.

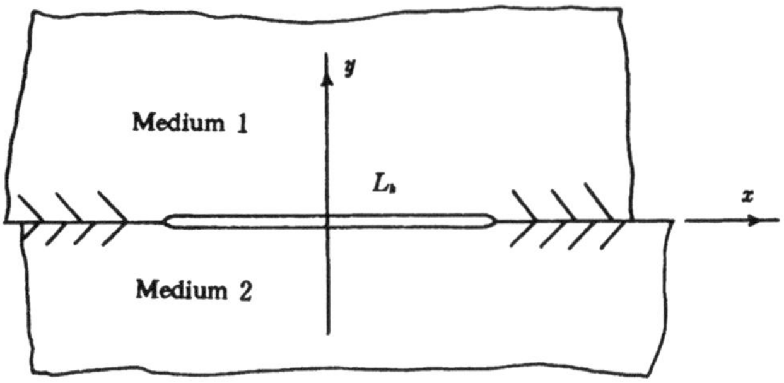

Fig. 1 The medium and the moving boundaries

For plane strain elastodynamic problems, the displacemt components can be expressed in terms of two scalar potentials, i. e.

$$u_1 = \partial F/\partial x + \partial G/\partial y, \qquad u_2 = \partial F/\partial y - \partial G/\partial x \tag{1}$$

where F, G satisfy the following differential equations

$$\nabla^2 F = \frac{1}{C_L^2}\ddot{F}, \qquad \nabla^2 G = \frac{1}{C_T^2}\ddot{G} \tag{2}$$

with ∇^2 being the Laplacian operator, C_L, C_T are dilatational and shear wave velocities, respectively, given by

$$C_L^2 = (\lambda + 2\mu)/\rho, \qquad C_T^2 = \mu/\rho \tag{3}$$

where λ, μ are Lame constants and ρ is the material density.

Let us make use of Fourier transform (with respect to x) and Laplace transform (with respect to t) (the definitions of them are given in the Appendix) to equations (2) and introduce the following two unknown functions

$$\varphi_k(s,p) = s[\bar{u}_k^*(s,o^+,p) - \bar{u}_k^*(s,o^-,p)] \qquad k = 1,2 \tag{4}$$

where $\bar{u}_k^*(s,y,p)$ are the transforms of displacement components. Then making use of the continuity conditions of stress and displacement on the interface and the condition that the stress should be finite when $y \to \infty$, the stress components on the interface can be expressed as

$$\bar{\sigma}_{k2}^*(s,o,p) = m_{kj}(s,p)\varphi_j(s,p) \qquad k,j = 1,2 \tag{5}$$

where $m_{kj}(s,p)$ are the basic solutions of the problem given in the Appendix, and the branches of

$$\alpha_l = (s^2 + p^2/C_{Ll}^2)^{1/2}, \qquad \beta_l = (s^2 + p^2/C_{Tl}^2)^{1/2} \tag{6}$$

are taken to be that α_l and β_l tend to positive real numbers when $s \to +\infty$ (where $l = 1,2$ refer to the upper and the lower half planes, respectively). Besides above equations, the following boundary conditions on $y = 0$ should be satisfied

$$\sigma_{k2}(x,o,t) = p_k(x,t) \qquad x \in L_k(t) \qquad k = 1,2 \tag{7}$$

$$u_k(x,o^+,t) = u_k(x,o^-,t) \qquad x \overline{\in} L_k(t) \qquad k = 1,2 \tag{8}$$

From equations (4) and (5), equations (7) and (8) become a set of simultaneous dual integral equations

$$L^{-1}F^{-1}[m_{kj}(s,p)\varphi_j(s,p)] = p_k(x,t), \qquad x \in L_k(t) \qquad k,j = 1,2 \tag{9}$$

$$L^{-1}F^{-1}[s^{-1}\varphi_k(s,p)] = 0, \qquad x \overline{\in} L_k(t), \qquad k = 1,2 \tag{10}$$

where L^{-1} and F^{-1} denote the inversion of Laplace and Fourier transforms, and p_k ($k = 1,2$) are the known stresses acting on the moving boundaries.

Introducing

$$\bar{\varphi}_k(x,t) = \frac{1}{2\pi}F^{-1}L^{-1}\varphi_k(s,p) \qquad k = 1,2 \tag{11}$$

then from the definition of φ_k, following relation exists

$$\bar{\varphi}_k(x,t) = \frac{i}{2\pi}\frac{\partial}{\partial x}[u_k(x,o^+,t) - u_k(x,o^-,t)] \qquad k = 1,2 \tag{12}$$

Making use of (8), φ_k can be expressed as

$$\varphi_k(s,p) = L\int_{L_k}\bar{\varphi}_k(x,t)exp(isx)dx, \qquad k = 1,2 \tag{13}$$

In fact, functions $m_{kj}(s,p)$ have the following limits at infinit

$$\lim_{s\to+\infty} m_{kj}(s,p) = a_{kj} \qquad k,j = 1,2 \tag{14}$$

where a_{kj} are constants given in the Appendix. From the Appendix, it can be seen that m_{11}, m_{22} and m_{12}, m_{21} are odd and even functions of s, respectively. Consider now the following functions

$$\begin{aligned} n_{kj}(s,p) &= m_{kj}(s,p) - a_{kj} \qquad k \neq j \\ n_{kj}(s,p) &= m_{kj}(sp) - a_{kj}\text{Sgn}(s) \qquad k = j \end{aligned} \tag{15}$$

where

$$\text{Sgn}(s) = \begin{cases} 1 & s > 0 \\ -1 & s < 0 \end{cases} \tag{16}$$

Obviously the following relation exists

$$\lim_{s\to\infty} n_{kj}(s,p) = 0 \qquad k,j = 1,2 \tag{17}$$

Substituting the above equations into equations(9), (10), the following singular integral equations can be obtained for zero initial conditions

$$a_{12}\bar{\varphi}_2(x,t) + \frac{ia_{11}}{\pi}\int_{L_1(t)}\frac{\bar{\varphi}_1(u,t)}{u-x}du$$

$$+\int_0^\infty\left\{\int_{\bar{L}_1}\bar{\varphi}_1(u,\tau)\bar{n}_{11}du + \int_{\bar{L}_2}\bar{\varphi}_2(u,\tau)\bar{n}_{12}du\right\}d\tau = p_1(x,t)/2\pi \qquad x \in L_1(t) \tag{18}$$

$$a_{21}\bar{\varphi}_1(x,t) + \frac{ia_{22}}{\pi}\int_{L_2(t)}\frac{\bar{\varphi}_2(u,t)}{u-x}du$$

$$+\int_0^\infty\{\int_{\bar{L}_1}\bar{\varphi}_1(u,\tau)\bar{n}_{21}du + \int_{\bar{L}_2}\bar{\varphi}_2(u,\tau)\bar{n}_{22}du\}d\tau = p_2(x,t)/2\pi \qquad x \in L_2(t) \tag{19}$$

$$\int_{L_1(t)}\bar{\varphi}_1(x,t)dx = 0, \qquad \int_{L_2(t)}\bar{\varphi}_2(x,t)dx = 0 \tag{20}$$

where $\bar{L}_1 = L_1(\tau), \bar{L}_2 = L_2(\tau)$ and $\bar{n}_{kj}(k,j=1,2)$ are Fredholm kernals of the above equations defined by

$$\bar{n}_{kj} = \bar{n}_{kj}(t-\tau,x,u) \qquad k,j = 1,2 \tag{21}$$

$$\bar{n}_{kj}(t,x,u) = \frac{1}{2\pi}L^{-1}\int_{-\infty}^{\infty} n_{kj}(s,p)exp[is(u-x)]ds \tag{22}$$

Equations(18)—(22) form the governing equations of the problem in which arbitrary moving boundaries $L_1(t)$ and $L_2(t)$ are involved.

INVERSION OF THE INTEGRAL TRANSFORMS

In the integral equations (18) and (19), it remains to invert the Laplace and Fourier transforms of $n_{kj}(s,p)$. To evaluate (22), consider first the function I_{kj} defined by

$$I_{kj} = \int_{-\infty}^{\infty} n_{kj}(s,p)exp[is(u-x)]ds, \qquad k,j = 1,2 \tag{23}$$

According to Lerch's theorem, as the Laplace transform of a function, the right side of (23) must be unique. Thus p can be taken to be real and positive in the following analysis. This idea had deen used by de-Hoop [9].

The object here is to transform the integral on the right side of (23) into a recognizable Laplace transform. It can be seen from the Appendix that m_{kj} are homogeneous for s and p, that is

$$m_{kj}(s,p) = m_{kj}(s/p,1) \qquad k,j = 1,2 \tag{24}$$

From (15), it can be seen that $n_{kj}(s,p)$ are also homogeneous for s and p. Changing the variable of integration to $s=p\eta$, (23) becomes

$$I_{kj} = p\int_{-\infty}^{\infty} n_{kj}(\eta,1)\exp[ip\eta(u-x)]d\eta, \qquad k,j = 1,2 \tag{25}$$

It can be seen from the Appendix that the branch points of $n_{kj}(\eta,1)$ in the η-plane are determined by

$$\bar{\alpha}_l = (\eta^2 + 1/C_{Ll}^2)^{1/2}, \qquad \bar{\beta}_l = (\eta^2 + 1/C_{Tl}^2)^{1/2}, \qquad l = 1,2 \tag{26}$$

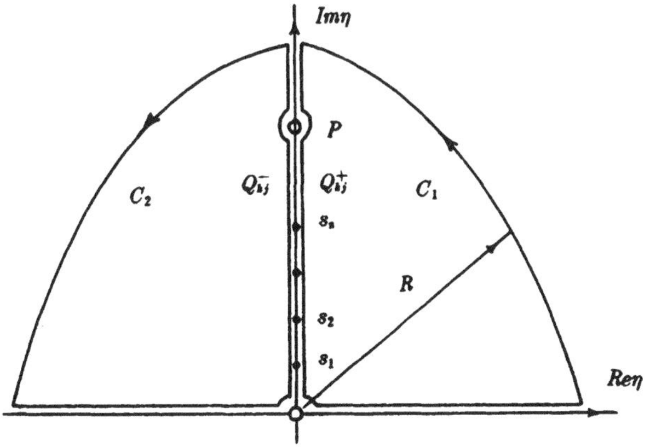

Fig. 2 The branches of Q_{kj} and the integral contours

To make the functions $n_{kj}(\eta,1)$ single-valued, cuts which pass through these points are made, as shown in Fig. 2. Denoting

$$Q_{kj}(\eta) = m_{kj}(\eta,1) \qquad k,j = 1,2 \tag{27}$$

the branches of $Q_{kj}(\eta)$ in η-plane can be expressed as in Fig. 2. In this figure, the superscripts '+' and '−' refer to the branches for $\text{Im}\bar{\alpha}_l \geqslant 0$, $\text{Im}\bar{\beta}_l \geqslant 0$ and $\text{Im}\bar{\alpha}_l \leqslant 0$, $\text{Im}\,\bar{\beta}_l \leqslant 0$, respectively. $s_n(n=1,2,3,4)$ denote the following branch points

$$s_n = i/C_{Ll}, \qquad i/C_{Tl}, \qquad l = 1,2 \tag{28}$$

and P is a pole of functions $Q_{kj}(\eta)$ which may or may not exist.

Since $m_{kj}(\eta,1)$ are single-valued and analytic, it can be seen from equation (14) that

$$\lim_{|\eta|\to\infty} Q_{kj}(\eta) = a_{kj}, \qquad k \neq j$$
$$\lim_{|\eta|\to\infty} Q_{kj}(\eta) = a_{kj}\mathrm{Sgn}(Re\eta), \quad k = j \tag{29}$$

To evaluate (23), for the case that $x-u>0$ and $k\neq j$, the following contour integrals are introduced in complex η-plane

$${}_1H_{kj} = \int_{C_1} [Q_{kj}(\eta) - a_{kj}]\exp[ip\eta(x-u)]d\eta \qquad k,j = 1,2 \tag{30}$$

$${}_2H_{kj} = \int_{C_2} [Q_{kj}(\eta) - a_{kj}]\exp[ip\eta(x-u)]d\eta \qquad k,j = 1,2 \tag{31}$$

where C_1, C_2 are the integral contours shown in Fig. 2. Letting $R\to\infty$ and making use of the asymptotic property of $Q_{kj}(\eta)$ $(m_{kj}(\eta,1))$ give by (14), Cauchy integral theorem and Jordan's lemma, the following result is obtained by adding (30) and (31)

$$I_{kj} = p\int_0^{\infty i} [Q^+_{kj}(\eta) - Q^-_{kj}(\eta)]\exp[ip\eta(x-u)]dy + I^0_{kj} \qquad k,j = 1,2 \tag{32}$$

where I^0_{kj} correspond to the pole of $Q_{kj}(\eta)$.

For the case that $x-u>0$ and $k=j$, $Q_{kj}(\eta)$ are singular at $\eta=0$ in the form $1/\eta$, that is

$$Q^0_{kj} = \lim_{\eta\to 0}[\eta Q_{kj}(\eta)] \qquad k = j = 1,2 \tag{33}$$

where Q^0_{kj} are constants given in the Appendix. Similarly, by the technique of contour integration, following result can be obtained

$$I_{kj} = p\int_0^{\infty i} [Q^-_{kj}(\eta) - Q^+_{kj}(\eta) + 2a_{kj}]\exp[ip\eta(x-u)]dy + i\pi Q^0_{kj}p + I^0_{kj}$$
$$k = j = 1,2 \tag{34}$$

To transform the integral given by (25) into a recognizable Laplace transform, made the change in variables

$$t = -i\eta(x-u) \tag{35}$$

Thus, for x>u, the value of the integral given by (25) can be expressed as

$$I_{kj} = \frac{ip}{x-u}\int_0^{\infty}\left[Q^+_{kj}\left(\frac{it}{x-u}\right) - Q^-_{kj}\left(\frac{it}{x-u}\right)\right]\exp(-pt)\,dt + I^0_{kj}$$
$$k \neq j, \quad k,j = 1,2 \tag{36}$$

$$I_{kj} = \frac{ip}{x-u}\int_0^{\infty}\left[Q^-_{kj}\left(\frac{it}{x-u}\right) - Q^+_{kj}\left(\frac{it}{x-u}\right) + 2a_{kj}\right]\exp(-pt)dt + i\pi pQ^0_{kj} + I^0_{kj}$$
$$k = j, \quad k,j = 1,2 \tag{37}$$

For the case $x<u$, the result can be obtained by simply exchanging $x-u$ for $u-x$ and exchanging Q^0_{kj} for $-Q^0_{kj}$.

Making use of (36), (37) and the properties of Laplace transform, the inversion of the Laplace and Fourier transforms in equation (22) can be obtained directly, that is

$$\bar{n}_{kj}(t,x,u) = \frac{-1}{2\pi(x-u)^2}\{R^+_{kj}(z) - R^-_{kj}(z)\}H(t) + R^0_{kj}, \quad k \neq j \tag{38}$$

$$\bar{n}_{kj}(t,x,u) = \frac{\mathrm{Sgn}(u-x)}{2\pi(x-u)^2}\{R^-_{kj}(z) - R^+_{kj}(z)\}H(t) + i\mathrm{Sgn}(x-u)Q^0_{kj}\,\delta'(t)/2 + R^0_{kj},$$
$$k = j \tag{39}$$

where $z=it/|x-u|$, $R_{ij}(z)=dQ_{ij}(z)/dz$, $R^0_{ij}=L^{-1}I^0_{ij}/2\pi$.

Substituting above result into eguations (18), (19), we have

$$a_{12}\bar{\varphi}_2(x,t)+\frac{ia_{11}}{\pi}\int_{L_1}\frac{\bar{\varphi}_1(u,t)}{u-x}du+\frac{iQ^0_{11}}{2}\int_{L_1}\dot{\bar{\varphi}}_1(u,t)\mathrm{Sgn}(x-u)du$$

$$+\int_0^t\left\{\int_{\bar{L}_1}\bar{\varphi}_1(u,\tau)\bar{n}^*_{11}du+\int_{\bar{L}_2}\bar{\varphi}_2(u,\tau)\bar{n}^*_{12}du\right\}d\tau=p_1(x,t)/2\pi,\qquad x\in L_1(t)\quad(40)$$

$$a_{21}\bar{\varphi}_1(x,t)+\frac{ia_{22}}{\pi}\int_{L_2}\frac{\bar{\varphi}_2(u,t)}{u-x}du+\frac{iQ^0_{22}}{2}\int_{L_2}\dot{\bar{\varphi}}_2(u,t)\mathrm{Sgn}(x-u)du$$

$$+\int_0^t\left\{\int_{\bar{L}_1}\bar{\varphi}_1(u,\tau)\bar{n}^*_{21}du+\int_{\bar{L}_2}\bar{\varphi}_2(u,\tau)\bar{n}^*_{22}du\right\}d\tau=p_2(x,t)/2\pi,\qquad x\in L_2(t)\quad(41)$$

where $\dot{(\ )}=\frac{\partial}{\partial t}(\ )$, $L_1=L_1(t)$, $L_2=L_2(t)$, $\bar{L}_1=L_1(\tau)$, $\bar{L}_2=L_2(t)$, $\bar{n}^*_{ij}$ are the remainders of $\bar{n}_{ij}$ from which the terms connected with Q^0_{ij} are excepted.

SOLUTION OF A KIND OF INTERFACE CRACK PROBLEM

Consider here an interface crack. The half length of it is assumed to be c(t). From known result [4], the interface crack is not fully open and its two faces are in contact with each other near the tips. Without any loss in generality, it is assumed that the crack faces are in frictionless contact in the interval $a(t)<|x|<c(t)$. Here a(t) is an unknown function of time. In such a case, we have

$$L_1(t)=(-c(t),\ c(t)),\qquad L_2(t)=(-a(t),\ a(t))\qquad(42)$$

From the definitions of $\bar{\varphi}_1$ and $\bar{\varphi}_2$, it can be concluded that $\bar{\varphi}_2$ is continous in $(-a,\ a)$ and bounded at $x=\pm a(t)$. Then equation (40) can be viewed as a Cauchy singular integral equation for the unknown function $\bar{\varphi}_1(x,t)$, so it can be easily proved that $\bar{\varphi}_1(x,t)$ has a square-root singularity at $x=\pm c(t)$. From (40) and (41), the stress components on the interface can be expressed as

$$\sigma_{12}(x,o,t)=2\pi a_{12}\varphi_2(x,t)[H(x+a)-H(x-a)]+2ia_{11}\int_{-c}^{c}\frac{\bar{\varphi}_1(u,t)}{u-x}du+O(1)\qquad(43)$$

$$\sigma_{22}(x,o,t)=2\pi a_{21}\varphi_1(x,t)[H(x+c)-H(x-c)]+2ia_{22}\int_{-a}^{a}\frac{\bar{\varphi}_2(u,t)}{u-x}du+O(1)\qquad(44)$$

From this result, the following conclusions can be obtained

1. There is no singularity for both σ_{22} and σ_{12} at $|x|=a$.
2. σ_{12} has square root singularity at $|x|=c^+(t)$.
3. σ_{22} has square root singularity at $|x|=c^-(t)$.

These results had been given by Comninou [4] for static problem. So $\bar{\varphi}_i(x,t)$ can be expressed as

$$\bar{\varphi}_1(x,t)=(1-x^2/c^2)^{-1/2}A_j(t)T_j(x/c)\qquad j=1,2,\cdots\qquad(45)$$

$$\bar{\varphi}_2(x,t)=(1-x^2/a^2)^{1/2}B_j(t)U_j(x/a)\qquad j=1,2,\cdots\qquad(46)$$

where A_j, B_j are unknown functions of time, T_j and U_j are first and second kinds of Chebyshev polynomials. Substitute the foregoing expressions into (40) and (41). From the orthogonality conditions, a system of non-singular derivative-integral equations are obtained

$$\frac{ia_{11}\pi}{2}A_{k+1}(t)+\bar{b}_{kj}(t)B_j(t)+e_{kj}\dot{A}_j(t)c(t)+\bar{e}_{kj}A_j(t)\dot{c}(t)+\int_0^t\{a_{kj}(t,\tau)A_j(\tau)$$

$$+b_{kj}(t,\tau)B_j(\tau)\}d\tau=F_{1k}(t)\qquad k=1,2,\cdots;\qquad j=1,2,\cdots\tag{47}$$

$$-\frac{ia_{22}\pi}{2}B_{k-1}(t)+\bar{c}_{kj}(t)A_j(t)+f_{kj}(t)\dot{B}_j(t)a(t)+\bar{f}_{kj}B_j(t)\dot{a}(t)+\int_0^t\{c_{kj}(t,\tau)A_j(\tau)$$

$$+d_{kj}(t,\tau)B_j(\tau)\}d\tau=F_{2k}(t)\qquad k=0,1,\cdots;\qquad j=1,2,\cdots\tag{48}$$

where $B_{-1}=0$. The details for the coefficients of the above equations are omitted here. It is easily seen that (47) and (48) are linear for A_j and B_j, but nonlinear for $a(t)$. It should be noted that, from the assumption that $\bar{\varphi}_2(x,t)$ is bounded at $|x|=a(t)$, equation (48) provides a additional equation (for $k=0$) for determining the unknown function $a(t)$.

Define stress intensity factors as

$$K_{\rm I}(t)=\lim_{x\to c^-}[(c-x)^{1/2}\sigma_{22}(x,o,k)]\tag{49}$$

$$K_{\rm II}(t)=\lim_{x\to c^+}[(x-c)^{1/2}\sigma_{12}(x,o,k)]\tag{50}$$

After the solution of (47) and (48) is obtained, the corresponding stress intensity factors can be expressed as

$$K_{\rm I}(t)=\sqrt{2}\,\pi a_{21}\sqrt{c}\sum_{j=1}^{\infty}A_j(t)\tag{51}$$

$$K_{\rm II}(t)=-\sqrt{2}\,\pi i a_{11}\sqrt{c}\sum_{j=1}^{\infty}A_j(t)\tag{52}$$

NUMERICAL RESULTS AND DISCUSSIONS

Denote equations (47)、(48) in the following manner

$$[\bar{K}_0(t)]\{(G(t)\}+[\bar{K}_1(t)]\{\dot{G}(t)\}+\int_0^t[\bar{K}_2(t-\tau)]]\{G(\tau)\}d\tau=\{F(t)\}\tag{53}$$

where $\{G\}=\{A_1,B_1,A_2,B_2\cdots\}^T$, $\{\dot{G}\}=\frac{d}{dt}\{G\}$, $[\bar{K}_0]$, $[\bar{K}_1]$, $[\bar{K}_2]$ are the corresponding coefficients matrixes, and $\{F\}$ is the matrix corresponding to p_1 and p_2. Equation (53) is solved numerically step by step in the form

$$[\bar{K}_1(t)\}\{\Delta G_j(t)\}=\{\{F(t)\}-[K_0(t)]\{G_j(t)\}-\int_0^{t_j}[\bar{K}_2(t-\tau)]\{(G(\tau)\}d\tau)\}\Delta t_j$$

$$j=0,1,2,\cdots\tag{54}$$

where $\{\Delta G_j(t)\}=\{G_{j+1}(t)\}-\{G_j(t)\}$, $\Delta t_j=t_{j+1}-t_j$. At each step, iteration must be made for a(t).

A series of interface crack problems are solved with the method given above. The load functions p_1 and p_2 are taken to be

$$p_1(x,t)=0,\qquad p_2(x,t)=-H(t)$$

which implies that the crack surface subjected to uniform pressure. The material constants are taken to be

$$a.\quad E_1 = 2.13 \times 10^{12} dym/cm^2, \nu_1 = 0.20, \rho_1 = 5.6 g/cm^2$$
$$b.\quad E_2 = 5.00 \times 10^{11} dyn/cm^2, \nu_2 = 0.33, \rho_2 = 7.6 g/cm^2$$
$$c.\quad E_2 = 2.70 \times 10^{11} dyn/cm^2, \nu_2 = 0.33, \rho_2 = 7.6 g/cm^2$$

1). Transient response of an interface crack

The half length of the crack is assumed to be one, that is, $L_1(t)=(-1,1), L_2(t)=(-a(t),a(t))$. The stress intensity factor (SIF) and the length of the contact region of the crack faces (LCR) for material combination a—b are as shown in Fig. 3. Both SIF and LCR reach maximum values after the initiation, then approach the corresponding static values. The maximum value of SIF is 30 percent higher than the static one. This result is somehow similar to that obtained by Sih[10] for the crack in a homogeneous medium. It is interesting to see that LCR reaches the maxima at the same time as that for SIF. Both SIF and LCR reach the maxima about $tC_{min}=2.0$, that is the time at which the elastic waves reflected by one crack tip arrive at the other one; here $C_{min}=\min\{C_{L1}, C_{T1}, C_{L2}, C_{T2}\}$.

2) Interface crack self—similar propagation

The crack is assumed to propagate with constant speed, V, from zero length, that is, $c(t)=Vt$. In such a case, $L_1(t)=(-Vt,Vt), L_2(t)=(-a(t),a(t))$. The SIF and LCR for material combination a—b are as shown in Fig. 4. The SIF decreases as the velocity of the crack increases. As V/C_{min} approaches 1.4, the SIF tends to zero. This result is analogous to that obtained by Broberg[11] for a homogeneous medium. The LCR has a property similar to SIF, that is, the LCR decreases as the velocity of the crack increases and tends to zero along with V/C_{min} approaching 1.4.

3) Interface crack arrest

In this case, $L_1(t)=(-c(t),c(t)), L_2(t)=(-a(t),a(t)), c(t)=Vt, Vt<1; c(t)=1, Vt\geqslant 1$, V is the velocity of crack propagation. The results for SIF and LCR for material combination $a-c$ are as shown in Fig. 5. The velocity of the crack tip is taken to be $V=0.8C_{min}$. It can be seen that both SIF and LCR reach maximum values after the crack arrest and then approach the corresponding static values. Since the maximum value of SIF is higher than that prior to the arrest and the corresponding static value, the arrested crack may reinitiate under some conditions. So, in some cases, additional resistance should be applied to the cracked structure to arrest the propagating crack.

In above results, LCR is of the order of $10^{-5}c(t)$. Although LCR is very small, it has a significant effect on the SIF, since small changes in the value of the LCR will cause large error of the SIF. It can be seen that the SIF, $K_I(t)$, for interface crack, is of the order of 1.0 for unit loading and unit half—length of the crack. It can be concluded that the spreading of the interface crack in dynamic tension field is intimately connected with failure in the shear on the interface. This behavior is similar to the interface crack under static loading [4, 8].

REFERENCES

[1] Erdogan, F., J. of Appl. Mech., Vol. 30, pp. 232—237, 1963.

[2] Erdogan, F., J. of Appl. Mech., Vol. 32, pp. 403—410, 1965.

[3] Lowengrub, M. and Sneddon, I. N. Int. J. of Engng Science, Vol. 11, pp. 1025—1034, 1973.

[4] Comnimou, M., J. of Appl. Mech., Vol. 44, pp. 631—636, 1977.

[5] Comnimou, M., J. of Appl. Mech., Vol. 45, pp. 287—290, 1978.

[6] Atkinson, C., Int. J. of Fracture, Vol. 18, pp. 161—177, 1982.

[7] Atkinson, C., Int. J. of Fracture, Vol. 19, pp. 131—138, 1982.

[8] Gautesen, A. K. and Dundurs, J., J. of Appl. Mech., Vol. 54, pp. 93—98, 1987

[9] de—Hoop, A. T., Doctoral Dissertation, Technische Hogeschool, Delft, 1958.

[10] Sin, G. C, et al., Int. J. Solids Structure., Vol. 7, pp. 977—993, 1972.

[11] Broberg, K. B., Arkiv for Fysik, Vol. 18, pp. 159—182, 1960

APPENDIX

The Laplace and Fourier transforms used in this paper are

$$f^*(p) = \int_0^\infty f(t)\exp(-pt)dt, \quad f(t) = \frac{1}{2\pi i}\int_{B_r} f^*(p)\exp(pt)dp \tag{55}$$

$$\bar{f}(s) = \frac{1}{2\pi}\int_{-\infty}^{\infty} f(x)\exp(isx)dx, \quad f(x) = \int_{-\infty}^{\infty} \bar{f}(s)\exp(-isx)ds \tag{56}$$

In equation (5), $m_{ij}(s,p)$ are given by

$$m_{11}(s,p) = \mu_1 s^{-1}[Gk_1^2(\alpha_2^{-1}r_2^2 - 4\beta_2 s^2) + k_2^2(\alpha_1^{-1}r_1^2 - 4\beta_1 s^2)]/F \tag{57}$$

$$\begin{aligned} m_{12}(s,p) = {} & i\mu_1[(\alpha_1^{-1}r_1 - 2\beta_1)G(\alpha_2^{-1}r_2^2 - 4\beta_1 s^2) \\ & - (\alpha_1^{-1}r_1^2 - 4\beta_1 s^2)(\alpha_2^{-1}r_2 - 2\beta_2)]/F \end{aligned} \tag{58}$$

$$m_{21}(s,p) = -m_{12}(s,p) \tag{59}$$

$$\begin{aligned} m_{22}(s,p) = {} & \mu_1 s^{-1}[\alpha_2^{-1}\beta_2 k_2^2(\alpha_1^{-1}r_1^2 - 4\beta_1 s^2) \\ & + \alpha_1^{-1}\beta_1 G k_1^2(\alpha_2^{-1}r_2^2 - 4s^2\beta_2)]/F \end{aligned} \tag{60}$$

where $F = s(\beta_1 s^{-1} - \alpha_1^{-1}s)G(\alpha_2^{-1}r_2^2 - 4\beta_2 s^2) + k_1^2 k_2^2(\beta_1\alpha_1^{-1} + \beta_2\alpha_2^{-1})$

$$+2s^2(\alpha_1^{-1}r - 2\beta_1)(\alpha_2^{-1}r - 2\beta_2) + s(\beta_2 s^{-1} - \alpha_2^{-1}s)(\alpha_1^{-1}r_1^2 - 4\beta_1 s^2)G^{-1}$$

$$\alpha_l^2 = s^2 + p^2/C_{Ll}^2, \quad \beta_l^2 = s^2 + p^2/C_{Tl}^2, \quad r_l = s^2 + \beta_l^2, \quad l = 1,2$$

$$G = \mu_2/\mu_1, \quad k_1^2 = p^2/C_{T1}^2, \quad k_2^2 = p^2/C_{T2}^2$$

In equation (14)

$$a_{11} = -2\mu_2[(\lambda_1 + \mu_1)(\lambda_2 + 2\mu_2) + (\lambda_2 + \mu_2)(\lambda_1 + 2\mu_1)G]/Co \tag{61}$$

$$a_{12} = -2i\mu_2^2[\lambda_1 + \mu_1 - (\lambda_2 + \mu_2)]/Co \tag{62}$$

$$a_{21} = -a_{12}, \quad a_{22} = a_{11} \tag{63}$$

$$Co = [(1+G)(\lambda_2+\mu_2)+2\mu_2][(1+G)(\lambda_1+\mu_1)+2\mu_2]$$

In equation (33)

$$Q^{\circ}_{11} = \mu_1(G\overline{K}_1\bar{k}_2^2 + \overline{K}_2\bar{k}_1^2)/(\bar{k}_1\overline{K}_1 + \bar{k}_2\overline{K}_2) \tag{64}$$

$$Q^{\circ}_{22} = \mu_1(\bar{k}_1 + G\bar{k}_2)\bar{k}_1\bar{k}_2/(\bar{k}_1\overline{K}_1 + \bar{k}_2\overline{K}_2) \tag{65}$$

where $\overline{K}_k = 1/C_{Lk}$, $\bar{k}_k = 1/C_{Tk}$, $k = 1, 2$.

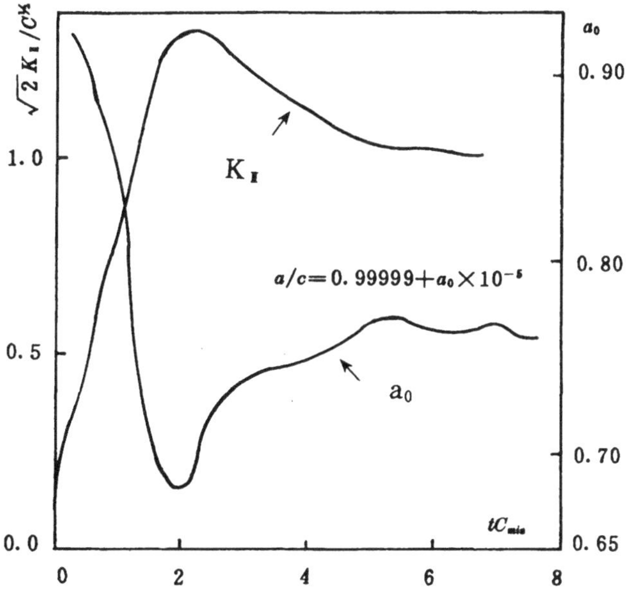

Fig. 3 SIF and LCR for non-propagating crack

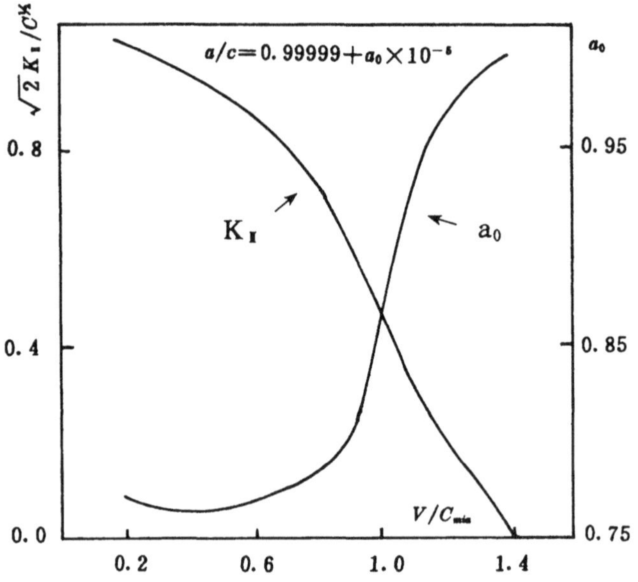

Fig. 4 SIF and LCR for self-similar propagating crack

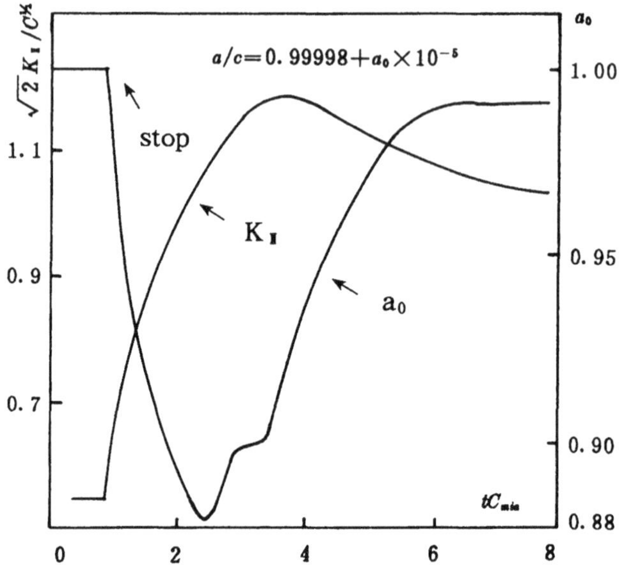

Fig. 5 SIF and LCR for arrest crack

Bei Fragen zur Produktsicherheit wenden Sie sich bitte an:
If you have any questions regarding product safety,
please contact:

Walter de Gruyter GmbH
Genthiner Straße 13
10785 Berlin
productsafety@degruyterbrill.com